AF358743

Advances in the Physics of Stars - in Memory of Prof. Yuri N. Gnedin

Advances in the Physics of Stars - in Memory of Prof. Yuri N. Gnedin

Editors

Nazar R. Ikhsanov
Galina L. Klimchitskaya
Vladimir M. Mostepanenko

MDPI • Basel • Beijing • Wuhan • Barcelona • Belgrade • Manchester • Tokyo • Cluj • Tianjin

Editors
Nazar R. Ikhsanov
Department of Astrophysics
Pulkovo Observatory of the
Russian Academy of Sciences
Saint Petersburg
Russia

Galina L. Klimchitskaya
Department of Astrophysics
Pulkovo Observatory of the
Russian Academy of Sciences
Saint Petersburg
Russia

Vladimir M. Mostepanenko
Department of Astrophysics
Pulkovo Observatory of the
Russian Academy of Sciences
Saint Petersburg
Russia

Editorial Office
MDPI
St. Alban-Anlage 66
4052 Basel, Switzerland

This is a reprint of articles from the Special Issue published online in the open access journal *Universe* (ISSN 2218-1997) (available at: www.mdpi.com/journal/universe/special_issues/star_memory).

For citation purposes, cite each article independently as indicated on the article page online and as indicated below:

LastName, A.A.; LastName, B.B.; LastName, C.C. Article Title. *Journal Name* **Year**, *Volume Number*, Page Range.

ISBN 978-3-0365-4074-0 (Hbk)
ISBN 978-3-0365-4073-3 (PDF)

Contents

Preface to "Advances in the Physics of Stars - in Memory of Prof. Yuri N. Gnedin"

This reprint collects articles devoted to different aspects of the physics of stars, their types, properties, stages of evolution, interstellar medium, dark matter and dark energy. It is devoted to the memory of outstanding scientist Prof. Dr. Yuri N. Gnedin, who not only organized the Department of Astrophysics at Pulkovo Astronomical Observatory of the Russian Academy of Sciences and led it for several decades but also initiated a number of prospective research projects in several other astrophysical Institutions. The scope of scientific interests and expertise of Prof. Gnedin were extraordinarily wide. He performed pioneering work in the theoretical description of polarized radiation transfer; generation of high-energy radiation in close binary star systems and galactic nuclei; cyclotron lines in the spectra of accreting neutron stars; determination of magnetic fields of cosmic sources from polarimetric observations; and the physics of intermediate-mass and supermassive black holes, magnetic white dwarfs, supernovae, exoplanets and dark satellites of stars, cosmic gamma-ray bursts, dark matter and dark energy. Some of these topics are covered in this reprint, which includes articles written by well-known experts in astrophysics and contains both research and review articles. It will be useful for both researches in the field of astrophysics and graduate students who are only beginning to choose the subject of their own research.

Nazar R. Ikhsanov, Galina L. Klimchitskaya, and Vladimir M. Mostepanenko

Editors

Editorial

Editorial to the Special Issue "Advances in the Physics of Stars—In Memory of Prof. Yuri N. Gnedin"

Nazar R. Ikhsanov [1], Galina L. Klimchitskaya [1,2] and Vladimir M. Mostepanenko [1,2,3,*]

1. Central Astronomical Observatory at Pulkovo of the Russian Academy of Sciences, 196140 Saint Petersburg, Russia; ikhsanov@gaoran.ru (N.R.I.); g.klimchitskaya@gmail.com (G.L.K.)
2. Peter the Great Saint Petersburg Polytechnic University, 195251 Saint Petersburg, Russia
3. Kazan Federal University, 420008 Kazan, Russia
* Correspondence: vmostepa@gmail.com

Citation: Ikhsanov, N.R.; Klimchitskaya, G.L.; Mostepanenko, V.M. Editorial to the Special Issue "Advances in the Physics of Stars—In Memory of Prof. Yuri N. Gnedin". *Universe* 2022, *8*, 239. https://doi.org/10.3390/universe8040239

Received: 30 March 2022
Accepted: 7 April 2022
Published: 13 April 2022

Publisher's Note: MDPI stays neutral with regard to jurisdictional claims in published maps and institutional affiliations.

This Special Issue collects articles devoted to various aspects of astrophysics which can be understood as a science investigating stars, galaxies, their types and properties, stages of their evolution, distribution in the Universe and the interstellar and intergalactic media. It is devoted to the memory of outstanding astrophysicist Prof. Dr. Yuri N. Gnedin whose scientific interests and expertise were extraordinarily wide by including the theoretical investigation of the polarized radiation transfer, generation of high-energy radiation in close binary star systems and galactic nuclei, and cyclotron lines in spectra of accreting neutron stars. Prof. Dr. Yuri N. Gnedin developed the pioneer method for the determination of magnetic fields of cosmic sources from polarimetric observations, contributed a great deal to physics of intermediate-mass and supermassive black holes and magnetic white dwarfs, supernovae, exoplanets and dark satellites of stars, cosmic gamma-ray bursts, dark matter and dark energy.

Yuri Gnedin was born on 13 August 1935 at the Russian city of Tula. After graduation from the Physico-Mechanical Department of Leningrad Polytechnic Institute (presently Peter the Great Saint Petersburg Polytechnic University) in 1959, he was employed by the famous A. F. Ioffe Physico-Technical Institute of the Russian Academy of Sciences, where he worked during 25 years in the field of theoretical physics and astrophysics. In 1966, Yu. N. Gnedin obtained his PhD and in 1979, after defending the Doctorate dissertation entitled "Propagation of polarized radiation in cosmic conditions", obtained the degree of Doctor of Physical and Mathematical Sciences (which is the second scientific degree in Russia similar to Habilitation in Germany which is significantly higher than PhD). In parallel with research activity at A. F. Ioffe Physico-Technical Institute, Yu. N. Gnedin delivered lectures in astrophysics for students of Leningrad Polytechnic Institute. In 1981, he obtained the academic status of Full Professor.

In 1984, Prof. Dr. Yu. Gnedin was invited to create the Department of Astrophysics and to lead the research work at Central Astronomical Observatory at Pulkovo of the Russian Academy of Sciences in the positions of the Head of Department and the Deputy Director of the Observatory on Science. It should be noted that Pulkovo Observatory was created in 1839. It is one of the oldest Institutions in Russia intended especially for scientific research. It is well known all over the world by many outstanding scientific results in the field of observational astronomy. However, by the end of 20th century, due to the great discoveries made in astrophysics and cosmology, astronomy already could not be further developed separately based mostly on the methods of mathematics. It should be mentioned, however, that in Russia astronomy was traditionally teached at the mathematical Departments of Universities and the most of members of the Pulkovo Observatory scientific staff graduated from the Department of Mathematics of Saint Petersburg (Leningrad) State University.

Because of this, the mission suggested to Prof. Gnedin was extremely complicated. He ought to become the part of old, well established and famous Institution, to create the new part of it consisting of researchers with quite a different background in physics and astrophysics, to organize the research activity in new scientific directions, and to convince everybody that the observational astronomy and astrophysics are not the hostile competitors but useful complements and assistants in obtaining new knowledge about our Universe. In order to cope with this task, one must be not only a brilliant scientist, but also an outstanding organizer. Prof. Dr. Gnedin accepted the offer and successfully solved all emerging scientific and organizational problems over the next more than 30 years.

Before reviewing the content of the present Special Issue, we briefly mention the research fields where important contributions by Prof. Dr. Yu. N. Gnedin are well known and recognized by the scientific community (in Figure 1 he is working in his office at Pulkovo Observatory [1]). These are the theory of radiation transfer, investigation of the polarized radiation emitted by various cosmic objects, X-rays and gamma-rays astronomy, physics of neutron stars and black holes. Specifically, Prof. Yu. N. Gnedin elaborated a new method for investigation of supermassive black holes based on polarimetric observations of active galactic nuclei. In collaboration with Prof. R. A. Sunyaev, Prof. Yu. N. Gnedin predicted an existence of cyclotron lines in the radiation of neutron stars. This discovery was used to develop the reliable method for measuring magnetic fields of neutron stars. Furthermore, Prof. Yu. N. Gnedin developed new method for measuring the magnetic fields of hot stars. This method is based on the phenomenon of Faraday rotation of the polarization plane of electromagnetic radiation scattered in their atmospheres. Prof. Yu. N. Gnedin initiated important investigations of several unusual cosmic phenomena using RT-32 radio telescopes. These include cosmic gamma-ray bursts, compact astrophysical objects, and active galactic nuclei. He also contributed a lot to constraining the parameters of axions and arions as possible constituents of dark matter by means of measuring different polarimetric effects in astrophysical processes. Prof. Yu. N. Gnedin is the author of 330 scientific papers and three books [2–4].

Figure 1. Prof. Dr. Yuri N. Gnedin (13 August 1935–27 March 2018) is working in his office at Pulkovo Observatory of the Russian Academy of Sciences.

An important part of Prof. Yu. N. Gnedin scientific life was his work as an outstanding organizer. He took part in many International research projects, was the member of International Astronomical Union, led the Program Committee of the 6-meter telescope of Special Astrophysical Observatory of the Russian Academy of Sciences, was the member of Bureau of the Scientific Council on Astronomy of the Russian Academy of Sciences. Prof. Yu. N. Gnedin was one of the main organizers and active participants of Alexander Friedmann International Seminars on Gravitation and Cosmology during the years 1993–2016. In honor of Prof. Yu. N. Gnedin, the small planet N 5084 is named "Gnedin".

Prof. Yu. N. Gnedin was a talented teacher whose lectures always created great interest among students. For several decades, he elaborated and delivered for students of the Peter the Great Saint Petersburg Polytechnic University the original lecture courses on modern problems of astrophysics, relativistic astrophysics, and general relativity theory, was the superwiser for many master's degree and PhD students.

Research articles and reviews published in this Special Issue, devoted to memory of Prof. Yu. N. Gnedin, touch many subjects of his wide scientific interests. Thus, article [5] outlines contributions of Prof. Gnedin to understanding polarization of radiation from neutron stars and presents new simulations of observations to be performed by NASA's Imaging X-ray Polarimetry Explorer (IXPE) launched in December 2021. These observations will provide the opportunity to test some models proposed by Prof. Gnedin and his collaborators.

Article [6] investigates unusual microwave and X-ray radiation from the ultracool dwarf TVLM 513-46546. According to the proposed model, the microwave radiation of this source comes from hundreds of magnetic loops quasi-uniformly distributed over its surface. This model suggests that the second population of magnetic loops is at the same time a source of X-ray emission.

Article [7] considers a new class of energetic transient sources called fast blue optical transients which possess moderately relativistic outflows in between the nonrelativistic and relativistic supernovae-related events. The kinetic particle-in-cell and Monte Carlo simulations are performed which allow to explain the observed broad band character of the radiation from these sources. Specifically, it is demonstrated that synchrotron radiation of accelerated relativistic electrons in the shock downstream may fit the observed radio fluxes. The suggested nonlinear Monte Carlo model predicts that protons and nuclei can be accelerated to petaelectronvolt energies and, therefore, the fast blue optical transients can be considered as one of the sources for galactic high energy cosmic rays.

In the article [8], the multi-component numerical model of hot Jupiter envelopes is suggested. By implementing the completed magnetohydrodynamic model of stellar wind, it becomes possible to calculate the structure of an extended envelope of hot Jupiter in both the super-Alfvén and sub-Alfvén regimes of the stellar wind flow as well in the trans-Alfvén regime. Using this approach, computations of changes in the chemical composition of hydrogen-helium envelopes of hot Jupiters are performed in different regimes. In particular, in the super-Alfvén flow regime, all the previously discovered types of extended gas-dynamic envelopes are reproduced. The dependence of the extended envelope on the magnitude of the wind magnetic field is investigated.

A new, interesting view on dark matter, dark energy and their relationship is suggested in the article [9]. According to this view, at the present epoch, the dark energy may have two components. The first of them is the standard one. It is determined by the cosmological constant in Einstein equations. Meanwhile, the second one, which is smaller in magnitude, is determined by remnants of the inflaton field which gives rise to the presently existing matter. In doing so, the second component of the dark energy becomes closely connected with the usual matter. Based on possible existence of the second component of dark energy, the so-called Hubble tension problem is discussed and its solution is suggested.

Thermal radiation spectra from a magnetized neutron star are further studied in [10] using the well known model assuming that the star is internally isothermal and possesses a dipole magnetic field in its outer layer. It is shown that the blackbody radiation emitted by any local surface element is nearly independent of the chemical composition of an envelope. The obtained theoretical results are found to be consistent with observations for certain neutron stars.

Constraints on the parameters of hypothetical interactions predicted by different extensions of the Standard Model following from the recent experiment on measuring the Casimir force at separations of a few micrometers are obtained in [11]. Specifically, the Yukawa-type corrections to Newton's law of gravitation and interactions of axions with nucleons are considered. As was mentioned above, both massive and massless axions (i.e., arions) were in the scope of scientific interests of Prof. Yu. N. Gnedin as the possible constituents of dark matter. It is shown that the obtained constraints are stronger than those following from other measurements of the Casimir interaction and gravitational experiments over certain interaction regions but weaker than the constraints obtained from the Casimir-less experiments where the Casimir force was completely nullified. The prospects of obtaining stronger constraints from experiments of different kinds are discussed.

The article [12] is devoted to estimations of the values of magnetic field at the event horizon of the supermassive black hole and the exponents of the power-law dependence of this magnetic field on the black hole radius. These estimations are obtained by using the polarimetric data of 33 Seyfert type 1 galaxies observed with the 6-meter telescope of Special Astrophysical Observatory of the Russian Academy of Sciences and the model of optically thick but geometrically thin Shakura-Sunyaev accretion disk. The obtained results are compared with those found by other methods.

A review on the progress in understanding of the first microquasar discovered in the Galaxy (SS433) is given in [13]. The stress is made on the results obtained by means of spectral and photometric observations at the Caucasian Mountain Observatory of the Sternberg Astronomical Institute of M. V. Lomonosov Moscow State University. According to common views, this mictoquasar is a massive eclipsing X-ray binary. The observations made at the Caucasian Mountain Observatory gave the possibility to determine the ellipticity of the microquasar orbit and to discover an increase in the orbital period of this system consisting of the relativistic object (black hole) and the optical star. Furthermore, a more precise constraint for the mass ratio of these objects was obtained.

The detection of magnetospheric accretion among Herbig Ae/Be stars of different kinds with accretion disks is reviewed in [14]. Specifically, the Herbig Ae star HD 101412 with a comparatively strong magnetic field, the early-type Herbig B6e star HD 259431 whose magnetosphere was recognized only recently, young binary system HD 104237 which includes a Herbig Ae star and a T Tauri star, and the Herbig Ae/Be star HD 37806. For the young binary system HD 104237, using the discovered periodic variations of equivalent widths of atmospheric lines in the spectrum of the primary, it is concluded that the star surface is spotted. The origin and nature of these spots are discussed.

A review of the most important results in investigation of magnetic fields of chemically peculiar stars obtained in the Special Astrophysical Observatory of the Russian Academy of Sciences using the 6-meter telescope is contained in the review [15]. Altogether more than 200 stars of this kind (i.e., more than 30% of their total number discovered elsewhere) were found at the Special Astrophysical Observatory. Objects with strong magnetic fields have been detected among stars with small rotation periods of years and decades. It was shown that for young chemically peculiar stars in Orion the probability of occurrence and strength of magnetic fields decreases sharply with age. An effective method of searching for magnetic stars was suggested.

The review [16] is devoted to the scattered radiation of circumstellar protoplanetary disks. In the majority of cases, this contribution to the total optical radiation of young stars is rather small. There are, however, two classes of stars in which the scattered radiation of circumstellar disks plays an important role. These are irregular variable stars with UX Ori as a typical example and highly embedded stars as well as stars with edge-on disks. The mechanism of variability of UX Ori and similar astrophysical objects was investigated using synchronous observations of linear polarization and brightness of their radiation. The highly embedded stars and stars with edge-on disks where examined by means of observations in the polarized light using the largest telescopes and a coronographic technique.

Although the articles included in this Special Issue cover a rather wide range of topics in modern astrophysics, they are incapable of embracing all the scope of scientific interests of Prof. Yu. N. Gnedin, whose breakthrough results and organizational and teaching activities were so important for the progress of this field of science.

Funding: N.R.I. acknowledges the financial support of the Ministry of Science and Higher Education (grant no. 075-15-2020-780 "Exoplanets-4"). The work of G.L.K. and V.M.M. was supported by the Peter the Great Saint Petersburg Polytechnic University in the framework of the Russian state assignment for basic research (Project N FSEG-2020-0024). The work of V.M.M. was also supported by the Kazan Federal University Strategic Academic Leadership Program.

Conflicts of Interest: The authors declare no conflict of interest.

References

1. Yuri Gnedin: "We Live in the Epoch of Revolution in Physics". Available online: https://naked-science.ru/article/interview/person-yurii-gnedin (accessed on 1 March 2022).
2. Dolginov, A.Z.; Gnedin, Y.N.; Silant'ev, N.A. *Propagation and Polarization of Radiation in Cosmic Media*; Gordon and Breach: Williston, ND, USA, 1995.
3. Gnedin, Y.N.; Silant'ev, N.A. *Basic Mechanism of Light Polarization in Cosmic Media*; Taylor & Francis: Cleveland, OH, USA, 1997.
4. Gnedin, Y.N.; Natsvlishvili, T.M. *Magnetic Fields of Stars: The Interaction Between Observations and Theory*; Harwood Academic Publishers: Reading, UK, 2000.
5. Heyl, J. Remembering Yury N. Gnedin at the Dawn of X-ray Polarimetry: Predictions of IXPE Observations of Neutron Stars. *Universe* **2022**, *8*, 84. [CrossRef]
6. Stepanov, A.; Zaitsev, V. On the Origin of Persistent Radio and X-ray Emission from Brown Dwarf TVLM 513-46546. *Universe* **2022**, *8*, 77. [CrossRef]
7. Bykov, A.; Romansky, V.; Osipov, S. Particle Acceleration in Mildly Relativistic Outflows of Fast Energetic Transient Sources. *Universe* **2022**, *8*, 32. [CrossRef]
8. Zhilkin, A.; Bisikalo, D. Multi-Component MHD Model of Hot Jupiter Envelopes. *Universe* **2021**, *7*, 422. [CrossRef]
9. Bisnovatyi-Kogan, G.S. Cosmological Model with Interconnection between Dark Energy and Matter. *Universe* **2021**, *7*, 412. [CrossRef]
10. Yakovlev, D. A Simple Model of Radiation from a Magnetized Neutron Star: Accreted Matter and Polar Hotspots. *Universe* **2021**, *7*, 395. [CrossRef]
11. Klimchitskaya, G.L.; Mostepanenko, V.M. Dark Matter Axions, Non-Newtonian Gravity and Constraints on Them from Recent Measurements of the Casimir Force in the Micrometer Separation Range. *Universe* **2021**, *7*, 343. [CrossRef]
12. Piotrovich, M.; Buliga, S.; Natsvlishvili, T. Determination of the Magnetic Field Strength and Geometry in the Accretion Disks of AGNs by Optical Spectropolarimetry. *Universe* **2021**, *7*, 202. [CrossRef]
13. Cherepashchuk, A. Progress in Understanding the Nature of SS433. *Universe* **2022**, *8*, 13. [CrossRef]
14. Pogodin, M.; Drake, N.; Beskrovnaya, N.; Pavlovskiy, S.; Hubrig, S.; Schöller, M.; Järvinen, S.; Kozlova, O.; Alekseev, I. Searching for Magnetospheres around Herbig Ae/Be Stars. *Universe* **2021**, *7*, 489. [CrossRef]
15. Romanyuk, I. Studies of Magnetic Chemically Peculiar Stars Using the 6-m Telescope at SAO RAS. *Universe* **2021**, *7*, 465. [CrossRef]
16. Grinin, V.P.; Tambovtseva, L.V. Scattered radiation of protoplanetary disks. *Universe* **2022**, *8*, 224. [CrossRef]

Article

Remembering Yury N. Gnedin at the Dawn of X-ray Polarimetry: Predictions of IXPE Observations of Neutron Stars

Jeremy Heyl

Department of Physics and Astronomy, University of British Columbia, Vancouver, BC V6T 1Z1, Canada; heyl@phas.ubc.ca

Abstract: NASA's Imaging X-ray Polarimetry Explorer (IXPE) was launched in December 2021. It is 100 times more sensitive to polarized X-ray emission than any preceding mission and it is opening a new observational window into high-energy astrophysics. I outline Yury N. Gnedin's many contributions to understanding polarization from neutron stars and present new simulations of observations that IXPE will perform of the X-ray pulsar Hercules X-1 and the magnetar 4U 0141+561 in February 2022. These observations highlight and test particular models that Gnedin and collaborators first proposed. I outline how IXPE will provide unique constraints on the structure and kinematics of the boundary region between the accretion flow and the neutron star surface of Hercules X-1 and how IXPE will verify the predictions of vacuum birefringence for the magnetar 4U 0142+561.

Keywords: magnetars stars; X-rays and stars; star atmospheres; polarization; plasmas; scattering; radiative transfer

Citation: Heyl, J. Remembering Yury N. Gnedin at the Dawn of X-ray Polarimetry: Predictions of IXPE Observations of Neutron Stars. *Universe* **2022**, *8*, 84. https://doi.org/ 10.3390/universe8020084

Academic Editors: Galina L. Klimchitskaya, Vladimir M. Mostepanenko and Nazar R. Ikhsanov

Received: 31 December 2021
Accepted: 21 January 2022
Published: 28 January 2022

Publisher's Note: MDPI stays neutral with regard to jurisdictional claims in published maps and institutional affiliations.

1. Introduction

I had the privilege of working with Yury Nikolaesh Gnedin (1935–2018) during the summer of 1992 at the Pulkovo Observatory. That summer, several students from the United States were visiting Pulkovo, but I was the only one who spoke Russian, so I acted as cruise director; further, I conducted a research project with Yury Nikolaesh on the compactness problem in AGN and, more fortunately for me, I got to know him well doing that short time. Working with him was always fun as we tried to unlock mysteries and he also tried to make my visit rewarding beyond science, during what was a very trying time for Russia and Pulkovo. He arranged for me to visit the astronomical outpost overlooking Urtsalanj, Armenia (Figure 1). As our group was waiting on the tarmac at Mineralnye Vody Airport until well after midnight to fly over Georgia into Yerevan, supposedly to avoid anti-aircraft fire, I realized that, if Yury Nikolaesh had explained in detail what the trip would entail, I might have chosen not to do it and the story of the adventure that followed in Armenia itself would easily fill a different sort of article.

Knowing Yury Nikolaesh fueled a sense of adventure both scientifically and otherwise that has remained with me since, but, beyond that, the project that I conducted with him was my first one in high-energy astrophysics and, in particular, astrophysical manifestations of QED, which is a theme that my career has followed since. In fact, I find that, throughout my career, I have been retracing much of his work, more so than that of any of my other mentors. Now, as I write this nearly thirty years later, just a few weeks after the launch of NASA's Imaging X-ray Polarimetry Explorer (IXPE), we are waiting with keen anticipation the observations of so many phenomena in X-ray polarization, which Yury N. Gnedin began to outline more than fifty years ago.

Figure 1. Top: wide view of the Ararat Scientific Site (taken by the author in August 1992). **Bottom**: inside cover of *150 Years of the Pulkovo Observatory*, a souvenir from 1992.

2. Polarized X-rays from Neutron Stars

Although Yury N. Gnedin started by working on comets, e.g., [1–3], the discovery of neutron stars and the beginnings of high-energy astrophysics quickly dominated his early work. I here focus on the polarized X-ray emission from accreting and isolated neutron stars and how Gnedin's work informs our expectations even today. On 9 December 2021, the IXPE observatory was launched from Cape Canaveral, Florida. In February 2022, IXPE will be almost completely devoted to the study of two objects, the X-ray pulsar Hercules X-1 and the magnetar 4U 0141+561. Gnedin devoted significant studies to the first object, e.g., [4–6], and the implications of his studies, e.g., [7–11] of vacuum polarization play an important role in our understanding of the second object.

The IXPE observatory measures the polarization of X-rays in 2–8 keV by measuring the direction of the initial photoelectron produced upon the absorption of an X-ray in a gas target. The total effective area of the three-mirror arrays is about $600\,\text{cm}^2$. Once the quantum efficiency of the detector is included, the total effective area is decreased to about $75\,\text{cm}^2$ at 2 keV and $8\,\text{cm}^2$ at 7 keV [12]. The interaction between an X-ray photon and the electron means that the direction of the photoelectron will be correlated with the polarization of the photon with a $\sin^2\theta$ dependence. The magnitude of the correlation for fully linearly polarized radiation is known as the modulation factor and, for IXPE, it ranges from 15% at 2 keV to 60% at 8 keV [12]; therefore, although the effective area drops quickly with the increase in energy, the increase in the modulation factor mitigates this for the detection of polarization. IXPE is not sensitive to circular polarization. The energy resolution of the detectors is typically about 20%, the angular resolution is about 20 arcseconds and the time resolution is one microsecond. All told, these specifications make IXPE a factor of 100 more capable than any X-ray polarimeter in space before it. The discovery potential is vast.

In the following sections, I describe what to expect from the first observations of X-ray polarization from Hercules X-1 and 4U 0142+561 in the context of two particular models for the emission from these sources. Although I discuss some other models, this is not meant to be an exhaustive review but rather a review of how these expectations are connected with Yury N. Gnedin's work; consequently, much of the literature remains uncited. The models that I discuss have been presented elsewhere [13,14], but the simulations for the planned

February 2022 IXPE observations of 4U 0142+561 (1Ms) and Hercules X-1 (about 400 ks in total) are presented here for the first time.

3. X-ray Pulsar Hercules X-1

The discovery of the pulsating X-ray sources Centaurus X-3 [15] and Hercules X-1 [16] with the Uhuru satellite in the early 1970s uncovered a new type of celestial object, the accreting X-ray pulsar. Gnedin and Sunyaev [4] soon thereafter outlined the key features of a model of X-ray pulsars that hold to this day. The magnetic field channels the in-falling material onto the polar regions with a typical radius one-tenth that of the neutron star. The mildly relativistic material slows dramatically through a strong shock that radiates the gravitational energy away. They argued that the radiation came from a superposition of high gyrocyclotron resonances with a net spectrum that resembled that of thermal bremsstrahlung. Furthermore, they argued that the radiation would be polarized and beamed perpendicular to the magnetic field [17], a knife beam. The magnetic field of Hercules X-1 was measured through a cyclotron feature at about 60 keV [18] to be about 100 times stronger than assumed by Gnedin and Sunyaev. Since the bulk of the emission lies below the cyclotron energy, the gyrocyclotron model of Gnedin and Sunyaev [4] does not hold. However, they argued, in a subsequent paper [17], that, if the bulk of the emission lies below the cyclotron resonance and the accretion shock lies close to the surface, the small opacity for photons traveling along the magnetic field results in a pencil beam along the field direction. This scenario is the basis for the slab models for X-ray polarization [19,20].

Additionally, because the magnetic field is stronger than that assumed in [4], the accretion flow is channeled into a much smaller spot (about 0.1 km vs. 1 km) and, for sufficiently high accretion rates, the flow piles up onto the surface in an accretion column (see Figure 2), where the accretion shock may lie well above the surface. In this situation, modern models for Hercules X-1, such as that of Becker and Wolff [21], find that the photon production is dominated by magnetic bremsstrahlung and photons scatter many times, gaining energy from the electron flow and ultimately emerge from the walls of the column, i.e., a knife beam (with polarization modeled for a static column in [22,23]). These models generally predicted a polarized fraction of about ten percent for the pencil beam and up to fifty percent for the knife beam in the range of 2–8 keV.

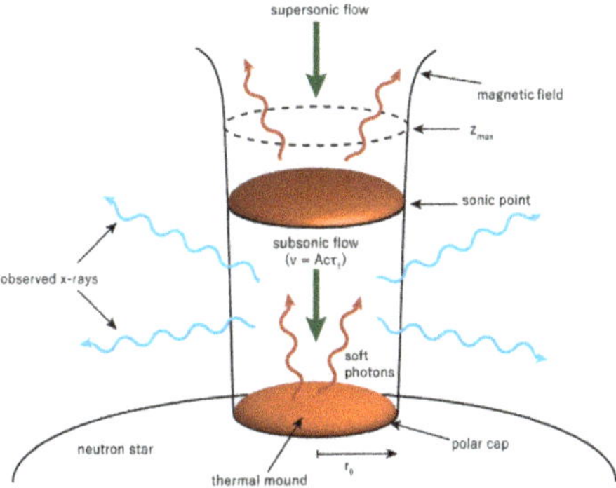

Figure 2. Schematic of the accretion flow near the surface of Her X-1; adapted from [21,24].

Although these models did not fit the broadband spectral energy distribution of X-ray pulsars such as Hercules X-1, it was nearly thirty-five years before new models for the polarization were developed [13,24]. Caiazzo and Heyl followed the spirit of Gnedin and Sunyaev [6,17] and focused on the scattering process itself to understand both the polarization and directional pattern of the emerging radiation. They used a matrix formalism [25] to find the asymptotic polarization state and radiation pattern within the plasma after many scatterings and study the final scattering before escape to derive the final polarization and emission pattern. To determine the final observed polarization and pulse profile, they included gravitational lensing, Doppler beaming and vacuum birefringence.

The results of two models that account for the observed pulse profile and cyclotron line of Hercules X-1 are depicted in Figure 3. The geometry of the two models is quite different. The model in the left-hand panels consists of emission from a single accretion column. The angle between the accretion column and the rotation axis of the neutron star is 86 degrees and the angle between the rotation axis and the line of sight is 83 degrees. It is nearly an orthogonal rotator and the bulk of the emission comes when the column is pointing away from the observer. Because the column is tall (seven kilometers) and the emission is beamed toward the surface, gravitational lensing dramatically amplifies emission in this configuration. The right-hand panels depict a geometry with two accretion columns. The columns make an angle of 42 degrees with respect to the rotation axis and the line of sight makes an angle of 50 degrees with respect to the line of sight. Again, the pulse is centered at the time when one of the accretion columns is pointing nearly directly away from the observer. The relativistic effects turn the typical rules for knife and pencil beams around. Pencil beams typically exhibit a minimum in the polarized fraction in the pulse accompanied with a rapid change in the polarization angle; on the other hand, a knife beam typically exhibits a peak in the polarized fraction and a slow change in the polarization angle during the pulse, e.g., [26]. The models presented in Figure 3 show a combination of these features. The polarized fraction reaches a peak in the pulse as for a non-relativistic knife beam, but the rate of change in the polarization angle is rapid through the pulse as in a pencil beam.

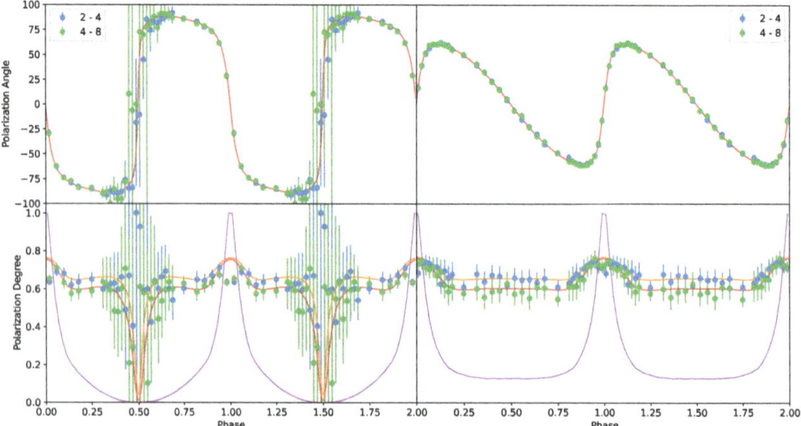

Figure 3. Predicted polarization of Her X-1 from the models of [13] with a 70 ks observation with IXPE. The purple curves in the lower panels depict the total counts from the accretion column in the model. The left panels show the result with a single accretion column and the right panels depict two accretion columns. The large error bars near phase 0.5 in the left panel result from the lack of flux from the accretion column at this phase. The slight discrepancy near phase 1.0 between the measured and model polarization fraction results from smearing the polarization direction over the bin and can be avoided with an unbinned analysis.

Figure 3 also depicts the anticipated observations and uncertainties for a 70 ks observation of Hercules X-1. In fact, IXPE will observe Hercules X-1 for a total of about 400 ks in February 2022 over six visits to span its 35-day precession period. The simulation depicts the results for a single visit. The instrument will be able to determine whether the relativistic model [13] accounts for the observations. It will be also able to verify the Becker and Wolff picture [21]. The results of the ixpeobssim simulation [27] are for a phase-binned calculation, so the rapid variation in the polarization angle through the pulse reduces the observed polarization in the center of the pulse for both geometries. However, because IXPE tags the photon arrivals with an accuracy of 1 microsecond, it will be straightforward to determine the polarization fraction reliably, even when the angle changes rapidly. The combination of a rapidly varying angle and a maximum in both the count rate and the polarization fraction is the unique signature of the model. A static accretion slab or column yields different signatures, as discussed above.

4. The Magnetar 4U 0142+561

Although Yury N. Gnedin only wrote a single paper about a magnetar per se [28], his work with George Pavlov, among others, outlined many of the crucial physical processes at play for the production and propagation of radiation for magnetars. In particular, he examined the interactions between photons and atoms in strong magnetic fields and highlighted that, even at the high temperatures of neutron star atmospheres, the material may have a significant neutral fraction [29]. Both this and the discussions on scattering in strong magnetic fields [6], when combined with the equations of radiative transfer in a strong magnetic field [7], form the basis for calculations of neutron star atmospheres, e.g., [30–34].

Additionally, Yury N. Gnedin and colleagues outlined the importance of vacuum birefringence from QED in the observations of neutron stars, e.g., [8–10], and, in particular, the spectral features that it can produce [11]. Following the spirit of the paper so far, the focus is X-ray polarization. Most important, for our discussion, is the argument of Novick et al. [35] whereby vacuum birefringence would generate a large phase difference between the normal modes and significant depolarization. Novick et al. wrote, "We conclude that it is extremely unlikely that polarized X-ray emission which arose from the surface of a magnetic neutron star would be observed". Gnedin et al. [9] countered the argument of Novick et al. by pointing out that a large phase retardation between the normal modes is a necessary but not sufficient condition for depolarization and, in fact, if the emission consists of an incoherent superposition of normal waves, no depolarization occurs, unless the properties of the birefringence vary too quickly. Birefringence would preserve the polarization along the path of the radiation. Furthermore, later work demonstrated that vacuum birefringence causes the direction of polarization in the X-rays for typical neutron stars to follow the direction of the magnetic field (the radiation remains in one of the normal modes) until a distance far from the surface of the neutron star, the polarization-limiting radius [36,37]. In particular, this QED effect makes the observed polarization many times larger than one would expect from emission coming from a large portion of the neutron star surface [38]. The result of this effect is shown in Figure 4. The left-hand panel shows the polarization map at the surface of the star, in particular, the direction of the polarization of the extraordinary mode. Across the stellar surface, the direction varies significantly, so, if one were to sum the total polarized radiation in this case, one would obtain a small polarized fraction. The right panel depicts the situation with vacuum birefringence. At a large distance from the surface of the star, the magnetic field follows a dipole pattern; its projection in the plane of the sky would be parallel to the magnetic axis, in this case, horizontal. The polarization vectors that still follow the extraordinary mode are all vertical and the total polarization integrated over the surface of the star can be large if the polarization fraction at emission is large.

Caiazzo et al. [14] presented a comprehensive suite of models to account for the broadband emission from the magnetar 4U 0142+561 [39]. In all cases, they assumed that vacuum birefringence existed at the level predicted by QED. Figure 5 shows the results for a particular model that assumes that all of the emission from 2 to 8 keV originates from a fully ionized hydrogen atmosphere [30], where the temperature and magnetic field follow a dipole distribution [40] with an additional polar hot spot. The figure shows a simulation of the planned 1 Ms IXPE observations for February 2022. In principle, other processes may contribute to this energy range, especially resonant cyclotron scattering [41–43]; however, for clarity, these are neglected here. With vacuum birefringence, the observed polarization fraction is high (nearly 100%) over the entire energy band. On the other hand, without vacuum birefringence, the observed polarized fraction is low at the low end of the IXPE band when the emission originates over the entire surface, so, from the left panel of Figure 4, the integrated polarization is small. At the high end of the IXPE band, the radiation comes from a very small portion of the stellar surface (the polar hot spot). Over this small region, the magnetic field in the plane of the sky is aligned in the horizontal direction; therefore, the resulting polarization is nearly vertical (i.e., in the extraordinary mode).

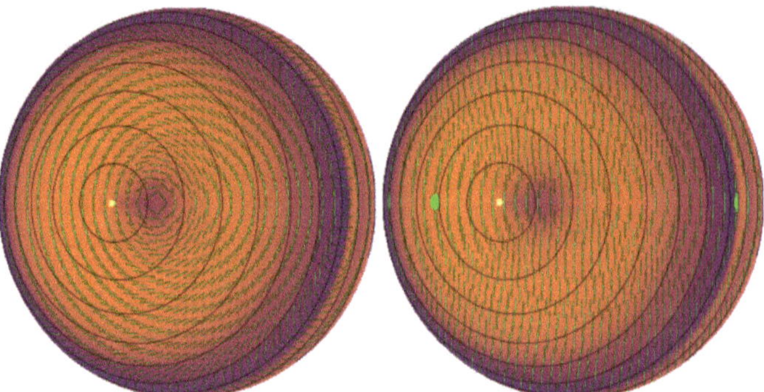

Figure 4. Surface polarization and emission map for the atmospheric model of 4U 0142+561; from [14].

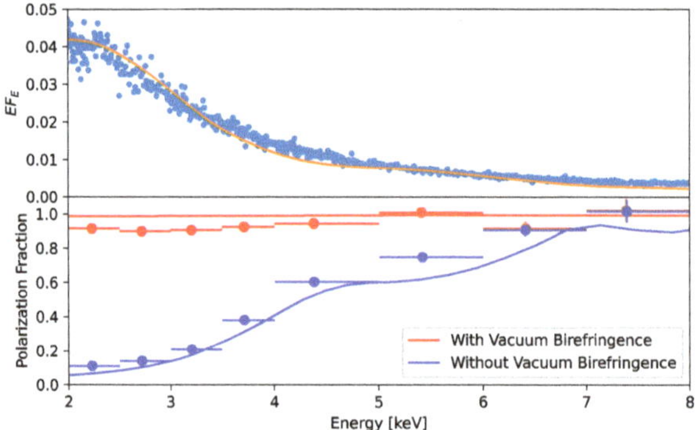

Figure 5. Predicted mean phase-resolved polarization of 4U 0142+61 and phase-averaged flux [39] from the atmospheric models of [14] with a simulated 1 Ms observation with IXPE.

If the emission in the low end of the IXPE band originates in a gaseous atmosphere, the effect of vacuum polarization will be dramatically verified by IXPE observations. On the other hand, if the surface is condensed, e.g., [44,45], the expected polarization will be small at the low end of the IXPE band [14,46]. However, at the high-energy end of the IXPE band, resonant cyclotron scattering begins to contribute, resulting in a larger polarization and another potential verification of vacuum birefringence.

5. Discussion

In February 2022, the observations of X-ray polarization from Hercules X-1 and 4U 0142+61 with the recent launched X-ray observatory IXPE will verify the models and predictions that Yury N. Gnedin made at the dawn of high-energy astrophysics. Beyond these predictions and models for particular sources, Yury N. Gnedin, with collaborators, developed the framework of radiative transfer in a strongly magnetized plasma where vacuum birefringence, incoherent scattering and absorption all play key roles that are crucial to understanding polarized X-ray emission, regardless of the source. The broad scientific goals of IXPE from AGN to axions dovetail with Yury N. Gnedin's many contributions to astrophysics over the years, e.g., [47,48]. As we venture into the undiscovered country of astrophysical X-ray polarization, let us remember the maps and tools that he gave us for the journey.

Funding: This research study was funded by NSERC Canada.

Data Availability Statement: Data are available upon request to the Author.

Acknowledgments: The calculations were performed on SciServer; the author would like to thank Ilaria Caiazzo and Denis González Caniulef for useful conversations.

Conflicts of Interest: The author declares no conflict of interest.

References

1. Gnedin, Y.N.; Dolginov, A.Z. Particle Distribution in a Comet Head. *Sov. Astron.* **1966**, *10*, 143.
2. Dolginov, A.Z.; Gnedin, Y.N. A theory of the atmosphere of a comet. *Icarus* **1966**, *5*, 64–74. [CrossRef]
3. Gnedin, Y.N. The Inverse Problem in the Theory of Cometary Forms. *Sov. Astron.* **1967**, *11*, 485.
4. Gnedin, Y.N.; Sunyaev, R.A. The Beaming of Radiation from an Accreting Magnetic Neutron Star and the X-ray Pulsars. *Astron. Astrophys.* **1973**, *25*, 233.
5. Gnedin, Y.N.; Sunyaev, R.A. Luminosity of thermal X-ray sources witha strong magnetic field. *Mon. Not. R. Astron. Soc.* **1973**, *162*, 53. [CrossRef]
6. Gnedin, Y.N.; Sunyaev, R.A. Scattering of radiation by thermal electrons in a magnetic field. *Sov. J. Exp. Theor. Phys.* **1974**, *38*, 51.
7. Gnedin, Y.N.; Pavlov, G.G. The transfer equations for normal waves and radiation polarization in an anisotropic medium. *Sov. J. Exp. Theor. Phys.* **1974**, *38*, 903–908.
8. Gnedin, Y.N.; Pavlov, G.G.; Shibanov, Y.A. Influence of polarization of vacuum in a magnetic field on the propagation of radiation in a plasma. *Sov. J. Exp. Theor. Phys. Lett.* **1978**, *27*, 305.
9. Gnedin, Y.N.; Pavlov, G.G.; Shibanov, Y.A. The effect of vacuum birefringence in a magnetic field on the polarization and beaming of X-ray pulsars. *Sov. Astron. Lett.* **1978**, *4*, 117–119.
10. Pavlov, G.G.; Gnedin, Y.N. Vacuum polarization by a magnetic field and its astrophysical appearances. In *Peculiar Stars, Strong Magnetic Fields and Gamma-ray Astronomy*; Syunyaev, R.A., Ed.; All-Union Institute of Scientific and Technical Information: Moscow, Russia, 1983; pp. 172–219.
11. Gnedin, I.N. Effect of vacuum polarization in a strong magnetic field and spectral features of X-ray source emission. In *IAU Colloq. 115: High Resolution X-ray Spectroscopy of Cosmic Plasmas*; Gorenstein, P., Zombeck, M., Eds.; Cambridge University Press: Cambridge, UK, 1990; pp. 78–84.
12. Ratheesh, A.; Rubini, A.; Marscher, A.; Manfreda, A.; Marrocchesi, A.; Brez, A.; Di Marco, A.; Paggi, A.; Profeti, A.; Nuti, A.; et al. The Imaging X-ray Polarimetry Explorer (IXPE): Pre-Launch. *arXiv* **2021**, arXiv:2112.01269.
13. Caiazzo, I.; Heyl, J. Polarization of accreting X-ray pulsars-II. Hercules X-1. *Mon. Not. R. Astron. Soc.* **2021**, *501*, 129–136. [CrossRef]
14. Caiazzo, I.; González-Caniulef, D.; Heyl, J.; Fernández, R. Probing magnetar emission mechanisms with spectropolarimetry. *arXiv* **2021**, arXiv:2112.03401.
15. Giacconi, R.; Gursky, H.; Kellogg, E.; Schreier, E.; Tananbaum, H. Discovery of Periodic X-ray Pulsations in Centaurus X-3 from UHURU. *Astrophys. J.* **1971**, *167*, L67. [CrossRef]
16. Tananbaum, H.; Gursky, H.; Kellogg, E.M.; Levinson, R.; Schreier, E.; Giacconi, R. Discovery of a Periodic Pulsating Binary X-ray Source in Hercules from UHURU. *Astrophys. J.* **1972**, *174*, L143. [CrossRef]
17. Gnedin, I.N.; Sunyaev, R.A. Polarization of optical and X-radiation from compact thermal sources with magnetic field. *Astron. Astrophys.* **1974**, *36*, 379–394.
18. Truemper, J.; Pietsch, W.; Reppin, C.; Voges, W.; Staubert, R.; Kendziorra, E. Evidence for strong cyclotron line emission in the hard X-ray spectrum of Hercules X-1. *Astrophys. J.* **1978**, *219*, L105–L110. [CrossRef]
19. Mészáros, P.; Nagel, W. X-ray pulsar models. I-Angle-dependent cyclotron line formation and comptonization. *Astrophys. J.* **1985**, *298*, 147–160. [CrossRef]
20. Kii, T.; Hayakawa, S.; Nagase, F.; Ikegami, T.; Kawai, N. Anisotropic X-ray transfer in a strongly magnetized plasma of the X-ray pulsar 4U 1626-67. *Publ. Astron. Soc. Jpn.* **1986**, *38*, 751–774.
21. Becker, P.A.; Wolff, M.T. Thermal and Bulk Comptonization in Accretion-powered X-ray Pulsars. *Astrophys. J.* **2007**, *654*, 435–457. [CrossRef]
22. Nagel, W. Radiative Transfer in a Strongly Magnetized Plasma—Part Two—Effects of Comptonization. *Astrophys. J.* **1981**, *251*, 288. [CrossRef]

23. Kii, T. X-ray polarizations from accreting strongly magnetized neutron stars—Case studies for the X-ray pulsars 4U 1626-67 and Hercules X-1. *Publ. Astron. Soc. Jpn.* **1987**, *39*, 781–800.
24. Caiazzo, I.; Heyl, J. Polarization of accreting X-ray pulsars. I. A new model. *Mon. Not. R. Astron. Soc.* **2021**, *501*, 109–128. [CrossRef]
25. Chou, C.K. Stokes Parameters for Thomson Scattering in a Strong Magnetic Field. *Astrophys. Space Sci.* **1986**, *121*, 333–344. [CrossRef]
26. Meszaros, P.; Novick, R.; Szentgyorgyi, A.; Chanan, G.A.; Weisskopf, M.C. Astrophysical implications and observational prospects of X-ray polarimetry. *Astrophys. J.* **1988**, *324*, 1056–1067. [CrossRef]
27. Pesce-Rollins, M.; Lalla, N.D.; Omodei, N.; Baldini, L. An observation-simulation and analysis framework for the Imaging X-ray Polarimetry Explorer (IXPE). *Nucl. Instrum. Methods Phys. Res. A* **2019**, *936*, 224–226. [CrossRef]
28. Gnedin, Y.N.; Ipatov, A.V.; Piotrovich, M.Y.; Finkel'Shtein, A.M.; Kharinov, M.A. Radio emission of the magnetar SGR 1806-20: Evolution of the magnetic field in the region of the radio afterglow. *Astron. Rep.* **2007**, *51*, 863–868. [CrossRef]
29. Gnedin, Y.N.; Pavlov, G.G.; Tsygan, A.I. Photoeffect in strong magnetic fields and X-ray emission from neutron stars. *Sov. J. Exp. Theor. Phys.* **1974**, *39*, 201.
30. Lloyd, D.A. Model atmospheres and thermal spectra of magnetized neutron stars. *arXiv* **2003**, arXiv:0303561.
31. Lloyd, D.A.; Perna, R.; Slane, P.; Nicastro, F.; Hernquist, L. A pulsar-atmosphere model for PSR 0656+14. *arXiv* **2003**, arXiv:0306235.
32. Lloyd, D.A.; Hernquist, L.; Heyl, J.S. Optical and X-ray Properties of Cooling Neutron Stars. *Astrophys. J.* **2003**, *593*, 1024–1031. [CrossRef]
33. Özel, F. Surface Emission Properties of Strongly Magnetic Neutron Stars. *Astrophys. J.* **2001**, *563*, 276–288. [CrossRef]
34. Ho, W.C.G.; Lai, D. Atmospheres and spectra of strongly magnetized neutron stars. *Mon. Not. R. Astron. Soc.* **2001**, *327*, 1081–1096. [CrossRef]
35. Novick, R.; Weisskopf, M.C.; Angel, J.R.P.; Sutherland, P.G. The effect of vacuum birefringence on the polarization of X-ray binaries and pulsars. *Astrophys. J.* **1977**, *215*, L117–L120. [CrossRef]
36. Shaviv, N.J.; Heyl, J.S.; Lithwick, Y. Magnetic lensing near ultramagnetized neutron stars. *Mon. Not. R. Astron. Soc.* **1999**, *306*, 333–347. [CrossRef]
37. Heyl, J.S.; Shaviv, N.J. Polarization evolution in strong magnetic fields. *Mon. Not. R. Astron. Soc.* **2000**, *311*, 555–564. [CrossRef]
38. Heyl, J.S.; Shaviv, N.J. QED and the high polarization of the thermal radiation from neutron stars. *Phys. Rev. D* **2002**, *66*, 023002. [CrossRef]
39. Tendulkar, S.P.; Hascöet, R.; Yang, C.; Kaspi, V.M.; Beloborodov, A.M.; An, H.; Bachetti, M.; Boggs, S.E.; Christensen, F.E.; Craig, W.W.; et al. Phase-resolved NuSTAR and Swift-XRT Observations of Magnetar 4U 0142+61. *Astrophys. J.* **2015**, *808*, 32. [CrossRef]
40. Heyl, J.S.; Hernquist, L. Almost analytic models of ultramagnetized neutron star envelopes. *Mon. Not. R. Astron. Soc.* **1998**, *300*, 599–615. [CrossRef]
41. Nobili, L.; Turolla, R.; Zane, S. X-ray spectra from magnetar candidates—I. Monte Carlo simulations in the non-relativistic regime. *Mon. Not. R. Astron. Soc.* **2008**, *386*, 1527–1542. [CrossRef]
42. Fernández, R.; Thompson, C. Resonant Cyclotron Scattering in Three Dimensions and the Quiescent Nonthermal X-ray Emission of Magnetars. *Astrophys. J.* **2007**, *660*, 615–640. [CrossRef]
43. Fernández, R.; Davis, S.W. The X-ray Polarization Signature of Quiescent Magnetars: Effect of Magnetospheric Scattering and Vacuum Polarization. *Astrophys. J.* **2011**, *730*, 131. [CrossRef]
44. van Adelsberg, M.; Lai, D.; Potekhin, A.Y.; Arras, P. Radiation from Condensed Surface of Magnetic Neutron Stars. *Astrophys. J.* **2005**, *628*, 902–913. [CrossRef]
45. Medin, Z.; Lai, D. Condensed surfaces of magnetic neutron stars, thermal surface emission, and particle acceleration above pulsar polar caps. *Mon. Not. R. Astron. Soc.* **2007**, *382*, 1833–1852. [CrossRef]
46. González Caniulef, D.; Zane, S.; Taverna, R.; Turolla, R.; Wu, K. Polarized thermal emission from X-ray dim isolated neutron stars: the case of RX J1856.5-3754. *Mon. Not. R. Astron. Soc.* **2016**, *459*, 3585–3595. [CrossRef]
47. Piotrovich, M.Y.; Gnedin, Y.N.; Buliga, S.D.; Natsvlishvili, T.M. Wavelength dependence of polarization and physical mechanisms of magnetic field generation in accretion disks around supermassive black holes in active galactic nuclei. *Astron. Lett.* **2014**, *40*, 459–463. [CrossRef]
48. Gnedin, Y.N.; Piotrovich, M.Y. New results in searching for axions by astronomical methods. *Int. J. Mod. Phys. A* **2016**, *31*, 1641019. [CrossRef]

Article

On the Origin of Persistent Radio and X-ray Emission from Brown Dwarf TVLM 513-46546

Alexander Stepanov [1,2,*] and **Valery Zaitsev [3]**

[1] Pulkovo Observatory, 196140 St. Petersburg, Russia
[2] Ioffe Institute, 194021 St. Petersburg, Russia
[3] Institute of Applied Physics, 603950 Nizhny Novgorod, Russia; za130@ipfran.ru
* Correspondence: astep44@mail.ru

Abstract: We study the origin of unusually persistent microwave and X-ray radiation from the ultracool dwarf TVLM 513-46546. It is shown that the source of ≈ 1 keV X-ray emission is not the entire corona of the brown dwarf, but a population of several hundreds of coronal magnetic loops, with 10 MK plasma heated upon dissipation of the electric current generated by the photospheric convection. Unlike models, which assume a large-scale magnetic structure of the microwave source, our model suggests that the microwave radiation comes from hundreds of magnetic loops quasi-uniformly distributed over the dwarf's surface. We propose a long-term operating mechanism of acceleration of electrons generating gyrosynchrotron radio emission caused by oscillations of electric current in the magnetic loops as an equivalent RLC circuit. The second population of magnetic loops—the sources of microwave radiation, is at the same time a source of softer (≈ 0.2 keV) X-ray emission.

Keywords: brown dwarf; X-ray emission; microwave radiation; magnetic loops; particle acceleration

Citation: Stepanov, A.; Zaitsev, V. On the Origin of Persistent Radio and X-ray Emission from Brown Dwarf TVLM 513-46546. *Universe* **2022**, *8*, 77. https://doi.org/10.3390/universe8020077

Academic Editors: Galina L. Klimchitskaya, Vladimir M. Mostepanenko and Nazar R. Ikhsanov

Received: 6 December 2021
Accepted: 25 January 2022
Published: 27 January 2022

Publisher's Note: MDPI stays neutral with regard to jurisdictional claims in published maps and institutional affiliations.

1. Introduction

The brown dwarf TVLM 513-46546 is currently one of the most studied among ultracool stars of spectral types M > 7, L, T, and Y. Radio, X-ray, ultraviolet, and optical spectroscopic observations have revealed rather unusual properties of TVLM 513-46546 radiation (see, for example, [1–3]). In addition to the surprising radio to X-ray luminosity ratio ($L_R/L_X \approx 3.3 \times 10^{-12}$ Hz^{-1}), by a factor of 10^4 larger than expected [2,4], a persistent (on a multiyear timescale) quiescent nonthermal radio emission from the brown dwarf TVLM 513-46546 is observed, which assumes a continuous source of accelerated electrons. The persistent quiescent radio emission is accompanied by long-term X-ray emission. By now, in low-mass dwarf stars, two thermal components of X-ray emission are observed: the "soft" one, with the temperature $T \approx (1–6) \times 10^6$ K, and the "hard" one, corresponding to the source temperature $T \approx 10^7$ K [5].

The current models of the microwave [1] and multiwavelength radiation sources in TVLM 513-46546 [2] provide no explanation for the nature of the accelerator of electrons producing the quiescent nonthermal gyrosynchrotron emission for several years. Moreover, the microwave radiation source was represented either by the dipole magnetic field of the dwarf [1] or as a large-scale magnetic structure with a "covering factor" of about 50% [2].

Here, we show that the sources of radio and X-ray emission of TVLM 513-46546 consist of a few $\times 100$ coronal magnetic loops quasi-uniformly distributed over the dwarf surface. We propose the heating mechanism of coronal loop plasma driven by dissipation of electric current generated by photospheric convection and the mechanism of long-term particle acceleration in the atmosphere of a brown dwarf supporting the persistent nature of the radio flux. Electric currents required for the heating of the loop up to $T \approx 10^7$ K are determined. Therefore, the magnetic loops responsible for the persistent microwave emission can also be the sources of the "soft" X-ray component.

2. Radio Emission from TVLM 513-46546

The M8.5V ultracool dwarf TVM 513-46546 (hereafter, TVLM 513) has an effective temperature of the photosphere $T_{\text{eff}} = 2200$ K, the mass $M_* = 0.07\,M_\odot$, the radius $R_* = 0.1\,R_\odot$, and is located at a distance of 10.6 pc from the Sun. The gravitational acceleration on the star surface is $g = 2 \times 10^5$ cm/s^2, and rotation velocity $v\,sin\,i \approx 60$ km/s. In the case of brown dwarfs, energy is transferred from the star core to its surface by convection. On the photosphere level, the convection velocity for the stars of M ≥ 7 spectral type varies from $V \approx 10^3$–10^4 cm/s [6] to $V \approx 1.4 \times 10^5$ cm/s [7]. The size of granulation cells for M8.5V stars approximately coincides with that of supergranulation cells, $D \approx 1.4 \times 10^7$ cm [8].

TVLM 513 displays two components of microwave radiation in the range of 4.8–8.5 GHz. The first one consists of periodic bursts with high brightness temperature $\geq 10^{11}$ K, highly beamed emission, and 100% circular polarization [3]. The periodicity presented in burst emission (≈ 1.96 h) is due to the rotation of the dwarf. The peculiarity of the microwave emission was explained in terms of the Electron Cyclotron Maser [9] or the plasma radiation mechanism [10]. The second component, which is called "quiescent", displays a brightness temperature of about 10^9 K and a small degree of circular polarization (<15%) [1,2]. The weak modulation of the quiescent component under the dwarf rotation means that the sources are distributed uniformly in the TVLM 513 atmosphere, and their total area is comparable with the area of the dwarf's surface. In addition, the respective sources of the radio emission should be supplied continuously by energetic particles, to compensate the losses caused by particles precipitation into the loss-cone and to provide the observed brightness temperature. Microwave observations of the TVLM 513 indicate that the quiescent component is stable on a multiyear timescale [2].

Observations with Very Large Array revealed the spectral index α in quiescent radio flux $F_\nu \sim \nu^\alpha$ for the wavelengths 20–3.6 cm [1]. It is positive between 20 and 6 cm and equal to $\alpha = 0.1 \pm 0.2$. For the wavelengths 6–3.6 cm, the spectral index is negative and equal to $\alpha = -0.4 \pm 0.1$. A change in the sign of the spectral index implies that the spectrum of quiescent emission has a maximum and the maximum frequency ν_{max} is in the range of $\nu_{\text{max}} = 1.4$–4.8 GHz (i.e., 20–6 cm) [1]. The change in the frequency index's sign in the gyrosynchrotron mechanism means that the radiation mode transits from optically thick to optically thin. In other words, the source of radiation in the range 4.8–8.4 GHz is most probably optically thin, which is important for further estimations.

We will use the information on the flux of the quiescent component at 8.4 GHz, i.e., where the emission source is optically thin. For the power-law distribution of energetic electrons $n_e(\varepsilon) \propto \varepsilon^{-\delta}$, the flux of gyrosynchrotron emission from an optically thin source at the frequency ν is [11]

$$F_\nu = 3.3 \times 10^{-24} \times 10^{-0.52\delta}\,(sin\vartheta)^{-0.43-0.65\delta}\left(\frac{\nu}{\nu_e}\right)^{1.22-0.9\delta}(n_e d)\,B\frac{S}{R^2}\ \text{ergs/cm}^2\ \text{s Hz} \quad (1)$$

Here, ϑ is the angle between the magnetic field B and the direction toward the observer, $\nu_e = 2.8 \times 10^6\,B$ is electron gyrofrequency, n_e is the number density of energetic electrons, d is the source thickness in the projection to the observer, S is the source area, and $R = 10.6$ pc is the distance to the source. For the maximum frequency in the gyrosynchrotron spectrum and the polarization degree, we use the following formulas [11]:

$$\nu_{peak} = 2.72 \times 10^3 \times 10^{0.27\delta}\,(sin\vartheta)^{0.41+0.03\delta}\,(n_e d)^{0.32-0.03\delta}\,B^{0.68+0.03\delta} \quad (2)$$

$$r_c = 1.26 \times 10^{0.035\delta} \times 10^{-0.071cos\vartheta}\left(\frac{\nu}{\nu_e}\right)^{-0.782+0.545cos\vartheta} \quad (3)$$

Equations (1)–(3) are true in the angle range $\vartheta = 20°–80°$ and for the optical thickness $\tau_v = \kappa_v d < 1$, where

$$\kappa_v = 1.4 \times 10^{-9-0.22\delta}(\sin\vartheta)^{-0.09+0.72\delta}\left(\frac{v}{v_e}\right)^{-1.30-0.98\delta}\frac{n_e}{B} \tag{4}$$

For the model of quiescent component, we determine the visible area of the source as $S \approx N_{loop}ld$, where N_{loop} is the number of elementary radiation sources (magnetic loops) on the visible hemisphere (Figure 1), and d and l are a typical thickness and length of a single loop.

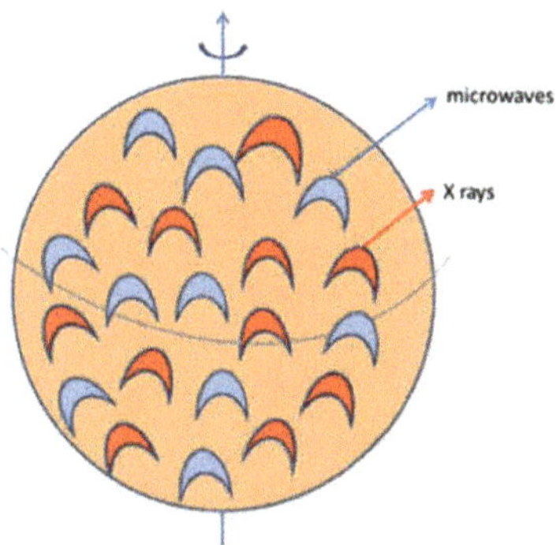

Figure 1. Sketch of brown dwarf disc with two populations of magnetic loops, the sources of "hard" X-ray emission (red), and both microwave and "soft" X-ray radiation (blue).

Using the relation $\alpha = 1.22 - 0.90\delta$ from the spectral index of the optically thin radio emission $\alpha = -0.4 \pm 0.1$, one can determine the spectral index of energetic electrons $\delta \approx 1.8 \pm 0.1$, indicating a rather hard spectrum of emitting particles. Equation (3) for the degree of circular polarization (15%) was used to estimate the magnetic field in coronal magnetic loops, the sources of gyrosynchrotron radiation of TVLM 513, $B \approx 100$ G [2].

It should be noted that one of the problems with multi-loop radio source is the estimation of the magnetic field using the observed degree of circular polarization, since various orientations of magnetic loops with different magnetic polarity should lead to a decrease in the Stokes V parameter. We assume here that despite this some predominant average polarity is retained in set of the loops, which gives a final degree of observed circular polarization (about 15%) that is rather low as compared to the polarization degree of gyrosynchrotron emission from solar loops (about 30%). The predominant average magnetic polarity can be associated with the large-scale magnetic field of the dwarf, which is partially fragmented into loops by the photospheric convection. A good example of the conservation of the magnetic field polarity in the loop set one can find from the observations of solar magnetic loops with the Solar and Heliospheric Observatory (SOHO) [12]. Among 30 loops recorded on 1999 August 30 in AR 7986, in 21 loops the negative magnetic polarity was retained. This example means that our estimates of the magnitude of the magnetic field from the observed polarization of microwave radiation from TVLM 513 give a lower average magnitude of the magnetic field in the loops.

The values of the magnetic field and the spectral index of energetic electrons, together with the condition of optically thin sources at 8.4 GHz, impose a limitation on the value $n_e d \leq 2 \times 10^{16}$ cm^{-2}, which determines the microwave flux. Assuming the loop thickness of the order of the granulation cell, $d \approx 10^7$ cm, $l \approx (2.5–5) \times 10^9$ cm, we obtain the area of all elementary sources $S \approx (2.5–5) \times 10^{16} N_{loop}$ cm^2; from the microwave flux (Equation (1))

$$F_{v=8.4} = 228\mu Jy \approx (2.28–5.6) \times 10^{-46}(n_e d)N_{loop}\text{ergs cm}^{-2}\text{s}^{-1}\text{Hz}^{-1} \tag{5}$$

one can estimate the number of sources of quiescent radio emission in the visible hemisphere required to explain the observed microwave flux, $N_{loop} \approx (0.4\text{--}1.0) \times 10^{19}/(n_e d)$ $\approx 200\text{--}500$. The filling factor of the stellar surface with the sources is about 3–16%. The number density of energetic particles is found from the condition of an optically thin source (Equations (1) and (4)) at 8.4 GHz, $n_e d \leq 2 \times 10^{16}$ cm^{-2}, which yields $n_e \leq 2 \times 10^9$ cm^{-3}.

3. Long-Term X-ray Emission of TVLM 513 and the Coronal Heating Problem

The information collected by Chandra ACIS-S3 shows that the quiescent radio emission from the brown dwarf TVLM 513 is accompanied by soft X-ray radiation in the energy range 0.3–2 keV [2]. To estimate the plasma temperature from 0.3–2 keV spectrum, the authors of [2] have shown that the best-fit temperature of TVLM 513 is 0.9 keV, or $T \approx 10^7$ K. The X-ray flux in the 0.3–2 keV range is $F_x \approx 6.3 \times 10^{-16}$ ergs/cm^2 s and the total luminosity $L_x = 4\pi R_*^2 F_x \approx 8.4 \times 10^{24}$ erg/s. This value of luminosity corresponds to the emission measure $EM_* = L_x/P(T)$, where [5]

$$P(T) = 2 \times 10^{-27}\sqrt{T} + 5 \times 10^{-25} exp\sqrt{2.8 + 10^6 \text{K}/T} \text{ ergs cm}^3\text{s}^{-1}. \tag{6}$$

putting $T = 10^7$ K in Equation (6), we obtain $EM_* \approx 9 \times 10^{47}$ cm^{-3}.

There is the problem of heating of the entire corona of TVLM 513. Indeed, if the source of the quasi-stationary X-ray radiation is a corona with a temperature of $T \approx 10^7$ K, then the emitting volume is $V \approx 4\pi R_*^2 H$, where $H = k_B T R_*^2/m_i G M_*$, k_B is the Boltzmann constant, $G = 6.67 \times 10^{-8}$cm^3/gs^2 is the gravitational constant. The X-ray luminosity from corona is

$$L_x = P(T)\bar{n}^2 V, \tag{7}$$

where $P(T)$ is determined by Equation (6), $\bar{n}$ is the average number density of electrons (ions) in the volume V. Using Equation (7), we determine the volume of the emitting corona V and the average values of the number density of particles in the corona from the observed values of the radiation luminosity and coronal temperature. For TVLM-513, we find $V \approx 2.5 \times 10^{30}$ cm^3, the height scale $H \approx 4 \times 10^9$ cm, and the average plasma density in the corona $\bar{n} \approx 6 \times 10^8$ cm^{-3}.

As a result of hot coronae, brown dwarfs noticeably increase their energy losses due to thermal conductivity. The radiation losses for TVLM-513 are $Q_r \approx 3 \times 10^{25}$ergs/s [13], while the energy losses due to thermal conductivity [14,15] $Q_T \approx 0.9 \times 10^{-6} T^{\frac{7}{2}} V H^{-2} \approx 4 \times 10^{29}$ergs/s significantly exceed similar losses in the solar corona, $Q_{T\odot} \approx 10^{28}$ergs/s. Therefore, to maintain the corona of TVLM 513 with $T \approx 10^7$ K, more powerful heating sources are required than those in the Sun. Let us assume that in the atmospheres of brown dwarfs, as is the case in the solar corona, there are hot Type II spicules, which in the last decade have been considered to be a probable source of solar corona heating [15–17]. Then the heat flux from the single spicule into the corona due to electron thermal conductivity along the magnetic field of the spicule is [15]

$$Q_{Tsp} \approx \frac{0.9 \times 10^{-6} T^{\frac{7}{2}}}{\Delta z} \pi r_0^2 \text{ ergs/s}. \tag{8}$$

For the radius of the spicule (magnetic flux tube) of brown dwarfs, the size of the granulation cell can be taken to be $r_0 \approx 10^7$ cm. Then for $T = 10^7$ K and a spicule length $\Delta z = 10^9$ cm, from Equation (8) we obtain $Q_{Tsp} \approx 10^{24}$ ergs/s. Therefore, to compensate for the energy losses from the TVLM 513 corona, ~4×10^5 hot spicules covered persistently the dwarf surface are required. Moreover, the total foot-points area of spicules should be of the order of the dwarf surface area (1.5×10^{20} cm^2), which significantly differs from the Sun, where spicules occupy $\approx$1% of the surface.

4. Generation of Electric Current and Plasma Heating in a Magnetic Loop: The Role of Photospheric Convection

Energy is transferred from the TVLM 513 core to the surface by the convection. At the photosphere level, the convection velocity is $V \approx 10^4$–10^5 cm/s [6,7]. The half-thickness of the magnetic loops is of the order of the size of a granulation cell $r_1 \approx D \approx 10^7$ cm [8]. When the radial component of the convection velocity V_r interacts with the azimuthal component of the magnetic field of the loop B_φ, an EMF arises, which generates an electric current flowing from one foot point of the loop through the coronal part to the other, and closing in the photosphere, where the inequality $\omega_e/\omega_i \nu_{ea}\nu_{ia} \ll 1$ is satisfied and the conductivity becomes isotropic. Here, ω_e and ω_i are gyrofrequencies of electrons and ions, ν_{ea} and ν_{ia} are electron-atom and ion-atom collision frequencies. Thus, the coronal magnetic loop with the photospheric current channel is an equivalent RLC-circuit [18–20]. In the self-consistent equation of the equivalent electric circuit, the resistance R and capacitance C are found to be dependent on the electric current [21]:

$$\frac{1}{c^2}L\frac{\partial^2 y}{\partial t^2} + \left[R(I) - \frac{|\overline{V}_r|l_1}{c^2 r_1}\right]\frac{\partial y}{\partial t} + \frac{1}{C(I)}y = 0, \tag{9}$$

where $y(t) = [\Delta I(t) - I]/I$, L is a loop inductance. EMF is in the photospheric foot point of the loop and is equal to

$$\Xi = \frac{l_1}{\pi c r_1^2}\int_0^{r_1} V_r B_\varphi 2\pi r dr \approx \frac{|\overline{V}_r|Il_1}{c^2 r_1} \tag{10}$$

where l_1 is the height of the RLC-circuit section in the area of EMF action, I is the longitudinal electric current, $B_\varphi \approx 2I/cr_1$. Estimates show that the main contribution to the loop resistance is made by the region of photospheric EMF, where the Cowling conductivity due to ion-atom collisions plays a decisive role. The resistance value in this case [22]

$$R(I) \approx \frac{1.5l_1 I^2 F^2}{\pi r_1^4 c^4 n m_i \nu'_{ia}(2 - F)}, \tag{11}$$

where $F = n_a/(n + n_a)$ is the relative density of neutrals, n is the density of electrons, $\nu'_{ia} \approx 1.6 \times 10^{-11} n_a \sqrt{T}$ Hz is the effective frequency of collisions of ions with neutrals. The steady-state value of the current in the loop is determined from the condition $R(I) = \Xi(I)$, hence at $l_1 \approx r_1$ we obtain

$$I \approx 0.8[|V_r|\pi r_1^3 c^2 n m_i \nu'_{ia}(2 - F)F^{-2}]^{1/2}\text{cgs} \tag{12}$$

For the brown dwarf considered, in the height interval $\Delta z \sim l_1$, the atomic density $n_a \approx 5 \times 10^{16} - 10^{17}$ cm^{-3}, $n \approx 5 \times 10^9 - 10^{10}$cm^{-3}, $T \approx 2200$ K. In this case, for the rate of photospheric convection $V_r \approx 10^4 - 10^5$ cm/s and $r_1 \approx 10^7$cm, we obtain the estimates of the current value: $I \approx 2.5 \times 10^9 \div 7 \times 10^{10}$ A. At the coronal level for an optically thin medium at temperatures $T > 2 \times 10^5$ K, hydrogen makes the main contribution to the neutral component, while the relative content of neutrals is determined by the formula [23]

$$F = 0.32 \times 10^{-3}\frac{1 + \frac{T}{6T_H}}{\left[\sqrt{\frac{T}{T_1}}\right]^{2-b}\sqrt{T}\left[1 + \sqrt{\frac{T}{T_1}}\right]^{1+b}}exp\frac{T_H}{T} \tag{13}$$

where $T_H = 1.58 \times 10^5$ K, $T_1 = 7.036 \times 10^5$ K, $b = 0.748$. In the temperature range of interest, 10^6–10^7 K, the approximation $F \approx 0.15/T$ is valid. The rate of Joule dissipation of the current per unit volume of the loop, taking into account Equation (11), is determined by the formula [22]

$$q_J = \frac{j_z^2}{\sigma} + \frac{F^2 B_\varphi^2 j_z^2}{(2-F)c^2 n m_i v'_{ia}} \approx 2.2 \times 10^{-9} \frac{I^4}{n^2 r_1^6 T^{2/3}} \ \text{ergs/cm}^3\text{s} \tag{14}$$

In the coronal part of the loop, the main contribution to the current dissipation is made by the second term in Equation (14). Despite the relatively low density of neutral atoms, $F \ll 1$, an effective dissipation channel associated with Cowling conductivity is switched on. This is due to a decrease in the effective plasma conductivity

$$\sigma_{eff} = \frac{\sigma}{1 + \frac{F^2 \omega_e \omega_i}{(2-F) v'_e v'_{ia}}} \tag{15}$$

because in the corona the second term in the denominator of Equation (15) is $\gg 1$. The heating of the magnetic loop is determined by the balance of Joule dissipation, thermal conductivity, and radiation losses:

$$\frac{d}{ds} \kappa_e T^{5/2} \frac{dT}{ds} = q_r - q_J. \tag{16}$$

Here, $\kappa_e = 0.9 \times 10^{-6}$, $q_r = \frac{\chi_0 p^2}{4 k_B^2} T^{-5/2}$, $q_J = \frac{10^{-8} k_B^2 I^4}{p^2 r_1^6} T^{1/2}$, s is the coordinate along the loop, $\chi_0 = 10^{-19}$, $p = 2 n k_B T$. Equation (15) is solved under the following boundary conditions at the foot point ($s = 0$) and at the loop top ($s = l/2$): $T = T_0$, $dT/ds = 0$ at $s = 0$ and $T = T_1$, $dT/ds = 0$ at $s = l/2$. It is assumed that $T_1 \gg T_0$. From Equation (16), we find the distribution of temperature and density along the loop [22]

$$T_1 \approx 2 \times 10^{-2} \frac{(IL)^{4/9}}{r_1^{2/3}} \ \text{K}, \quad n_1 = \frac{1}{3} \left(\frac{2\kappa_e}{\chi_0} \right)^{1/2} \frac{T_1^2}{L} \ \text{cm}^{-3}, \quad L = l/2 \tag{17}$$

It can be seen from Equation (17) that the temperature at the top of the loop reaches its maximum, and the plasma density its minimum. Both values vary slightly over most of the length of the loop, and at the foot points they reach the equilibrium values. Let us determine the number of hot loops in the corona of TVLM 513, which provide the observed measure of X-ray emission $EM_* \approx 9 \times 10^{47}$ cm^{-3}. By analogy with the solar corona, we assume that the thickness of the loops does not change with height, i.e., $r_1 \approx 10^7$ cm. Then from Equation (17) we find the electric current required to heat the loop plasma, $I \approx 6 \times 10^{10}$ A, the plasma density $n \approx 4.8 \times 10^{10}$ cm^{-3} and the measure of the emission of an individual loop $EM_{loop} \approx 4.6 \times 10^{45}$ cm^{-3}. To ensure the observed emission measure of "hard" X-ray radiation, it is necessary that $N_{loop} = EM_*/EM_{loop} \approx 200$ hot loops should be presented in the TVLM 513 corona. As shown in Section 2, to explain the persistent microwave emission from TVLM 513 as well as the "soft" X-ray radiation, about 200–500 loops are needed (Figure 1).

5. Mechanism of Long-Term Acceleration of Electrons in a Magnetic Loop—An Equivalent Electric Circuit

The persistent microwave emission generated in the TVLM 513 corona by a set of magnetic loops suggests that magnetic loops must be constantly replenished with energetic particles to compensate for the losses associated with the escape of particles into the loss-cone. In [24,25] a mechanism of long-term acceleration of electrons, caused by oscillations of the electric current in the loop, generated by photospheric convection, was proposed and developed. Let us show that even with a relatively small electric current, the acceleration of electrons by current oscillations in the loop is effective. From Equation (9) it follows that the current oscillations of the loop as an equivalent electric circuit are excited when the negative resistance of the photospheric EMF exceeds the loop resistance, $R(I) \leq |V_r| l_1 / (r_1 c^2)$. The eigen-frequency of the equivalent RLC-circuit depends on the loop radius r_1, density n, and length l of the coronal part of the loop [21,22]:

$$\nu_{RLC} = \frac{c}{2\pi\sqrt{LC(I_0)}} \approx \frac{1}{(2\pi)^{3/2}\sqrt{\Lambda}}\frac{I_0}{cr_1^2\sqrt{nm_i}}, \quad \Lambda = ln\left(\frac{4l}{\pi r_1}\right) - \frac{7}{4} \tag{18}$$

Assuming $r_1 \approx 10^7$ cm, $n \approx 10^{10}$ cm^{-3}, $I_0 \approx 10^8$ A, $l \approx 6 \times 10^9$ cm, from Equation (18) we obtain an estimate for the eigen-frequency of the electric circuit: $\nu_{RLC} \approx 2 \times 10^{-2}$ Hz (that is, a period of 50 s). Oscillations of the electric current are associated with those of the azimuthal component of the magnetic field in the magnetic loop, $B_\varphi(r,t) = 2rI_z(t)/cr_1^2$. These oscillations, in turn, according to the equation $rot\,\vec{E} = -(1/c)\partial\vec{B}_\varphi/\partial t$, lead to the generation of an electric field directed along the loop axis. Assuming $I(t) = I_0 + \Delta I sin(2\pi\nu_{RLC}t)$, we obtain the electric field averaged over the loop cross section [24]

$$\overline{E}_z = \frac{4\nu_{RLC}I_0}{3c^2}\frac{\Delta I}{I_0}. \tag{19}$$

Then from Equation (19) we obtain at $I_0 = 10^8$ A $= 3 \times 10^{17}$ cgs, $\Delta I/I_0 = 10^{-2}$ the value of the electric field: $\overline{E}_z \approx 8.9 \times 10^{-8}$ cgs $\approx 2.67 \times 10^{-5}$ V/cm. In such an electric field at a distance of $l = 6 \times 10^9$ cm, electrons can acquire energy $\varepsilon \approx 160$ keV. At $\Delta I/I_0 = 0.02$ more energetic electrons, $\varepsilon \approx 320$ keV, will be accelerated. For the plasma density $n = 10^{10}$ cm^{-3} and the temperature $T = 10^6$ K, the Dreicer field is [26]

$$E_D = 6 \times 10^{-8}(n/T)\text{ V/cm} \approx 6 \times 10^{-4}\text{ V/cm} \tag{20}$$

and the ratio of the Dreicer field to the accelerating field is $x = E_D/\overline{E}_z$. The kinetic theory gives the rate of electron acceleration per unit volume [27]:

$$\dot{n}_e = 0.35 n\nu_{ei}x^{3/8}exp\left(-\sqrt{2x} - x/4\right) \tag{21}$$

where $\nu_{ei} \approx 60nT^{-3/2}$ s^{-1}. Substituting the values of n and T into Equation (21), we find the acceleration rate $\dot{n}_e \approx 3 \times 10^7$ el/cm^{3}s. The condition for the generation of microwave radiation (Section 2) requires $n_e d < 2 \times 10^{16}$ cm^{-2}, which for $d = 2r_1 = 2 \times 10^7$ cm gives $n_e < 10^9$ cm^{-3}. The developing small-scale turbulence (whistlers for example) prevents the free escape of particles from the magnetic loop and contributes to the accumulation of accelerated particles in coronal magnetic loops.

The regimes of pitch-angle diffusion of accelerated particles into the loss-cone depend on the ratio of the following time scales [28]: the diffusion time t_D, which corresponds to the mean time for the change in the pitch-angle of a fast electron at $\pi/2$ due to the wave-particle interaction, the time of filling the loss-cone t_D/σ, and the time of flight when escaping into the loss-cone, $t_0 = l/v \approx 0.3$ s. Here, $\sigma = B_{foot}/B_{top}$ is the mirror ratio of a magnetic loop. In the case of intermediate diffusion, when the condition $t_0 < t_D < \sigma t_0$ is satisfied, the loss-cone is filled faster than the particles escape into the loss-cone. In this case, the lifetime of a particle in a magnetic trap is of the order of σt_0 and the number density of accelerated electrons is [29]

$$n_e \approx \frac{\dot{n}_e(x)\sigma l}{\sqrt{2\varepsilon/m}} \tag{22}$$

Substituting into Equation (22) $\dot{n}_e = 3 \times 10^7$ el/cm^{3}s, $\sigma = 3$, $l = 6 \times 10^9$ cm, and $\varepsilon = 160$ keV, we obtain $n_e \approx 2.3 \times 10^7$ cm^{-3}. In the case of strong diffusion, when the condition $t_D < t_0$ is satisfied, the density of accelerated electrons n_e is by about an order of magnitude higher. Therefore, the proposed acceleration mechanism will provide the persistent microwave emission.

6. Discussion

In the problem of the origin of X-ray radiation and heating of stellar coronae, the question arises: what is emitting? Is it the entire hot corona or its local regions, such as coronal magnetic loops, and what are the constant sources of the heating for the corona or the local regions? We have shown that plasma heating to the temperature T~10 MK provided by electric current in several hundred magnetic loops quasi-uniformly distributed over the surface of the brown dwarf TVLM 513, which explains the X-ray emission luminosity of ~1 keV, is energetically more favorable than heating the entire corona. The second population of colder, with T~1 MK, coronal loops is a source of persistent nonthermal microwave radiation explained by gyrosynchrotron radiation of ~150–300 keV electrons in a magnetic field of ~100 G. The proposed mechanism of long-term acceleration of electrons by induced electric fields is based on small oscillations of electric current in the coronal loops supported by photospheric convection.

In the model proposed by Osten et al. [1], the source of microwave radiation is represented by the dipole magnetic field of the dwarf. Berger et al. [2] suggested that the source of microwave radiation is a large-scale magnetic structure with a "covering factor" of about 50%. Both models cannot provide a long-term supply of sub-relativistic electrons to the radiation sources. In addition, the Berger's model does not identify the X-ray source in the brown dwarf corona.

It follows from Equation (17) that the plasma heating temperature is proportional to $T\sim I^{4/9}$. For a current $I = 10^8$ A, at which the acceleration of electrons is sufficiently efficient, the plasma in the loop heats up only to a temperature of 4×10^5 K, and at $I = 10^9$ A up to 10^6 K, i.e., the "microwave" loops are also sources of a softer X-ray component with a maximum in the spectrum of ≈ 0.2 keV. With an increase in the current value, the lumped-loop approach used by us is invalid, since the period of the RLC oscillations becomes shorter than the propagation time of the Alfvén disturbance along the loop $t_A = l/V_A \approx 30$ s. During this time, the current oscillations change direction many times and the acceleration of electrons is ineffective. This circumstance can explain the absence of noticeable nonthermal radio emission from hot coronal loops with the plasma temperature $T = 10$ MK. In this work, we did not consider the origin of the intense, with the brightness temperature $T_b \geq 10^{11}$ K, flare component of microwave radiation from TVLM 513, the nature of which was discussed in [9,10]

7. Conclusions

Photospheric convection, interacting with the magnetic field at the foot points of the magnetic loops, leads to generation of an electric current, $I \approx 10^9 - 7 \times 10^{10}$ A, the dissipation of which under the conditions of partially ionized plasma causes the loop plasma heating up to a temperature of $T \approx 10^6 - 10^7$ K. Hot plasma of ~200 loops, quasi-uniformly distributed over the dwarf's surface, is a source of ~1 keV X-ray emission from TVLM 513.

Photospheric convection maintains continuous oscillations of electric current in the magnetic loops as an equivalent RLC circuit, generating the induced electric fields that accelerate electrons up to 150–300 keV in magnetic loops distributed over the dwarf's surface. This provides the persistent quiescent microwave radiation of a few $\times 100$ loops of TVLM 513 by the gyrosynchrotron mechanism.

Effective acceleration of electrons by the electric current oscillations is possible even at relatively low values of electric current, $I \geq 10^8$ A. At currents $I \approx 3 \times 10^8 - 10^9$ A, the loop plasma heats up only to temperature $T \approx 10^6$ K, i.e., magnetic loops emitting microwave radiation can simultaneously be the sources of ~0.2 keV X-ray emission.

Thus, photospheric convection not only forms numerous magnetic loops in the corona of TVLM 513, but also generates an electric current leading to heating of the loop plasma and acceleration of electrons. As a result, two populations of loops quasi-distributed over the TVLM 513 surface are formed, one of which is a source of "hard" X-ray radiation, and the other is simultaneously a source of microwave and "soft" X-ray radiation.

Author Contributions: Investigation, A.S. and V.Z.; writing, A.S. and V.Z. All authors have read and agreed to the published version of the manuscript.

Funding: This research was partially funded by the Russian Science Foundation, grant No. 20-12-00268, the RFBR, grant No. 20-02-00108, and the State Assignment Nos. 0030-2021-0002, 1021032422589-5, and 0040-2019-0025.

Institutional Review Board Statement: Not applicable.

Informed Consent Statement: Not applicable.

Data Availability Statement: All necessary data are contained in this paper.

Conflicts of Interest: The authors declare no conflict of interest.

References

1. Osten, R.A.; Hawley, S.L.; Bastian, T.S.; Reid, I.N. The Radio Spectrum of TVLM 513-46546: Constraints on the Coronal Properties of a Late M Dwarf. *Astrophys. J.* **2006**, *637*, 518–521. [CrossRef]
2. Berger, E.; Gizis, J.E.; Giampapa, M.S.; Rutledge, R.E.; Liebert, J.; Martín, E.; Basri, G.; Fleming, T.A.; Johns-Krull, C.M.; Phan-Bao, N.; et al. Simultaneous Multiwavelength Observations of Magnetic Activity in Ultracool Dwarfs. I. The Complex Behavior of the M8.5 Dwarf TVLM 513-46546. *Astrophys. J.* **2008**, *673*, 1080–1087. [CrossRef]
3. Hallinan, G.; Bourke, S.; Lane, C.; Antonova, A.; Zavala, R.T.; Brisken, W.F.; Boyle, R.P.; Vrba, F.J.; Doyle, J.G.; Golden, A. Periodic Bursts of Coherent Radio Emission from an Ultracool Dwarf. *Astrophys. J. Lett.* **2007**, *663*, L25–L28. [CrossRef]
4. Güdel, M.; Benz, A.O. X-Ray/Microwave Relation of Different Types of Active Stars. *Astrophys. J. Lett.* **1993**, *405*, L63–L66. [CrossRef]
5. Giampapa, M.S.; Rosner, R.; Kashyap, V.; Fleming, T.A.; Schmitt, J.H.M.M.; Bookbinder, J.A. The Coronae of Low-Mass Dwarf Stars. *Astrophys. J.* **1996**, *463*, 707–725. [CrossRef]
6. Mohanty, S.; Basri, G.; Shu, F.; Allard, F.; Chabrier, G. Activity in Very Cool Stars: Magnetic Dissipation in Late M and L Dwarf Atmospheres. *Astrophys. J.* **2002**, *571*, 469–486. [CrossRef]
7. Osterbrock, D.T. The Internal Structure of Red Dwarf Stars. *Astrophys. J.* **1953**, *118*, 529–546. [CrossRef]
8. Rucinski, S.M. Sizes of spots in spotted stars. *Acta Astron.* **1979**, *29*, 203–209.
9. Yu, S.; Hallinan, G.; MacKinnon, A.L.; Antonova, A.; Kuznetsov, A.; Golden, A.; Zhang, Z.H. Modelling the radio pulses of an ultracool dwarf. *Astron. Astrophys.* **2011**, *525*, A39–A49. [CrossRef]
10. Zaitsev, V.V.; Stepanov, A.V. On the Origin of Intense Radio Emission from the Brown Dwarfs. *Radiophys. Quantum Electron.* **2017**, *59*, 867–875. [CrossRef]
11. Dulk, G.A. Radio emission from the Sun and stars. *Annu. Rev. Astron. Astrophys.* **1985**, *23*, 169–224. [CrossRef]
12. Aschwanden, M.J.; Newmark, J.S.; Delaboudinière, J.-P.; Neupert, W.M.; Klimchuk, J.A.; Gary, G.A.; Portier-Fozzani, F.; Zucker, A. Three-dimensional Stereoscopic Analysis of Solar Active Region Loops. I. SOHO/EIT Observations at Temperatures of (1.0–1.5) $\times 10^6$ K. *Astrophys. J.* **1999**, *515*, 842–867. [CrossRef]
13. Zaitsev, V.V.; Stepanov, A.V. X-ray Emission from Ultracool Stars. *Radiophys. Quantum Electron.* **2021**, *64*, 419–429. (In Russian) [CrossRef]
14. Priest, E.R. *Solar Magnetohydrodynamics*; Springer: Berlin/Heidelberg, Germany, 1982.
15. Zaitsev, V.V.; Stepanov, A.V.; Kronshtadtov, P.V. On the Possibility of Heating the Solar Corona by Heat Fluxes from Coronal Magnetic Structures. *Sol. Phys.* **2020**, *295*, 166–181. [CrossRef]
16. De Pontieu, B.; McIntosh, S.W.; Hansteen, V.H.; Schrijver, C.J. Observing the Roots of Solar Coronal Heating—In the Chromosphere. *Astrophys. J. Lett.* **2009**, *701*, L1–L6. [CrossRef]
17. De Pontieu, B.; McIntosh, S.W.; Carlsson, M.; Hansteen, V.H.; Tarbell, T.D.; Boerner, P.; Martínez-Sykora, J.; Schrijver, C.J.; Title, A.M. The Origins of Hot Plasma in the Solar Corona. *Science* **2011**, *331*, 55–58. [CrossRef]
18. Alfven, H.; Carlquist, P. Currents in the Solar Atmosphere and a Theory of Solar Flares. *Sol. Phys.* **1967**, *1*, 220–228. [CrossRef]
19. Zaitsev, V.V.; Stepanov, A.V. Towards the circuit theory of solar flares. *Sol. Phys.* **1992**, *139*, 343–356. [CrossRef]
20. Zaitsev, V.V.; Stepanov, A.V.; Urpo, S.; Pohjolainen, S. LRC-circuit analog of current-currying magnetic loop: Diagnostics of electric parameters. *Astron. Astrophys.* **1998**, *337*, 887–896.
21. Khodachenko, M.L.; Zaitsev, V.V.; Kislyakov, A.G.; Stepanov, A.V. Equivalent Electric Circuit Models of Coronal Magnetic Loops and Related Oscillatory Phenomena on the Sun. *Space Sci. Rev.* **2009**, *149*, 83–117. [CrossRef]
22. Stepanov, A.V.; Zaitsev, V.V.; Nakariakov, V.M. *Coronal Seismology: Waves and Oscillations in Stellar Coronae*; WILEY-VCH Verlag: Karlsruhe, Germany, 2012.
23. Verner, D.A.; Ferland, G.J. Atomic Data for Astrophysics. I. Radiative Recombination Rates for H-like, He-like, Li-like, and Na-like Ions over a Broad Range of Temperature. *Astrophys. J. Suppl.* **1996**, *103*, 467–473. [CrossRef]
24. Zaitsev, V.V.; Stepanov, A.V.; Kaufmann, P. On the Origin of Pulsations of Sub-THz Emission from Solar Flares. *Sol. Phys.* **2014**, *289*, 3017–3032. [CrossRef]

25. Zaitsev, V.V.; Stepanov, A.V. Acceleration and Storage of Energetic Electrons in Magnetic Loops in the Course of Electric Current Oscillations. *Sol. Phys.* **2017**, *292*, 141–152. [CrossRef]
26. Benz, A. *Plasma Astrophysics. Kinetic Processes in Solar and Stellar Coronae*; Kluwer Academic Publishers: Dordrech, The Netherlands; Boston, MA, USA; London, UK, 1993.
27. Knoepfel, H.; Spong, D.A. Runaway electrons in toroidal discharges. *Nucl. Fusion* **1979**, *19*, 785–825. [CrossRef]
28. Trakhtengerts, V.Y.; Rycroft, M.J. *Whistlers and Alfven Mode Cyclotron Masers in Space*; Cambridge University Press: Cambridge, UK, 2008.
29. Zaitsev, V.V.; Stepanov, A.V. Particle Acceleration by Induced Electric Fields in Course of Electric Current Oscillations in Coronal Magnetic Loops. *Geomagn. Aeron.* **2018**, *58*, 831–840. [CrossRef]

Article

A Simple Model of Radiation from a Magnetized Neutron Star: Accreted Matter and Polar Hotspots

Dmitry Yakovlev

Ioffe Institute, Politekhnicheskaya 26, 194021 St. Petersburg, Russia; yak.astro@mail.ioffe.ru

Abstract: A simple and well known model for thermal radiation spectra from a magnetized neutron star is further studied. The model assumes that the star is internally isothermal and possesses a dipole magnetic field ($B \lesssim 10^{14}$ G) in the outer heat-insulating layer. The heat transport through this layer makes the surface temperature distribution anisotropic; any local surface element is assumed to emit a blackbody (BB) radiation with a local effective temperature. It is shown that this thermal emission is nearly independent of the chemical composition of insulating envelope (at the same taken averaged effective surface temperature). Adding a slight extra heating of magnetic poles allows one to be qualitatively consistent with observations of some isolated neutron stars.

Keywords: neutron stars; radiation transfer; magnetic fields

Citation: Yakovlev, D. A Simple Model of Radiation from a Magnetized Neutron Star: Accreted Matter and Polar Hotspots. *Universe* **2021**, *7*, 395. https://doi.org/10.3390/universe7110395

Academic Editors: Nazar R. Ikhsanov, Galina L. Klimchitskaya and Vladimir M. Mostepanenko

Received: 30 September 2021
Accepted: 18 October 2021
Published: 21 October 2021

Publisher's Note: MDPI stays neutral with regard to jurisdictional claims in published maps and institutional affiliations.

1. Yury N. Gnedin

This paper is dedicated to the memory of Yury N. Gnedin (1935–2018) Figure 1. He was my senior colleague at Theoretical Astrophysics Department of Ioffe Insitute from 1971 to 1984, and we often met later, when he moved to Pulkovo Observatory. He was always interested in many scientific fields. For instance, his pet subjects were radiation transfer, neutron stars and magnetic fields (although he never lost scientific interest in a great amount of other things). When neutron stars were discovered in 1967, he was extremely enthusiastic about them and initiated ultrafast spreading of interest to these objects among colleagues. He was the first author of the first publication [1] on neutron stars at the Ioffe Institute after the discovery of these stars. He wrote—with different co-authors—plenty of (now) classical papers on neutron star physics. In particular, they include a basic formulation of the radiation transfer problem in a strongly magnetized plasma [2], a theoretical prediction of electron cyclotron lines in the spectra of radiation from magnetized neutron stars [3], the first realistic studies of ionization in the magnetized hydrogen atmospheres of neutron stars [4]. He was also greatly fond of history.

Although we did not work much together, I, as many others, respected Yury Gnedin very much. When he could help, he really did, and he was a Real Gentleman—friendly and open to any one, from a prominent scientist to a hopeless student. I will always remember the wonderful atmosphere, created by Yury Gnedin, and warm hospitality of his family and home. It is amazing that both his sons have grown up into excellent scientists.

I present a small piece of neutron star physics on thermal radiation from magnetized neutron stars. Hopefully, he would like it.

Figure 1. Yury N. Gnedin. (**Left**): A rare moment of leisure on his birthday, 13 August 2007, near his summer cottage. (**Right**): At a conference (June 2004). Courtesy to Oleg Gnedin.

2. Introduction

I will mostly consider thermal X-ray radiation from the surface of a middle-aged ($t \lesssim 10^6$ yr) isolated magnetized neutron star. The radiation is supposed to emerge from the warm neutron star interior and is emitted from a thin atmosphere or condensed surface (e.g., [5]), where the radiation spectrum is formed. The interiors of the star are thought to be nearly isothermal because of the high thermal conductivity of dense matter. Nevertheless, the interiors are thermally insulated from the surface by a thin heat blanketing envelope, where the thermal conduction is poorer and more sensitive to the chemical composition and magnetic field. The field redirects the emerging heat flow and results in anisotropic surface temperature distribution. The thermal surface emission can also become anisotropic.

The anisotropy of observed surface radiation is used to infer magnetic field strength and geometry, the composition of the surface layers, global parameters of the star (such as mass and radius), as well as important parameters of superdense matter in stellar interiors (see, e.g., [5,6] and references therein).

Here, I study a simple model of thermal emission from magnetized neutron stars with isothermal interiors, assuming the star emits a blackbody (BB) radiation with a local effective surface temperature T_s, which varies over the stellar surface under the effects of dipole surface magnetic field. This model has been invented long ago by Greenstein and Hartke [7], elaborated in the literature (e.g., [8–13]) and reviewed in [6,14]. Here I point out some properties of the surface emission, which, to the best of my knowledge, have not been studied in the literature.

Logically, this paper continues the previous one [15] (hereafter Paper I), which shows that the simple thermal spectra of magnetized neutron stars can be accurately approximated by two-BB (2BB) models. Section 3 outlines theoretical formalism. Section 4 is devoted to the effects of chemical composition of the blanketing envelopes and Section 5 extends the solution to the case, in which magnetic poles contain additional hot spots. The conclusions are formulated in Section 6.

It is important to mention more complicated neutron star emission models. They include sophisticated magnetic field effects on thermal emission, leading to specific spectral, polarization and angular properties of radiation (see, e.g., [6,16–18]), which are not reproduced by the given model.

3. Simple Model Spectra

3.1. General Formalism

The calculation of radiation spectral flux from a spherical neutron star under formulated assumptions is simple. Here is the summary using the notations of Paper I. Any small surface temperature element emits like a BB, and the flux is obtained by intergating over a visible part of the surface, taking into account that emitted quanta propagate in Schwarzschild space-time and demonstrate the gravitational redshift of photon energies and light bending. The effects of General Relativity are specified by the compactness parameter $x_g = r_g/R$, where $r_g = 2GM/c^2$ is the Schwartschield radius, M is the gravitational star's mass, R is the circumferential radius, G is the gravitational constant and c is the velocity of light. One deals either with local quantities at the stellar surface (e.g., the local surface temperature T_s) or with the quantities detected by a distant observer. The letter ones will often be marked by the symbol '∞'. For instance, $T_s^\infty = T_s\sqrt{1-x_g}$ is the redshifted surface temperature. As an exclusion, we will denote local (non-redshifted) photon energy by E_0, and the redshifted energy by $E \equiv E_\infty = E_0\sqrt{1-x_g}$.

Let F_E^∞ be a radiative spectral flux density [erg cm^{-2} s^{-1} keV^{-1}] detected at a distance $D \gg R$. It is customary to express F_E^∞ as:

$$F_E^\infty = \frac{R^2}{D^2}\,H_E^\infty,\tag{1}$$

where H_E^∞ is the effective flux, that is formally independent of D. It can be calculated as an integral of the radiating surface flux over the visible part of the surface,

$$H_E^\infty = \frac{15\sigma_{\rm SB}}{16\pi^5 k_{\rm B}^4}\int_{\rm viz} d\Omega_s\,\frac{(1-x_g)^{-1}\,\mathcal{P}E^3}{\exp(E/k_{\rm B}T_s^\infty)-1},\tag{2}$$

where $\sigma_{\rm SB}$ is the Stefan–Boltzmann constant, $k_{\rm B}$ is the Boltzmann constant, $d\Omega_s$ is a surface solid angle element, and $\mathcal{P}$ is the light bending function (e.g., [5,19–21]).

It is convenient to integrate over the star's surface and calculate the bolometric luminosity of the star, L_s, as well as the averaged non-redshifted effective surface temperature $T_{\rm eff}$ (e.g., [10]),

$$L_s = \sigma_{\rm SB}R^2\int_{4\pi} d\Omega_s\,T_s^4 \equiv 4\pi\sigma_{\rm SB}R^2 T_{\rm eff}^4.\tag{3}$$

For a uniform surface temperature ($T_s = T_{\rm eff}$), one immediately gets the standard BB flux,

$$H_E^{\rm BB\infty} = \frac{15\sigma_{\rm SB}}{4\pi^4 k_{\rm B}^4}\,\frac{(1-x_g)^{-1}\,E^3}{\exp(E/k_{\rm B}T_{\rm eff}^\infty)-1},\tag{4}$$

and the bolometric effective flux,

$$H_{\rm bol}^{\rm BB\infty} = \int_0^\infty H_E^{\rm BB\infty}\,dE = \frac{\sigma_{\rm SB}T_{\rm eff}^{\infty 4}}{1-x_g}.\tag{5}$$

3.2. Input Parameters

Equation (2) allows one to compute thermal spectra for any given temperature distribution T_s over the neutron star surface. We focus on the distribution created by a dipole magnetic field (with the field strength $B_{\rm pole}$ at magnetic poles) due to anisotropic heat transport in a thin (maximum a few hundreds meters) heat blanketing envelope. The input parameters are M, R, chemical composition of the blanketing envelope, and a nonredshifted temperature T_b at its bottom (at density $\rho_b \sim 10^{10}$ g cm^{-3}; see, for example, [14]). The local T_s is usually determined by solving local quasistationary radial heat transport within the envelope mediated by an effective radial thermal conductivity. For studying the thermal surface emission, it is profitable to use $T_{\rm eff}$ [see Equation (3)] instead of T_b.

3.3. 2BB Representation

According to Paper I, the spectral fluxes H_E^∞, computed from Equation (2) for iron heat blankets, are accurately fitted by a familiar 2BB model,

$$H_E^\infty = s_c H_E^{\mathrm{BB}\infty}(T_{\mathrm{effc}}) + s_h H_E^{\mathrm{BB}\infty}(T_{\mathrm{effh}}). \tag{6}$$

Here, $H_E^{\mathrm{BB}\infty}(T_{\mathrm{eff}})$ is given by Equation (4); 'c' and 'h' refer, respectively, to colder and hotter BB components. Any fit contains four parameters, which are two effective temperatures T_{effc} and T_{effh}, and two fractions of effective radiating surface areas, s_c and s_h. Instead of T_{effc} and T_{effh}, it is convenient to introduce two dimensionless parameters $p_c = T_{\mathrm{effc}}/T_{\mathrm{eff}}$ and $p_h = T_{\mathrm{effh}}/T_{\mathrm{eff}}$, with $p_c < p_h$. In Paper I the fits have been done for a number of representative values of M, R, $\log T_{\mathrm{eff}}[K]$ (from 5.5 to 6.8), $\log B_{\mathrm{pole}}[G]$ (from 11 to 14), photon energies ($0.064 < E \lesssim 40$ keV, removing those E at which the fluxes are negligibly small) and inclination angles i (between line of sight and the magnetic axis). Typical relative fit errors have not exceeded a few percent, meaning the fits were very good, providing excellent analytic representation of the original computed data. Therefore, thermal X-ray spectral fluxes of magnetized neutron stars within the given model are nearly identical to those of 2BB spectral models.

Moreover, as shown in Paper I, it is sufficient to calculate the fluxes H_E^∞ for two observation directions which are (i) pole observation, $i = 0$, to be denoted as $H_E^{\parallel\infty}$, and (ii) equator observations, $i = 90°$, to be denoted as $H_E^{\perp\infty}$. If these fluxes are known, the radiation flux in any direction i is accurately approximated as:

$$H_E^{i\infty} = H_E^{\parallel\infty} \cos^2 i + H_E^{\perp\infty} \sin^2 i. \tag{7}$$

We will see (Section 5) that this approximation is more restrictive than Equation (6).

This accurate representation of numerically calculated $H_E^{i\infty}$ by a map of four fit parameters (p_c, p_h, s_c, s_h) in the space of input parameters has to be taken with the grain of salt. The problem is that the fluxes $H_E^{i\infty}$ are close to 1BB fluxes (4), which leads to some degeneracy of fit parameters (with respect to a fit procedure and a choice of grid points). On the other hand, the assumed surface temperature distribution model and the local BB emission model are definitely approximate by themselves, so that a too-accurate fitting of $H_E^{i\infty}$ is actually purely academic. However, these results can be helpful for the interpretation of observations, especially because the 2BB model is often used by observers.

4. Iron and Accreted Heat Blankets

Paper I considered heat blankets made of iron. Here, we employ a more general model [10,11,22], in which a heat blanket consists of shells (from top to bottom) of hydrogen, helium, carbon and iron. Light elements (H, He, C) are thought to appear because of accretion of H and/or He and further burning into C. Iron is either formed at neutron star birth or is a final result of carbon burning. The mass of lighter elements ΔM in the heat blanket is treated as a free parameter. The heat blanketing envelope may fully consist of iron (Paper I) or contain some mass $\Delta M \lesssim 10^{-6}\, M_\odot$ of lighter elements. Lighter elements affect the surface temperature distribution and thermal surface spectral fluxes.

To simplify the task, we will study fully accreted blankets and conclude on the partly accreted ones in the end of Section 4.2.

4.1. T_s-Distributions

Our surface temperature distribution T_s is axially symmetric and symmetric with respect to the magnetic equator. For illustration, Figure 2 presents calculated T_s for a star with $M = 1.4\, M_\odot$ and $R = 12$ km [x_g=0.344 and the surface gravity $g_s = GM/(R^2\sqrt{1-x_g}) = 1.59 \times 10^{14}$ cm s^{-2}]. The effective surface temperatures are shown versus colatitude ϑ, which varies from $\vartheta = 0$ at the north pole, to $\vartheta = 90°$ at the magnetic equator, and then to $\vartheta = 180°$ at the south pole. Three panels (a–c) correspond to three values of the magnetic field $B_{\mathrm{pole}} = 10^{12}$, 10^{13} and 10^{14} G at the pole. Each panel displays $T_s(\vartheta)$ for six values of $\log T_{\mathrm{eff}}$ [K]= 5.8, 6.0, ... 6.8. Solid lines are plotted for the heat blankets made of iron, while dashed lines are for the fully accreted blankets. The iron and accreted matter have different thermal conductivities and, hence, different $T_s(\vartheta)$ distributions at the same T_{eff}. The dotted lines correspond to a non-magnetic star to guide the eye (they give $T_s = T_{\mathrm{eff}}$).

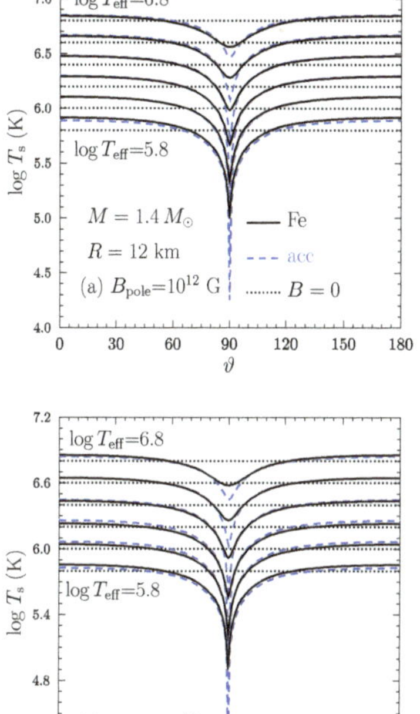
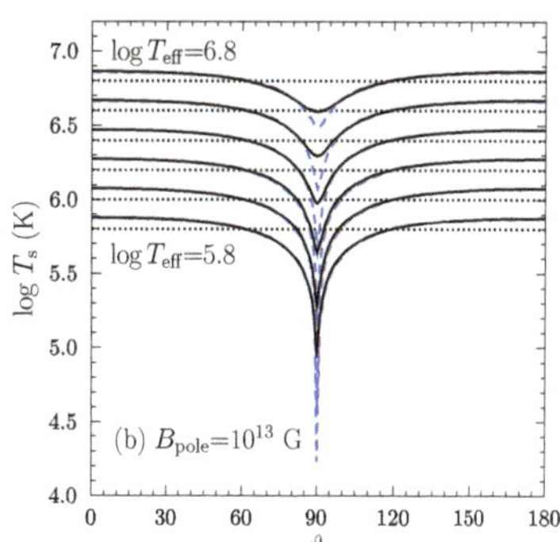

Figure 2. Effective surface temperature T_s versus colatitude ϑ, from magnetic pole ($\vartheta = 0$) to equator ($\vartheta = 90°$) and the opposite magnetic pole ($\vartheta = 180°$) of a neutron star with $M = 1.4\,M_\odot$ and $R = 12$ km at three values of $B_{\mathrm{pole}} = 10^{12}$, 10^{13} and 10^{14} G [panels (**a**–**c**)]. The curves are plotted for six effective surface temperatures, $\log T_{\mathrm{eff}}$ [K] = 5.8, 6.0 ... 6.8. Solid curves refer to the heat-blanket model made of iron, dashed cures refer to the heat blanket of fully accreted matter; dotted curves correspond to the field-free star.

Heat conduction is mainly radiative in the outer non-degenerate layer. In deeper layers, where the electrons become degenerate, heat is mostly transported by electrons. The field affects also thermodynamic properties of the matter, for instance, the pressure (e.g., [23]). The magnetic field effects, which regulate the anisotropy of the surface temperature distribution, are twofold.

The effects of the first type are the classical effects of electron rotation about magnetic field lines. They are especially important near the equator, where the heat, emergent from the stellar interiors, propagates mainly across field lines and is thus suppressed. This may strongly enhance thermal insulation of the equatorial part of the heat blanketing envelope, producing rather narrow equatorial dips of $T_s(\vartheta)$ (the cold equatorial belt, Figure 2). The dips for the accreted matter are seen to be much stronger than for the iron matter, especially at lower T_{eff}.

The effects of the second type are associated with the Landau quantization of electron motion across B-lines. They become more efficient at higher B near magnetic poles, where they make the heat insulating layers more heat transparent and tend to increase T_s. These relatively warmer polar 'caps' are wide (Figure 2); T_s increases inside them, if we approach a pole, but not dramatically.

As long as B_{pole} is rather weak ($B_{\mathrm{pole}} \lesssim 10^{11}$ G), the classical effects stay more important, producing a pronounced colder equatorial belt, but weakly affecting the polar zones. The surface becomes overall colder, compared with the field-free star with the same internal temperature T_b (e.g., see Figures 22 and 23 in [14]). With increasing B_{pole}, the quantum effects become dominant, creating hot polar zones and making the surface overall hotter (at the same T_b). At $B_{\mathrm{pole}} \gtrsim 10^{12}$ G, the surface becomes hotter than at $B_{\mathrm{pole}} = 0$; hot polar zones make the equatorial belt less significant.

It is remarkable (Figure 2) that, outside the equatorial belt, at fixed T_{eff}, the surface temperature profiles $T_s(\vartheta)$ for iron and accreted heat blankets are mainly close. Moreover, these profiles almost coincide at $B_{\mathrm{pole}} \approx 10^{13}$ G. The latter conclusion is independent of specific values of M and R.

4.2. Spectral Fluxes

Despite the dramatic interplay of the opposite effects and substantial temperature anisotropy, the surface-integrated spectral fluxes H_E^∞ of thermal radiation behave smoothly under variations of B_{pole} and/or T_{eff}. This is because of strong effects of General Relativity, which smooth out photon propagation from the surface to the distant observer [8,10,11]. As noted in the cited papers, General Relativity 'hides' magnetic effects on thermal surface emission. Very large magnetic fields, with $B_{\mathrm{pole}} \gtrsim 10^{14}$ (typical of magnetars), start to affect the surface emission much more strongly, but such fields are not considered here. They deserve a special study. In particular, their heat-insulating envelopes can become thick and the 1D approach to calculating the heat insulation within them can be questionable.

We have checked that the spectral fluxes, calculated for the accreted envelopes, possess the same properties (Section 3.3) as for the iron envelopes. In particular, the flux remains to be close to that calculated in the 1BB approximation at uniform surface temperature $T_s = T_{\mathrm{eff}}$ (see, e.g., Figure 1 in Paper I). It is disappointing, meaning that such a flux, taken at T_{eff} as an input parameter, carries little information on the magnetic field. Then, one should study deviations from the 1BB approximation. It is important that if one fixes T_b and varies B_{pole}, one obtains noticeable T_s and flux variations (as demonstrated, for instance, in Figures 23 and 24 of [14]) but these variations are accompanied by variations of T_{eff}. This would be a good theoretical construction, if one knew T_b. However, the observer can measure T_{eff} and wishes to infer T_b, which is not easy because at fixed T_{eff}, the theoretical spectral flux is slightly sensitive to T_b.

Figure 3 shows thermal spectral flux densities $H_E^{\|\infty}$ for polar observations of the same star as in Figure 2 (at seven values of T_{eff} and three values of B_{pole}). The fluxes are plotted in logarithmic scale versus the decimal logarithm of the redshifted photon energy E. For each value of T_{eff} we present three curves. The solid curves give the fluxes radiated from the star with iron heat blanket. The dashed curves present similar fluxes for accreted heat blankets. The dotted curves demonstrate the 1BB model, Equation (4), with given constant $T_s = T_{\mathrm{eff}}$ (as if the magnetic field is absent).

All three fluxes for each T_{eff} look close in the logarithmic format of Figure 3. However, they do not coincide (as clearly seen from Figure 3 of Paper I). Note that the difference between the fluxes emitted from the star with accreted and iron envelopes is noticeably smaller than the difference between these fluxes and 1BB ones. The difference from 1BB fluxes monotonically increases with growing B_{pole} and E, which is quite understandable. Higher B_{pole} produces stronger anisotropy of T_s distribution. Radiation at higher energies is predominantly emitted from hotter places of the surface.

Let us remark that at $B_{\mathrm{pole}} = 10^{13}$ G the spectral fluxes from the star with iron and accreted heat blankets are almost indistinguishable (but sufficiently different from the 1BB flux). This is because the T_s distributions for iron and accreted blankets are very close (Figure 2, Section 4.1). They differ only in cold equatorial belts, but the contribution into the fluxes from cold narrow belts appears almost negligible (also see Paper I). In contrast, the contribution from hotter surface regions can be important (Section 5). At $B_{\rm pole} = 10^{14}$ G the situation remains nearly the same as at 10^{13} G.

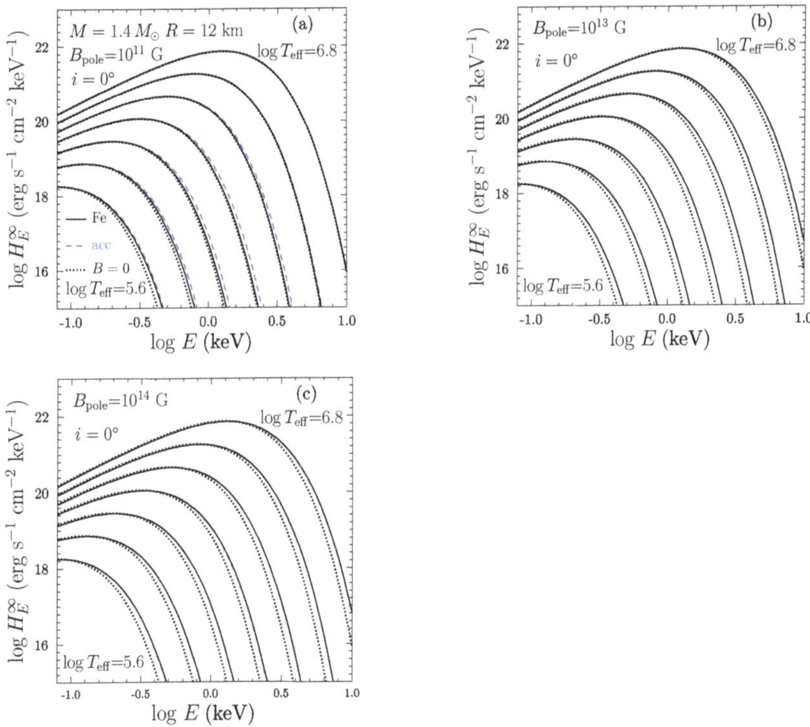

Figure 3. Thermal spectral fluxes observed from magnetic poles for a neutron star with $M = 1.4\,M_\odot$ and $R = 12$ km at three values of $B_{\rm pole} = 10^{11}$ G (a), 10^{13} G (b), and 10^{14} G (c) at $\log T_{\rm eff} = 5.6$, 5.8, ..., 6.6 assuming iron (solid lines) or accreted (dashed lines) heat blankets. The dotted lines refer to non-magnetic star to guide the eye.

Figure 4 shows similar spectral fluxes but for equator observations (and only for $B_{\rm pole} = 10^{11}$ and 10^{13} G). In this case, the situation is similar to that for pole observations.

The main outcome of Figures 3 and 4 is that the difference of the fluxes emitted from stars with iron and accreted heat blanketing envelopes is rather small and can be ignored in many applications, especially taking into account approximate nature of heat blanketing models. The results for partly accreted heat blankets would be similar. Accordingly, one can neglect chemical composition of heat blankets for calculating thermal emission from magnetized neutron stars in the adopted model of heat blankets. Therefore, the chemical composition of heat blankets does affect cooling of the neutron stars but almost does not affect thermal emission (under formulated assumptions).

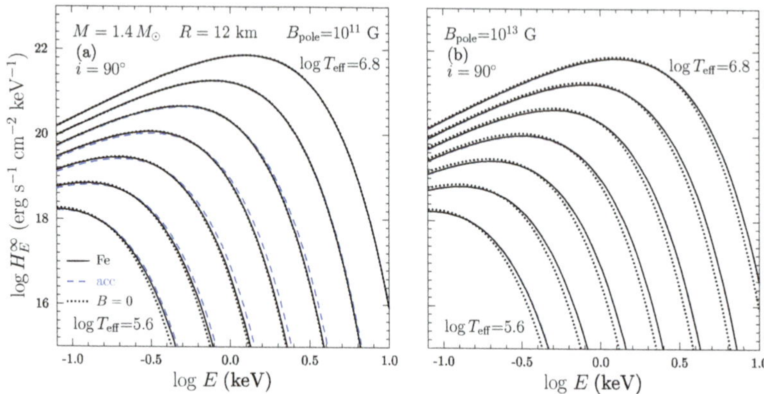

Figure 4. Same as in Figure 3 but for equator observations at $B_{\rm pole} = 10^{11}$ G (**a**) and 10^{13} G (**b**).

4.3. Phase-Space Maps

As discussed in Section 3.3 (see also Paper I), thermal spectral fluxes can be presented by maps of fit parameters ($p_{\rm c}$, $p_{\rm h}$, $s_{\rm c}$, $s_{\rm h}$) as functions of input parameters. For completeness, Figure 5 shows these 2BB maps versus $\log T_{\rm eff}$ at $B_{\rm pole} = 10^{11}$ G [panels (a) and (e)], 10^{12} G [panels (b) and (f)], 10^{13} G [(c) and (g)] and 10^{14} G [(d) and (h)]. Panels (a)–(d) refer to pole observations, while (e)–(h) to equator observations. Solid lines correspond to the iron heat blankets, while dashed lines correspond to the accreted ones.

One can see that the dependence of fit parameters on $T_{\rm eff}$ is smooth, without any specific features. The parameters of both effective BB components are of the same order of magnitude. Most importantly, the temperature $T_{\rm effh}$ of the hotter BB component exceeds the temperature $T_{\rm effc}$ of the colder component by less than 20%. The effective surface fraction $s_{\rm c}$ of the colder component slightly exceeds the fraction $s_{\rm h}$ of the hotter component.

The maps in Figure 5 represent the same thermal fluxes, which are plotted in Figures 3 and 4. The main conclusion of Section 4.2, based on Figures 3 and 4, is that the fluxes for iron and accreted envelopes are nearly the same. If they were identical, then respective solid and dashed curves in Figure 5 should have been identical as well. However, these curves differ. The difference is especially visible for $B_{\rm pole} = 10^{11}$ and 10^{12} G, but less visible at higher $B_{\rm pole}$. The difference is mainly explained by two things. First, weak variations of 2BB fit parameters lead to weaker variations of the fluxes. Secondly, the 2BB approximation is accurate but not exact by itself, as discussed in the previous section.

It is important that we fit *unabsorbed* calculated spectral fluxes using *unabsorbed* 2BB models. Accordingly, as argued in Paper I, one can compare 2BB parameters, inferred from observations (corrected for interstellar absorption and non-thermal radiation component), with the theoretical parameters of unabsorbed 2BB models.

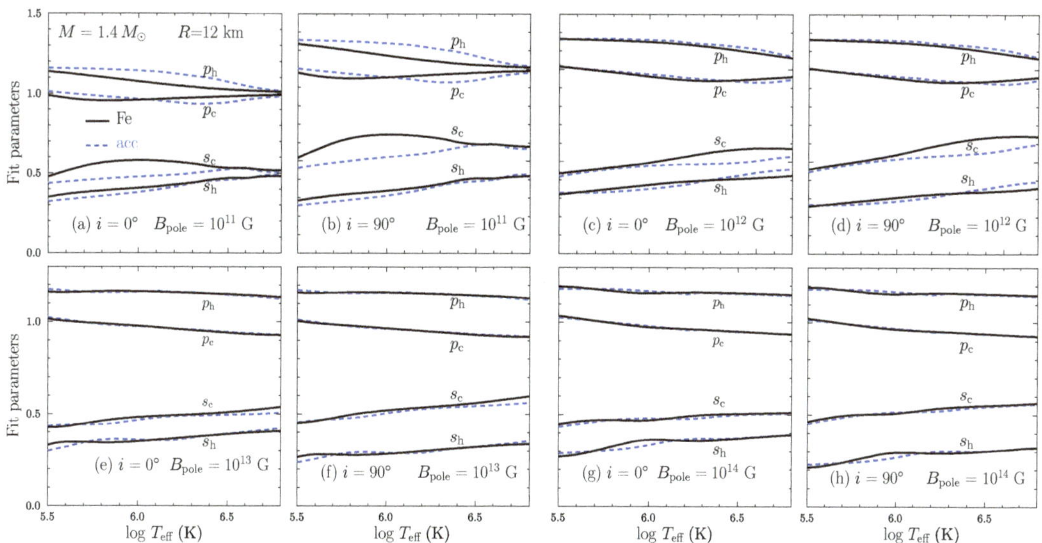

Figure 5. The 2BB fit parameters versus $\log T_{\rm eff}$ for a 1.4 $M_\odot$ neutron star with $R = 12$ km and $B_{\rm pole} = 10^{11}$, 10^{12}, 10^{13}, or 10^{14} G [panels (**a–h**)] for polar ($i = 0$) and equator ($i = 90°$) observations. Solid lines refer to the star with iron heat blanket, while dashed ones are for accreted heat blanket.

5. Extra Heating of Magnetic Poles

Section 4 describes the thermal emission of cooling neutron stars with dipole magnetic field near their surfaces, using the $T_{\rm s}$ model calculated in [10,11]. According to the theory, radiative spectral fluxes should be almost insensitive to the chemical composition of the heat blanketing envelopes; 2BB fits to their thermal X-ray spectral fluxes do not give the difference between the temperatures $T_{\rm effh}$ and $T_{\rm effc}$ of the hotter and colder BB components higher than 20% (for $B_{\rm pole} \lesssim 10^{14}$ G).

A brief comparison with observations of isolated middle-aged neutron stars in Paper I shows that there are no reliable candidates for such objects at the moment; the difference between $T_{\rm effh}$ and $T_{\rm effc}$ is actually larger, at least $\sim$50% for RX J1856.5$-$3754 [24,25], as a promising example.

Here, we propose a possible phenomenological extension of the discussed model, which may simplify the interpretation of observations of some sources. Specifically, let us assume the presence of additional heating of magnetic poles, which produces polar hotspots and raises the pole temperature. Such an extension has been used, for instance, in Paper I but assuming heating of both poles. Now we consider heating either of both poles or one of them. For simplicity, this heating is assumed to be axially symmetric. It does not violate the axial symmetry of the $T_{\rm s}$ distribution. Note that two equivalent hotspots do not destroy the initial symmetry of the north and south hemispheres, whereas one hotspot does destroy it. Possible physical justifications for extra heating are mentioned in Section 6.

Let $T_{\rm s0}(\vartheta)$ be the basic effective surface temperature used in Section 3. We introduce a small phenomenological angle ϑ_0 that detemines the size of a hotspot. Following Paper I we assume that at $\vartheta < \vartheta_0$ (in the nothern hotspot),

$$T_{\rm s}(\vartheta) = T_{\rm s0}(\vartheta) \left[1 + \delta \, \cos^2\left(\frac{\pi \vartheta}{2\vartheta_0} \right) \right] \quad \text{at } \vartheta \leq \vartheta_0, \tag{8}$$

and $T_{\rm s}(\vartheta) = T_{\rm s0}(\vartheta)$ outside the hotspot. The parameter δ specifies an extra temperature enhancement at the magnetic pole; the enhancement smoothly disappears as $\vartheta \to \vartheta_0$. This will be our *model with one hotspot*. Assuming a similar temperature enhancement near the second magnetic pole, we will get the *model with two hot spots*. The presence of spots renormalizes the total effective temperature T_{eff}, Equation (3).

Otherwise, the computation of spectral fluxes is the same as in Section 3. These fluxes can also be approximated by 2BB fits (6), and can be analyzed via phase-space maps. In some cases presented below the relative the fit accuracy becomes worse (reaching sometimes $\sim$10%) but the fit remains robust. We will again take the star with $M = 1.4\,M_\odot$ and $R = 12$ km. For certainty, we use the model of iron heat blanket and $\vartheta_0 = 10°$.

5.1. Extra Hotspots on Both Poles

The left panel of Figure 6 demonstrates calculated spectral fluxes for $B_{\text{pole}} = 10^{14}$ G and seven values of $\log T_{\text{eff}}[\text{K}] = 5.6, 5.8, \ldots, 6.8$ at $\delta = 0.3$. Since these hotspots are identical, observations of the first and second poles ($i = 0$ and $180°$) give the same fluxes. Such fluxes are shown by dashed lines. For comparison, the solid lines are calculated at $\delta = 0$, in which case the extra hotspots are absent and the results of Section 3 apply. The dotted lines show the fluxes for non-magnetic star. One can see that the presence of hotspots enhances the spectral fluxes at high photon energies. This is expected: enhanced surface temperature of hotspots intensifies generation of high-energy radiation. The higher the T_{eff}, the larger the photon energies are affected. The solid and dotted lines are the same as in Figure 3c.

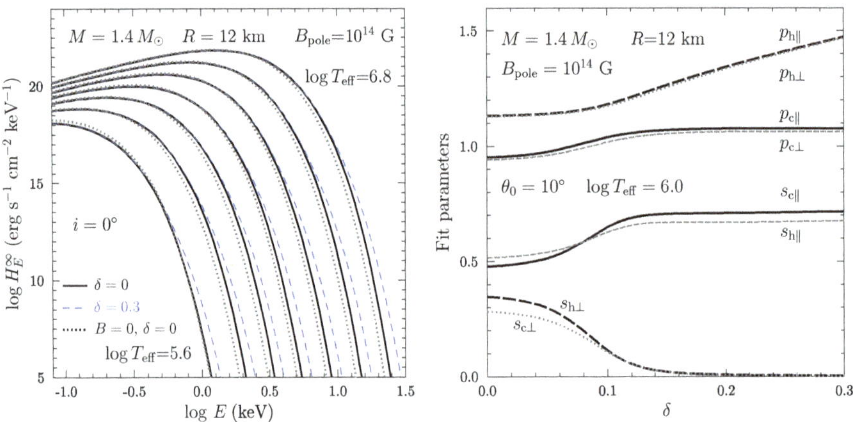

Figure 6. (Left): Thermal spectral fluxes observed from magnetic poles of a neutron star with $M = 1.4\,M_\odot$ and $R = 12$ km at $B_{\text{pole}} = 10^{14}$ G and $\log T_{\text{eff}}$ [K]= 5.6, 5.8,…,6.6. The solid lines are obtained without extra heating, while the dashed lines are for two extra hotspots of angular size $\vartheta_0 = 10°$ on magnetic poles at the heating parameter $\delta = 0.3$. Dots show 1BB model for a non-magnetic star to guide the eye. **(Right)**: 2BB fit parameters for the same star with $T_{\text{eff}} = 1$ MK possessing two polar hotspots of angular size $\vartheta_0 = 10°$ on magnetic poles [see Equation (8)] versus extra relative surface temperature increase δ at the pole.

The right panel of Figure 6 shows the 2BB fit parameters versus δ for the same star with $T_{\text{eff}}=1$ MK. At $\delta = 0$, the results naturally coincide with those in Figures 5g,h. However, with increasing δ, the fit parameters become different. The temperature T_{effh} of the hotter BB component and the effective emission surface area s_c of the colder component noticeably increase, whereas the emission surface area of the hotter component dramatically falls down. Even with really small hotspot temperature enhancement $\delta \leq 0.3$ (which gives the fraction of extra luminosity $\leq$1.4%), one obtains an absolutely new phase-space portrait with $s_h \ll s_c$. Such 2BB fits have been often inferred from observations of cooling isolated neutron stars (see, e.g., Paper I); these sources are usually interpreted as neutron stars, which have small hotspots with noticeably enhanced temperature.

Therefore, the theory with hotspots predicts (e.g., Paper I) two types of neutron stars, whose spectra can be approximated by the 2BB models. The first ones are those with smooth surface temperature distributions, created by nonuniform surface magnetic fields (as considered in Section 3) to be called spectral 2BB models *of smooth magnetic atmospheres*. The second sources are those with distinct hotspot BB component to be called 2BB *with hotspots*. Naturally, there is a smooth transition between them (for instance, by increasing δ in the right panel of Figure 6). It seems that the observations do not provide good candidates for the sources of the first type (Paper I), but there are some candidates for the sources of the second and intermediate types as we will outline later.

5.2. Phase Resolved Spectroscopy in Case of Two Hotspots

Let us present a few calculated lightcurves to illustrate an important problem of pulse fraction. For simplicity, we consider the star as an orthogonal rotator with the spin axis perpendicular to the line of sight. Figure 7 presents several lightcurves for a star with $B_{\mathrm{pole}} = 10^{14}$ G versus phase angle θ. The lightcurves are normalized by $H_E(0)$ (the spectral flux for pole observations at given energy E). The displayed ratio of redshifted or non-redshifted fluxes is the same at a given E, so that we drop the symbol ∞. Note that the normalization flux $H_E(0)$ strongly depends on energy by itself. Panel (a) corresponds to $\delta = 0$, in which case the hotspots are actually absent and we have the emission from the smooth magnetic atmosphere. In case (b) the hotspots with $\delta = 0.3$ are available and strongly affect the lightcurves.

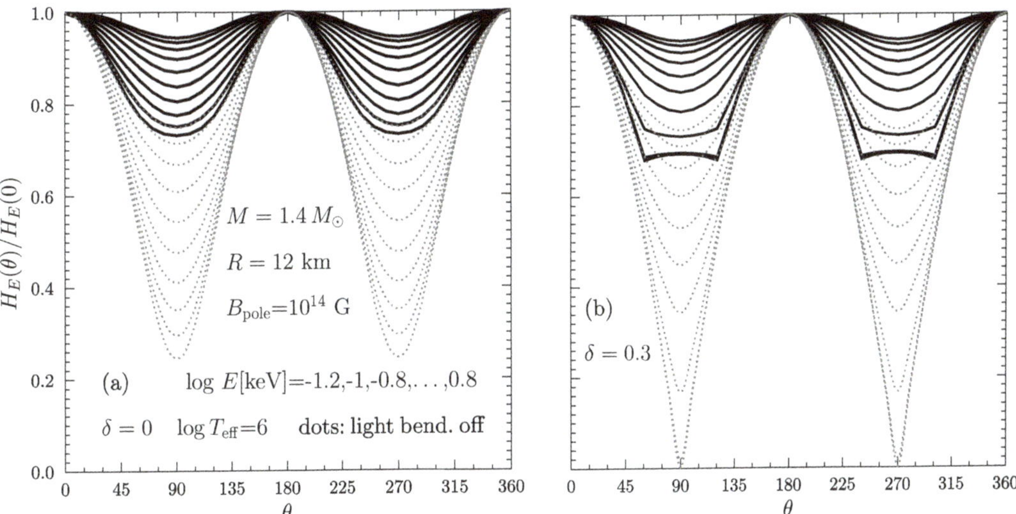

Figure 7. Theoretical lightcurves, normalized to pole observations, versus phase angle θ for a neutron star with T_{eff} = 1 MK and $B_{\mathrm{pole}} = 10^{14}$ G; (a) $\delta = 0$ (no extra pole heating) and (b) $\delta = 0.3$. Solid curves are for photon energies $\log E[\mathrm{keV}] = -1, -0.8, \ldots, 0.8$ (from top to bottom). The two lowest curves on panel (b) (with highest E) almost coincide. Dotted curves show the same but neglecting gravitational light bending. The star is assumed to be orthogonal rotator with the spin axis perperndicular to the line of sight.

Solid lines on each panel in Figure 7 present the lightcurves at 10 energies (from top to bottom), $\log E[\mathrm{keV}] = -1, -0.8, \ldots, 0.8$. The higher E, the stronger phase variations. The last two lines on panel (b) almost coincide. The lightcurves on panel (a) are seen to be smooth. The curves on panel (b) at $E \lesssim 1$ keV are although smooth and resemble those on panel (a). However, at higher energies the lightcurves (b) have shapes with erased dips at nearly equator observations [and then Equation (7) becomes inaccurate although

Equation (6) works well]. These curves are typical for lightcurves produced by antipodal point sources on the neutron star surface; see, e.g., Figure 4 in [19] or Figure 6 in [21]. Clearly, at high energies the star emits from the poles, the hottest places on the surface. Nevertheless, the pulse fraction remains not too high ($\leq 30\%$) in both cases (a) and (b) even at the highest taken energy $E \approx 6.3$ keV, at which thermal spectral flux becomes typically very low by itself. Therefore, the enhancement of thermal emission by extra heating of two magnetic poles does not lead to sizeable pulse fractions.

The dotted lines on Figure 7 show the same lightcurves as the solid lines but calculated neglecting gravitational bending of light rays. One sees that without the light bending the pulse fraction would be much stronger (as is well known; see, for example, [8]). The difference of solid and dotted curves on Figure 7 has simple explanation. Without light bending, the equator observations ($\theta = 90°$ or $270°$) would collect more light emitted from the cold equator. This would lower the observed emission and produce stronger dips of the lightcurves.

5.3. Extra Hotspot on One Pole

Finally, consider the case of a single hotspot, which we put at the north pole. This case is different from the previous one.

The left panel of Figure 8 presents spectral fluxes calculated for the same values of T_{eff} and B_{eff} as in Figure 6. The dashed lines are for the north-pole observations ($i = 0$) at $\delta = 0.3$. These fluxes are the same as in Figure 6: the observer sees a large fraction of the surface (mainly the northern hemisphere); the southern polar region is unseen.

The solid line in the left panel of Figure 8 refers to the same case of $\delta = 0.3$ but the star is observed from the south pole ($i = 180°$). Then the observer cannot see the hotspot, so the line exactly coincides with the solid line in Figure 3c. The dotted line in Figure 8 is for a non-magnetic star (as in Figures 3c and 6). Now, with only one extra heated pole, the difference between observations of the north and south poles is pronounced.

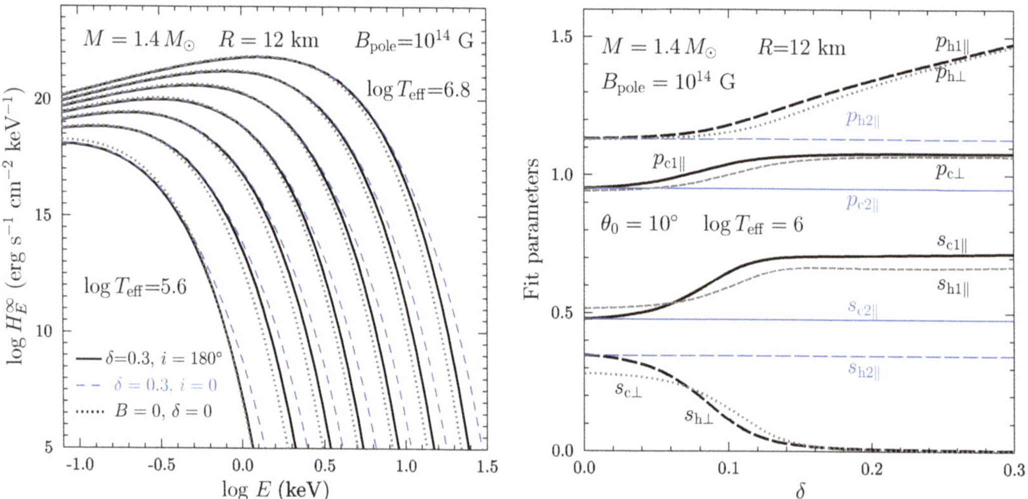

Figure 8. (Left): Theoretical spectral fluxes for a neutron star with $M = 1.4\,M_\odot$, $R = 12$ km, $B_{\mathrm{pole}} = 10^{14}$ G, different T_{eff} and one extra hotspot on one magnetic pole at the extra heating parameter $\delta = 0.3$. The solid and dashed curves are for non-heated ($i = 180°$) and heated pole ($i = 0$) observations, respectively. **(Right):** Maps of fit parameters versus δ for this star at $T_{\mathrm{eff}} = 1$ MK. The thick solid and gray curves are for the north-pole and equator observations, respectively. Thin horizontal curves are for the south-pole observations.

The right panel of Figure 8 presents the maps of 2BB fit parameters versus δ for the same star at $T_{\text{eff}} = 1$ MK for the north-pole, south-pole, and equator observations (to be compared with Figure 6). The maps the for north-pole and equator observations are nearly the same as those in Figure 6, but the maps for the south-pole observations are plain. Corresponding fit parameters are independent of δ just because the observer cannot see the emission from the hotspot (which is the only emission which increases with δ).

5.4. Phase Resolved Spectroscopy in Case of One Hotspot

Phase-resolved lightcurves are plotted in Figures 9 and 10. They are naturally different from the case of two extra heated magnetic poles. Figure 9 is analogous to Figure 7 and shows the light curves at the same T_{eff}, B_{pole}, and extra heating parameter $\delta = 0.3$ (a) and $\delta = 0.14$ (b). With one hotspot, the fractions of extra luminosities are 0.7% (a) and 0.3% (b). The phase-space variations are similar to those in Figure 7 only at low photon energies, in which case the extra polar heating is almost unnoticeable and the lightcurves have two shallow dips over one rotation period. However with growing E, the extra heating becomes important and increases the depth and width of the dips (such shapes are known to be produced by point sources on neutron star surface, e.g., Figure 4 in [19]). At $E \gtrsim 4$ keV in Figure 9a the pulse fraction becomes close to 100%. In that case the observer would see only one, but very pronounced dip over one rotation, because the extra hotspot would be poorly visible in a wide range of θ with the center at $\theta = 180°$. In Figure 9b the heating is weaker, the dips are smaller, but nevertheless they are much larger than for two heated spots at $\delta = 0.3$ in Figure 7. Evidently, one extra-heated magnetic pole is much more profitable than two, for producing large pulse fractions.

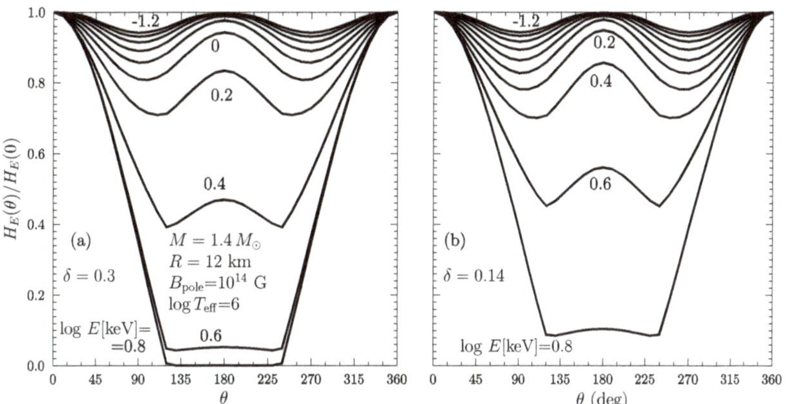

Figure 9. Theoretical lightcurves for a neutron star with $M = 1.4\,M_\odot$, $R = 12$ km, $T_{\text{eff}} = 1$ MK and $B_{\text{pole}} = 10^{14}$ G with an extra hotspot of angular size $\vartheta_0 = 10°$ on one magnetic pole versus phase angle θ at extra heating parameter $\delta = 0.3$ (**a**) and 0.14 (**b**). The curves are for photon energies $\log E$ [keV]$= -1, -0.8, \ldots, 0.8$. See the text for details.

Figure 10 is plotted for $B_{\text{pole}} = 10^{11}$ G to illustrate the effects of B_{pole} on the pulse fraction. Figure 10a presents the normalized lightcurves for the star with $\delta = 0.3$. Now the pulse fraction is lower than at $B_{\text{pole}} = 10^{14}$ G. Nevertheless, it remains much higher than it would be if the additional heating was switched on at both poles. Figure 10b demonstrates that at lower $\delta = 0.14$ at $B_{\text{pole}} = 10^{11}$ G the pulse fraction is weaker affected by the extra heating but nevertheless reaches about 70% at $E \sim 6$ keV.

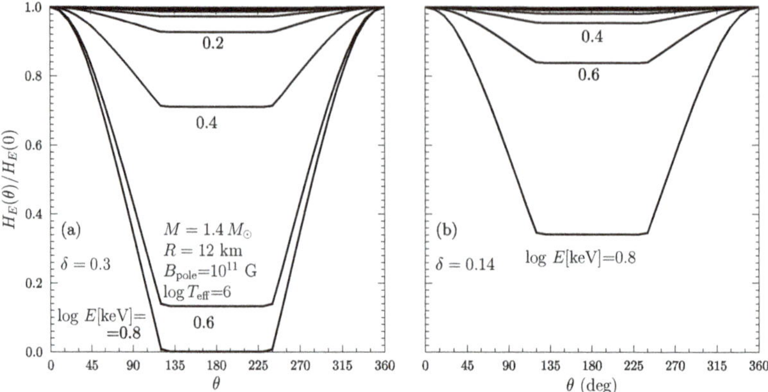

Figure 10. Same as in Figure 9 but for $B_{\mathrm{pole}} = 10^{11}$ G.

6. Discussion and Conclusions

Several isolated middle-aged neutron stars which radiation spectra have been (or can be) fitted by 2BB models have been selected in Paper I, based on the recent catalog [26] presented also on the Web: www.ioffe.ru/astro/NSG/thermal/cooldat.html, accessed on 17 October 2021.

These can be cooling neutron stars with dipole magnetic fields and additionally heated polar caps. This selection is illustrative (does not pretend to be complete).

1. XMMU J172054.5−372652 is a neutron star that is probably associated with the SNR G350.1−0.3 [27]; there is no direct evidence of pulsations. Potekhin et al. [26] used archival *Chandra* data and fitted the X-ray spectrum with a neutron star (NSX) atmosphere model assuming $M = 1.4M_\odot$ and $R = 13$ km. They obtained $T_{\mathrm{eff}} \approx 2$ MK, but did not perform two-component fits which would be interesting.
2. PSR B1055−52 (J1057−5226) is a moderately magnetized middle-aged pulsar. Its characteristic effective magnetic field, reported in the ATNF pulsar catalog [28], is $B_{\mathrm{eff}} = 1.1 \times 10^{12}$ G. Potekhin et al. [26] present the value $k_B T_{\mathrm{eff}}^\infty \approx 70$ eV ($T_{\mathrm{eff}}^\infty \approx 0.8$ MK) based on the 2BB spectral fit by De Luca et al. [29]; the fit includes also a power-law (PL) non-thermal radiation component. Potekhin et al. [26] have corrected the results of [29] for the distance estimate to the source made by Mignani et al. [30] and reported $T_{\mathrm{effh}}/T_{\mathrm{effc}} \sim 2.3$ with $s_h \ll s_c$. It seems worth to try to explain these data with a 2BB model containing hotspots.
3. PSR J1740+1000 possesses the magnetic field $B_{\mathrm{eff}} = 1.8 \times 10^{12}$ G. The 2BB spectral fit was performed in [31]. With the same version of the fit as selected in [26], one has $k_B T_{\mathrm{eff}}^\infty \approx 70$ eV ($T_{\mathrm{eff}}^\infty \approx 0.8$ MK), $T_{\mathrm{effh}}/T_{\mathrm{effc}} \sim 2.8$ and $s_h \ll s_c$, with the same conclusion as for the PSR B1055−52.
4. PSR B1823−13 (J1826−1334), located in the SNR G18.0−00.7, has $B_{\mathrm{eff}} = 2.8 \times 10^{12}$ G. Its X-ray emission is mainly non-thermal [32,33] but has some thermal component. The 1BB+PL fit gives the radius of thermally emitting region $R_{\mathrm{eff}}^\infty \approx 5$ km, smaller than the expected radius of a neutron star. Adding the second BB component is statistically insignificant with the present data, but might be possible in the future.
5. RX J1605.3+3249 (RBS 1556) is a neutron star studied by many authors (e.g., [34−38]) with contradictory conclusions on its properties (as detailed in [26]). There is no solid evidence of stellar rotation [37]. Its X-ray emission has been analyzed using BB fits and neutron star atmosphere models. Recently Pires et al. [37] analysed the *XMM-Newton* observations and Malacaria et al. [38] jointly analysed the *NICER* and *XMM-Newton* data. These teams improved 2BB fits by adding a broad Gaussian absorption line, in which case they got $k_B T_{\mathrm{effc}}^\infty \sim 60$ eV ($T_{\mathrm{effc}}^\infty \sim 0.7$ MK), and $T_{\mathrm{effh}}/T_{\mathrm{effc}} \sim 2$, a good opportunity to assume a dipole magnetic field and additionally heated poles.

6. RX J1856.5−3754 is a neutron star with nearly thermal spectrum. It was discovered by Walter et al. [39]. It is rotating with the spin period of $\sim$7 s; the effective magnetic field is $B_{\mathrm{eff}}\sim$1.5 $\times$ 10^{13} G; magnetic properties are highly debated (e.g., [18,40] and references therein). The spectrum has been measured in a wide energy range, including X-rays, optical and radio, and interpreted with many spectral models, particularly, with the model of thin partially ionized magnetized hydrogen atmosphere on top of solidified iron surface (see, e.g., [5,41]). Note alternative 2BB, 2BB+PL, and 3BB fits constructed by [24] and [25]. The 2BB fits give $k_B T^{\infty}_{\mathrm{effc}}\sim$40 eV ($T^{\infty}_{\mathrm{effc}} \approx 0.46$ MK), $T_{\mathrm{effh}}/T_{\mathrm{effc}}\sim$1.6, $R_{\mathrm{effh}}\sim$0.5 R and $R_{\mathrm{effc}}\sim R$; they seem closer to the 2BB spectral models with smooth magnetic atmosphere, than 2BB fits for other sources. Extra heating of magnetic poles would simplify this interpretation.

This brief analysis does not reveal any good candidate which would belong to the family of neutron stars with dipole magnetic fields and smooth surface temperature distribution (Section 4). This does not mean that such neutron stars do not exist; it can be difficult to identify them because their spectra are close to 1BB spectrum. To interpret the above sources one needs to complicate the model, for instance, by introducing extra heating of magnetic poles.

Now let us summarize our results. Following Paper I we have studied simple models (Section 3) of thermal spectra emitted from surfaces of isolated neutron stars with dipole surface magnetic fields $10^{11} \lesssim B_{\mathrm{pole}} \lesssim 10^{14}$ G. Such fields make the surface temperature distribution noticeably non-uniform. The model assumes BB emission with a local temperature from any surface element.

In Section 4 we have shown that the spectral X-ray fluxes emitted from such neutron stars are almost the same for heat blanketing envelopes composed of iron and fully accreted matter (at the same average effective surface temperatures T_{eff}). By measuring the spectral fluxes and T_{eff}, it would be difficult to infer composition of their heat blanketing envelopes, although this composition can noticeably affect neutron star cooling (e.g., [14] and references therein). In the presence of fully accreted matter, the fluxes are accurately fitted by 2BB models (as in Paper I for the iron heat blankets). At $M\sim$1.4 $M_{\odot}$ and $R\sim$12 km, the ratio $T_{\mathrm{effh}}/T_{\mathrm{effc}}$ of effective temperatures of the hotter to colder BB components cannot be essentially larger than $\sim$1.2. For less compact stars, with smaller M/R, this value can be somewhat higher (see e.g., [8], Paper I). If this ratio is larger from observations, then the model of Sections 3 and 4 cannot explain the data.

In Section 5, following Paper I, we have extended the model by assuming sufficiently weak extra heating of one or both magnetic poles. Increasing extra relative temperature rise δ at the pole(s), does not greatly violate the accuracy of the 2BB spectral approximation, but allows one to obtain larger $T_{\mathrm{effh}}/T_{\mathrm{effc}}$. Also, it affects the pulse fraction, which increases with B_{pole} and photon energy E. The pulse fraction can be very high if one magnetic pole is additionally heated, while the other is not.

Let us stress, that there have been many other studies of thermal emission from magnetized neutron stars. The emission produced by anisotropic surface temperature distribution of the star with dipole magnetic fields was analyzed by Page [8]. Quadrupole magnetic field components were added in [9]. A possible presence of toroidal field component in the neutron star crust was studied in [12,13]. Also, there were studies of thermal emission during magnetic field evolution in neutron stars (e.g., [18,40]). These works give a variety of magnetic fields configurations, surface temperature distributions, and phase-resolved spectra of neutron stars to be compared with observations and determine simultaneously both, the surface temperature distribution and magnetic field geometry.

Introducing some extra heating of magnetic poles, we are actually doing the same. Note that hot spots can be produced by pulsar mechanism. Also, the toroidal crustal magnetic fields can significantly widen the cold equatorial belt, creating relatively stronger heated poles [12], which can mimic hotspots. At the first step it might be simpler to introduce phenomenological polar heating, that can be different on the two poles, and to determine the parameters of the extra heating (the T_s distribution) by varying the parameters like δ and extra temperature profile [Equation (8)] at both poles. This would be a simple and physically transparent model of surface temperature distribution. At the next step, one could infer (constrain) the magnetic field geometry. This method can be easily extended to different T_s distributions, not necessarily axially symmetric.

Another very important direction would be to go beyond the approximation of local BB emission from any surface patch, and use more realistic emission models (with spectral features and anisotropic radiation, as well as with account of polarization effects, for example, [42,43]).

Funding: This research was partly (in analyzing the case of accreted matter) supported by the Russian Science Foundation, grant 19-12-00133.

Institutional Review Board Statement: Not applicable.

Informed Consent Statement: Not applicable.

Data Availability Statement: The data underlying this article will be shared on reasonable request to the corresponding author.

Acknowledgments: I am grateful to Oleg Gnedin for providing me photos of his Father.

Conflicts of Interest: The author declares no conflict of interest.

Abbreviations

The following abbreviations are used in this manuscript:

BB   blackbody

References

1. Gnedin, Y.N.; Dolginov, A.Z.; Tsygan, A.I. Possible mechanism of formation of the spectra of cosmic X-ray sources. *Sov. J. Exp. Theor. Phys. Lett.* **1969**, *10*, 283–285.
2. Gnedin, Y.N.; Pavlov, G.G. The transfer equations for normal waves and radiation polarization in an anisotropic medium. *Sov. J. Exp. Theor. Phys.* **1974**, *38*, 903–908.
3. Gnedin, Y.N.; Sunyaev, R.A. Polarization of optical and X-radiation from compact thermal sources with magnetic field. *Astron. Astrophys.* **1974**, *36*, 379–394.
4. Gnedin, Y.N.; Pavlov, G.G.; Tsygan, A.I. Photoeffect in strong magnetic fields and x-ray emission from neutron stars. *Sov. J. Exp. Theor. Phys.* **1974**, *39*, 201.
5. Potekhin, A.Y. Atmospheres and radiating surfaces of neutron stars. *Phys. Uspekhi* **2014**, *57*, 735. [CrossRef]
6. Potekhin, A.Y.; De Luca, A.; Pons, J.A. Neutron Stars — Thermal Emitters. *Space Sci. Rev.* **2015**, *191*, 171–206. [CrossRef]
7. Greenstein, G.; Hartke, G.J. Pulselike character of blackbody radiation from neutron stars. *Astrophys. J.* **1983**, *271*, 283–293. [CrossRef]
8. Page, D. Surface temperature of a magnetized neutron star and interpretation of the ROSAT data. 1: Dipole fields. *Astrophys. J.* **1995**, *442*, 273–285. [CrossRef]
9. Page, D.; Sarmiento, A. Surface Temperature of a Magnetized Neutron Star and Interpretation of the ROSAT Data. II. *Astrophys. J.* **1996**, *473*, 1067. [CrossRef]
10. Potekhin, A.Y.; Yakovlev, D.G. Thermal structure and cooling of neutron stars with magnetized envelopes. *Astron. Astrophys.* **2001**, *374*, 213–226. [CrossRef]
11. Potekhin, A.Y.; Yakovlev, D.G.; Chabrier, G.; Gnedin, O.Y. Thermal Structure and Cooling of Superfluid Neutron Stars with Accreted Magnetized Envelopes. *Astrophys. J.* **2003**, *594*, 404–418. [CrossRef]
12. Geppert, U.; Küker, M.; Page, D. Temperature distribution in magnetized neutron star crusts. II. The effect of a strong toroidal component. *Astron. Astrophys.* **2006**, *457*, 937–947. [CrossRef]
13. Zane, S.; Turolla, R. Unveiling the thermal and magnetic map of neutron star surfaces though their X-ray emission: Method and light-curve analysis. *MNRAS* **2006**, *366*, 727–738. [CrossRef]
14. Beznogov, M.V.; Potekhin, A.Y.; Yakovlev D.G. Heat blanketing envelopes of neutron stars. *Phys. Rep.* **2021**, *919*, 1–68. [CrossRef]
15. Yakovlev, D.G. Two-blackbody portraits of radiation from magnetized neutron stars. *MNRAS* **2021**, *506*, 4593–4602. [CrossRef]
16. Zavlin, V.E.; Pavlov, G.G. Modeling Neutron Star Atmospheres. In *Neutron Stars, Pulsars, and Supernova Remnants*; Becker, W., Lesch, H., Trümper, J., Eds., Max-Plank-Institut für extraterrestrische Physik: Garching bei München, Germany, 2002; p. 263.
17. González Caniulef, D.; Zane, S.; Taverna, R.; Turolla, R.; Wu, K. Polarized thermal emission from X-ray dim isolated neutron stars: the case of RX J1856.5-3754. *MNRAS* **2016**, *459*, 3585–3595. [CrossRef]

18. De Grandis, D.; Taverna, R.; Turolla, R.; Gnarini, A.; Popov, S.B.; Zane, S.; Wood, T.S. X-ray Emission from Isolated Neutron Stars revisited: 3D magnetothermal simulations. *Astrophys. J.* **2021**, *914*, 118. [CrossRef]
19. Beloborodov, A.M. Gravitational Bending of Light Near Compact Objects. *Astrophys. J. Lett.* **2002**, *566*, L85–L88. [CrossRef]
20. Poutanen, J.; Gierliński, M. On the nature of the X-ray emission from the accreting millisecond pulsar SAX J1808.4-3658. *MNRAS* **2003**, *343*, 1301–1311. [CrossRef]
21. Poutanen, J. Accurate analytic formula for light bending in Schwarzschild metric. *Astron. Astrophys.* **2020**, *640*, A24. [CrossRef]
22. Potekhin, A.Y.; Chabrier, G.; Yakovlev, D.G. Internal temperatures and cooling of neutron stars with accreted envelopes. *Astron. Astrophys.* **1997**, *323*, 415–428.
23. Haensel, P.; Potekhin, A.Y.; Yakovlev, D.G. Neutron Stars. 1. Equation of State and Structure; *Astrophysics and Space Science Library*; Springer: New York, NY, USA, 2007; Volume 326.
24. Sartore, N.; Tiengo, A.; Mereghetti, S.; De Luca, A.; Turolla, R.; Haberl, F. Spectral monitoring of RX J1856.5-3754 with XMM-Newton. Analysis of EPIC-pn data. *Astron. Astrophys.* **2012**, *541*, A66. [CrossRef]
25. Yoneyama, T.; Hayashida, K.; Nakajima, H.; Inoue, S.; Tsunemi, H. Discovery of a keV-X-ray excess in RX J1856.5-3754. *Publ. Astron. Soc. Jpn.* **2017**, *69*, 50. [CrossRef]
26. Potekhin, A.Y.; Zyuzin, D.A.; Yakovlev, D.G.; Beznogov, M.V.; Shibanov, Y.A. Thermal luminosities of cooling neutron stars. *MNRAS* **2020**, *496*, 5052–5071. [CrossRef]
27. Gaensler, B.M.; Tanna, A.; Slane, P.O.; Brogan, C.L.; Gelfand, J.D.; McClure-Griffiths, N.M.; Camilo, F.; Ng, C.Y.; Miller, J.M. The (Re-)Discovery of G350.1-0.3: A Young, Luminous Supernova Remnant and Its Neutron Star. *Astrophys. J. Lett.* **2008**, *680*, L37. [CrossRef]
28. Manchester, R.N.; Hobbs, G.B.; Teoh, A.; Hobbs, M. The Australia Telescope National Facility Pulsar Catalogue. *Astron. J.* **2005**, *129*, 1993–2006. [CrossRef]
29. De Luca, A.; Caraveo, P.A.; Mereghetti, S.; Negroni, M.; Bignami, G.F. On the Polar Caps of the Three Musketeers. *Astrophys. J.* **2005**, *623*, 1051–1069. [CrossRef]
30. Mignani, R.P.; Pavlov, G.G.; Kargaltsev, O. Optical-Ultraviolet Spectrum and Proper Motion of the Middle-aged Pulsar B1055-52. *Astrophys. J.* **2010**, *720*, 1635–1643. [CrossRef]
31. Kargaltsev, O.; Durant, M.; Misanovic, Z.; Pavlov, G.G. Absorption Features in the X-ray Spectrum of an Ordinary Radio Pulsar. *Science* **2012**, *337*, 946. [CrossRef]
32. Pavlov, G.G.; Kargaltsev, O.; Brisken, W.F. Chandra Observation of PSR B1823-13 and Its Pulsar Wind Nebula. *Astrophys. J.* **2008**, *675*, 683–694. [CrossRef]
33. Zhu, W.W.; Kaspi, V.M.; McLaughlin, M.A.; Pavlov, G.G.; Ng, C.Y.; Manchester, R.N.; Gaensler, B.M.; Woods, P.M. Chandra Observations of the High-magnetic-field Radio Pulsar J1718-3718. *Astrophys. J.* **2011**, *734*, 44. [CrossRef]
34. Motch, C.; Sekiguchi, K.; Haberl, F.; Zavlin, V.E.; Schwope, A.; Pakull, M.W. The proper motion of the isolated neutron star J1605.3+3249 . *Astron. Astrophys.* **2005**, *429*, 257–265. [CrossRef]
35. Posselt, B.; Popov, S.B.; Haberl, F.; Trümper, J.; Turolla, R.; Neuhäuser, R. The Magnificent Seven in the dusty prairie. *Astrophys. Space Sci.* **2007**, *308*, 171–179. [CrossRef]
36. Tetzlaff, N.; Schmidt, J.G.; Hohle, M.M.; Neuhäuser, R. Neutron Stars From Young Nearby Associations: The Origin of RX J1605.3+3249. *Publ. Astron. Soc. Aust.* **2012**, *29*, 98–108. [CrossRef]
37. Pires, A.M.; Schwope, A.D.; Haberl, F.; Zavlin, V.E.; Motch, C.; Zane, S. A deep XMM-Newton look on the thermally emitting isolated neutron star RX J1605.3+3249. *Astron. Astrophys.* **2019**, *623*, A73. [CrossRef]
38. Malacaria, C.; Bogdanov, S.; Ho, W.C.G.; Enoto, T.; Ray, P.S.; Arzoumanian, Z.; Cazeau, T.; Gendreau, K.C.; Guillot, S.; Güver, T.; et al. A Joint NICER and XMM-Newton View of the "Magnificent" Thermally Emitting X-Ray Isolated Neutron Star RX J1605.3+3249. *Astrophys. J.* **2019**, *880*, 74. [CrossRef]
39. Walter, F.M.; Wolk, S.J.; Neuhäuser, R. Discovery of a nearby isolated neutron star. *Nature* **1996**, *379*, 233–235. [CrossRef]
40. Popov, S.B.; Taverna, R.; Turolla, R. Probing the surface magnetic field structure in RX J1856.5-3754. *MNRAS* **2017**, *464*, 4390–4398. [CrossRef]
41. Ho, W.C.G.; Kaplan, D.L.; Chang, P.; van Adelsberg, M.; Potekhin, A.Y. Magnetic hydrogen atmosphere models and the neutron star RX J1856.5-3754. *MNRAS* **2007**, *375*, 821–830. [CrossRef]
42. Ho, W.C.G.; Potekhin, A.Y.; Chabrier, G. Model X-Ray Spectra of Magnetic Neutron Stars with Hydrogen Atmospheres. *Astrophys. J. Suppl.* **2008**, *178*, 102–109. [CrossRef]
43. Zyuzin, D.A.; Karpova, A.V.; Shibanov, Y.A.; Potekhin, A.Y.; Suleimanov, V.F. Middle aged γ-ray pulsar J1957+5033 in X-rays: pulsations, thermal emission, and nebula. *MNRAS* **2021**, *501*, 4998–5011. [CrossRef]

Review

Progress in Understanding the Nature of SS433

Anatol Cherepashchuk

P. K. Sternberg State Astronomical Institute, M. V. Lomonosov Moscow State University, Universitetskij Prospect, 13, 119234 Moscow, Russia; cherepashchuk@gmail.com

Abstract: SS433 is the first example of a microquasar discovered in the Galaxy. It is a natural laboratory for studies of extraordinarily interesting physical processes that are very important for the relativistic astrophysics, cosmic gas dynamics and theory of evolution of stars. The object has been studied for over 40 years in the optical, X-ray and radio bands. By now, it is generally accepted that SS433 is a massive eclipsing X-ray binary in an advanced stage of evolution in the supercritical regime of accretion on the relativistic object. Intensive spectral and photometric observations of SS433 at the Caucasian Mountain Observatory of the P. K. Sternberg Astronomical Institute of M. V. Lomonosov Moscow State University made it possible to find the ellipticity of the SS433 orbit and to discover an increase in the system's orbital period. These results shed light on a number of unresolved issues related to SS433. In particular, a refined estimate of the mass ratio $\frac{M_x}{M_v} > 0.8$ was obtained (M_x and M_v are the masses of the relativistic object and optical star). Based on these estimates, the relativistic object in the SS433 system is the black hole; its mass is $>8M_\odot$. The ellipticity of the orbit is consistent with the "slaved" accretion disc model. The results obtained made it possible to understand why SS433 evolves as the semi-detached binary instead of the common envelope system.

Keywords: close binaries; black holes; evolution of binary stars

Citation: Cherepashchuk, A. Progress in Understanding the Nature of SS433. *Universe* **2021**, *8*, 13. https://doi.org/10.3390/universe8010013

Academic Editors: Nazar R. Ikhsanov, Galina L. Klimchitskaya and Vladimir M. Mostepanenko

Received: 26 October 2021
Accepted: 24 December 2021
Published: 27 December 2021

Publisher's Note: MDPI stays neutral with regard to jurisdictional claims in published maps and institutional affiliations.

1. Introduction

The unique object SS433 shows in its optical spectrum [1], in addition to the standard emission lines of hydrogen and helium, moving emissions, which move along the spectrum with a period of 162.3 d [2]. The amplitude of these displacements reaches an enormous value of ~1000 Å, which amounts up to tens of thousands of km s^{-1} on the scale of velocities. Before the discovery of SS433 in the 1970s [2,3], astronomers had never seen anything like this. Therefore, the nature of object SS433 seemed mysterious; some scientists called SS433 "the enigma of the century". In the earliest press releases, there were even mentions that it was possible that we were observing signals from an extraterrestrial intelligence which was shining on us with a powerful re-configurable laser. Fortunately, this hypothesis did not manage to appear in scientific publications since, in 1979, it turned out [4,5] that the moving lines in the spectrum of SS433 are formed in relativistic ($v \simeq 0.26c$) collimated jets that are ejected from some central source and precess with a period of $\approx$164 d. The opening angle of the precession cone is $\approx$20°, and the precession axis is inclined with respect to the line of sight by an angle of $\approx$79°. Although the nature of the central source remained unclear, speculation arose that all of these phenomena were enacted in a binary system [6,7]. The binary system model was proposed in 1980 by [8], who measured the Doppler shifts of the narrow components of the stationary hydrogen lines and found a period of $\approx$13.1 d, which testified in favor of the close binary system model. The authors from the analysis of their data, assuming that the radial velocity curve measured by them reflects the orbital motion of the components, came to the conclusion that SS433 is a low-mass binary system consisting of a low-mass optical star ($1M_\odot$) and a neutron star. As it turned out later, according to our research, this conclusion turned out to be incorrect.

In 1980, we made photometric observations of SS433 and found that SS433 is an eclipsing binary system [9]. The discovery of optical eclipses in the SS433 system made it possible to establish that the radial velocity curve constructed by [8] does not reflect the orbital motion of the components but describes the motion of gas in a gas stream flowing from the Lagrange point L1 onto a relativistic object. Analysis of optical eclipses in SS433, as well as studies of the system's extra eclipse brightness depending on the phase of the precessional period [9], made it possible to construct a self-consistent model of SS433 as a massive X-ray eclipsing binary system at an advanced stage of evolution with an optically bright supercritical accretion disc predicted in [10]. Relativistic jets are ejected from the central parts of the supercritical disc perpendicular to its plane. The plane of the disc is inclined to the orbital plane of the SS433 system at an angle of 20°, the disc precesses with a period of 162.3 d and relativistic jets track the disc precession. The moving emissions are formed in the outer parts of the jets.

The question arose about the reasons for the disc precession. Back in 1973, Katz proposed a precessing disc model to explain the 35-day X-ray cycle in the Her X-1 system [11]. In 1974, Roberts proposed a slaved precession model for an accretion disc in the Her X-1 system [12]. Even earlier, the idea of the slaved disc was expressed by Shakura [13]. The slaved disc model for SS433 was proposed in 1980 by van den Heuvel et al. [14]. In this model, the axis of rotation of the optical star precesses and the disc tracks this precession. The non-perpendicularity of the axis of rotation of the optical star to the orbital plane may be associated with an asymmetric supernova explosion, which inclined the orbital plane of the binary system relative to the axis of rotation of the star [12,15]. To explain the high and stable velocity of matter in the SS433 jets, a model of radiative acceleration was proposed (the so-called line-looking effect [16,17]).

Thus, by the early 1980s, the main features of SS433 as a massive close binary X-ray system in the supercritical accretion regime had been clarified. However, many problems remained unresolved. The massive binary system model for SS433 [9] was confirmed by [18], who plotted the radial velocity curve from the stationary HeII 4686 line and measured the mass function of a relativistic object (the lower limit of the mass of an optical star), which turned out to be $\approx 10 M_\odot$.

2. Unsolved Problems

Over the past 40 years, the object SS433 has been studied in detail in the optical, IR, X-ray and radio ranges (see, for example, reviews [6,7] where all the necessary references are given). Many new and interesting results were obtained on the object SS433, which turned out to be the first example of a microquasar in our Galaxy. However, the main problems associated with this unique object have been unresolved until recently. Let us list some of them.

1. The nature of the precessional variability of SS433. Which model is applicable to SS433: slaved precessing disc model or slaved disc model?
2. The nature of the accreting relativistic object: a neutron star or a black hole?
3. Why has a common envelope not formed in the SS433 system, which undergoes secondary mass exchange in the thermal evolution time scale, and the system evolves as a semi-detached one?

Recently, new observational data have been obtained on the object SS433, which bring us closer to solving these problems.

3. Discovery of the Ellipticity of SS433's Orbit: Strong Support for a Slaved Accretion Disc Model

The slaved accretion disc model proposed in [14] seems to be very attractive since a natural physical justification can be formulated for it. Indeed, a small (~1%) asymmetry of a supernova explosion, which accompanied the formation of a relativistic object in a binary system, can turn the orbital plane of the binary system relative to the axis of rotation of the optical star by a significant angle [12,15]. A supernova explosion could also significantly increase the eccentricity of the system's orbit. Under the action of tidal interaction from the side of the resulting relativistic object, the axis of rotation of the optical star begins to precess, which leads to the formation of an accretion disc that does not lie in the plane of the system's orbit and precesses with the precession period of the optical star. Since a precessing star has a large mass and a huge rotational moment, its precession is very stable and, as a consequence, the observed precession of the accretion disc should also be stable on average. Our long-term (24 years) spectral observations of SS433 [19,20] using published data indicate the stability of the parameters of the kinematic model SS433 over 40 years. This stability can be considered as an argument in favor of the model of a slaved accretion disc in the SS433 system.

However, until recently, there was one fundamental difficulty in applying the model of a slaved accretion disc to the SS433 system. The orbit of SS433 was considered circular since the secondary minimum on the eclipsing light curve, within the errors associated with significant physical variability of the system, was located approximately in the middle between the two primary minima.

The presence of a circular orbit in the SS433 system during asynchronous rotation of the optical star (the star's rotation axis is not perpendicular to the orbital plane) contradicts the theory of tidal synchronization of stars in binary systems. According to this theory [21,22], the synchronization of axial and orbital rotation due to the dissipation of the energy of orbital motion in dynamic tides should occur before the circularization of the orbit. Therefore, if the orbit of SS433 is circular, then the optical star must be synchronized with the orbital revolution and its rotation axis must be perpendicular to the orbital plane. In this case, the accretion disc should lie in the plane of the system's orbit. The slaved disc model is not applicable in this case. As noted above, a significant initial eccentricity of the orbit could appear in a binary system after a supernova explosion, so the assumption of an initial circular orbit seems unnatural. In this regard, the problem of finding direct signs of the ellipticity of the SS433 orbit arises. Indirect evidence of the ellipticity of the orbit was obtained in [19,20,23]. In these works, it was found that the velocity of matter in relativistic jets SS433 is modulated by the orbital period $P_{orb} = 13.082$ d. In this regard, a hypothesis was put forward about the possible ellipticity of the SS433 orbit: changes in the distance between the components in an elliptical orbit cause a change in the rate of influx of stellar matter into the accretion disc, which can lead to orbital modulation of the velocity of matter in the jets. Other indirect indications of the ellipticity of the SS433 orbit were obtained in [24,25]. In this regard, it is of interest to search for direct evidence of the ellipticity of the SS433 orbit.

During 2018–2021, at the Caucasian Mountain Observatory (KGO) of the GAISH MSU, we carried out intensive photometric monitoring of SS433. Several hundred brightness estimates of SS433 were obtained in BVRI filters during 276 observation nights. These data allowed us, in combination with published photometric observations, to construct a high-precision (including 909 individual observations) mean SS433 light curve in the V band in the phases of the precessional period $P_{prec} = 163.3$ d, corresponding to the maximum opening of the accretion disc with respect to the observer [26]. Figure 1 shows this curve. It can be seen that the secondary minimum has a noticeable shift relative to the middle between the two primary minima. Analysis of this displacement and the study of the difference in the widths of the primary and secondary minima performed in [26] made it possible to establish that the eccentricity of the SS433 orbit is $e = 0.050 \pm 0.007$, and the longitude of the periastron for the epoch of $\approx$1990 is $\omega = 40° \pm 20°$. The ellipticity of the SS433 orbit discovered by us is in good agreement with the model of a slaved accretion disc. Both observational facts—the long-term stability of the kinematic model parameters and the ellipticity of the SS433 orbit—strongly support the model of a slaved accretion disc in the SS433 system [14], as well as the hypothesis that the reason that the disc does not lie in the plane orbit is an asymmetric supernova explosion in the system [12,15].

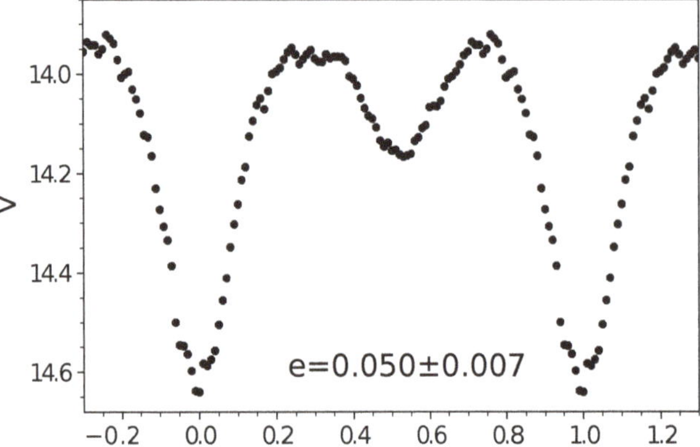

Figure 1. Average orbital light curve of SS433 derived from 909 individual measurements taken during 1978–2012. The observed points were selected in the phases of maximum opening of the accretion disc with respect to the observer. From the paper [26].

4. The Discovery of an Increase in the Orbital Period of the SS433 System: A Strong Argument for the Presence of a Black Hole

Despite more than forty years of research, the nature of the relativistic object in the SS433 system remained completely unclear until recently (see discussions on this topic in [6,7,20]). In [19,20], using information on the constancy of the orbital period of SS433 over 28 years [27,28], the authors estimated the ratio of the masses of the components in the SS433 $q = \dfrac{M_x}{M_v} = 0.6$ (M_x and M_v are the masses of a relativistic object and an optical star, respectively). The constancy of the orbital period for the SS433 system is very surprising; after all, the rate of mass loss from the system in the form of wind from the supercritical accretion disc is very high, $\sim 10^{-4} M_\odot$ year^{-1} (see, for example, [9]). In [20,29], the balance equations for the loss of angular momentum of the SS433 system were derived by various mechanisms: the transfer of mass and angular momentum during the flow of matter of an optical star through the internal Lagrange point L_1 to a relativistic object, Jeans mass loss in a symmetric high-velocity outflow matter in the form of wind from the disc as well as the outflow of matter outside the binary system through the outer Lagrange point L_2. The first mechanism, when flowing from a more massive optical star to a relativistic object, tends to bring the components closer together (and, accordingly, to decrease the orbital period). The second mechanism seeks to increase the distance between the components, that is, to increase the orbital period. The third mechanism leads to a decrease in the distance between the components, i.e., to a decrease in the orbital period. It turns out that, at a huge rate of mass loss in the form of a wind $\sim 10^{-4} M_\odot$ year^{-1}, an equilibrium between these three mechanisms is established at a very large mass ratio $q \simeq 0.6$. With an optical star mass of $\approx 10 M_\odot$ (see, for example, [18]), it follows that the mass of a relativistic object is $6 M_\odot$, which is typical of a black hole. In this regard, it is very important to check with additional observations the nature of the constancy or change in the orbital period of SS433. As noted above, in 2018–2021, on the automated 60-cm reflector of the Caucasian Mount Observatory of P. K. Sternberg Astronomical Institute of M. V. Lomonosov Moscow State University, we obtained dense series of photometric observations of SS433 in *BVRI* filters. With the involvement of all published photometric observational data (since 1979), we discovered an increase in the orbital period of SS433 with time [26]. Figure 2 shows the corresponding plots $O - C$—the differences between the observed moments of the primary minimum of the light curve of SS433 and theoretical moments calculated with linear elements [28]. Here, the parabola corresponds to an increase in the period, and the inclined line corresponds to a constant period (the slope of this line to the abscissa axis determines the correction to a constant period). It can be seen that the parabolic approximation of the points in Figure 2 is much more preferable than the linear approximation. Thus, it is shown that the orbital period of SS433 increases. The rate of this increase is $\dot{P} = (1.0 \pm 0.3) \times 10^{-7}$ s per s. It is also shown that the model of the third body in the SS433 system, which can also lead to an increase in the orbital period, is rejected, since the mass of the third body must be more than $16 M_\odot$. The lines of such a massive star would be visible in the total spectrum of the SS433 system, which is not observed.

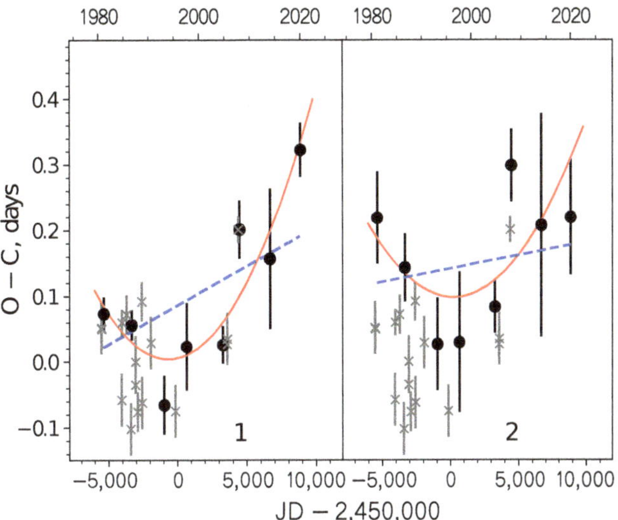

Figure 2. $O - C$ values for SS433, calculated using the ephemeris with a constant orbital period $P_{\text{orb}} = 13.08223$ d. Right and left panels: $O - C$ calculated by different methods. The parabola corresponds to an increase in the orbital period. The inclined straight line corresponds to a constant period with a ΔP correction. From the paper [26].

Using the equation to balance the loss of angular momentum and taking into account that the orbital period of SS433 increases, an improved estimate of the mass ratio $q > 0.8$ is given in [26]. With the mass of the optical star $M_v = 10 M_\odot$, an estimate is obtained for the mass of the black hole in SS433 $M_x > 8 M_\odot$. Thus, it has been shown that the SS433 system contains a black hole with a mass close to the average mass of stellar black holes in X-ray binaries ($\approx 8 \div 10 M_\odot$).

The increase in the orbital period of SS433 found by us allows us to reliably reject the presence of a neutron star in the system since, in this case, the orbital period of SS433 should not increase, but should decrease with time, which contradicts observations.

5. Other Estimates of the Mass Ratio

The supercritical optically bright accretion disc in the SS433 system makes spectroscopic investigations of the donor star more difficult due to the disc's powerful radiation. Nevertheless, there is a significant progress in this problem by now.

Careful spectroscopic investigations of SS443 made by [30–32] let us estimate the spectral type of the optical star as A7I and to obtain (using absorption lines) the radial velocity curve of this star with the semi-amplitude $K_v = 58.2 \pm 3.1$ km s^{-1}. It corresponds to the optical star's mass function $f_v(m) = 0.268 M_\odot$. Similar results were obtained by [33]. In the recent paper by [34], the axial rotation velocity $v_{\text{rot}} = 140 \pm 20$ km s^{-1} was measured using the Doppler widening of absorption lines of the optical star. Using this value, the corresponding masses and mass ratio of the components of SS433 were estimated:

$q = \dfrac{M_x}{Mv} = 0.37 \pm 0.04$, the mass of the relativistic object $M_x = 4.2 \pm 0.4 M_\odot$ and the mass of the optical star $M_v = 11.3 \pm 0.6 M_\odot$.

There are reasons to suppose that the estimates of the mass of the relativistic object in SS433 based on the spectroscopic data cannot be assumed as reasonably reliable, because the SS433 radiation travel through the intercomponent moving gas medium and also through the rotating circumbinary gas envelope. For the gigantic mass loss rate $\dot{M} \simeq 10^{-4}$ year^{-1}, the density of matter in these structures is very high. Therefore, the selective absorption of the optical star's light in moving dense gas structures can significantly distort the orbital radial velocity curve of the donor star measured using absorption lines, see, e.g., [35–37].

The situation with emission lines is also uncertain. For example, the mass function of the relativistic object calculated using the He II 4686 emission line changed from $10 M_\odot$ [18] to $2 M_\odot$ [38]; it depends on the phase of the precessional period. This can be related with the complicated structure of the wind from the precessing accretion disc whose shape can be asymmetric and twisted.

Because of these problems, great hopes were pinned on studies of X-ray eclipses in the SS433 system. Since the orbital inclination of the system $i = 79°$ is known from the analysis of displacements of moving emission lines, and the optical star fills or even overfills its inner critical Roche lobe, the analysis of the duration of the X-ray eclipse (when the star hides X-ray structures) gives a principal possibility to estimate the mass ratio of the components.

First such estimate was made by [39–41]. From the analysis of X-ray eclipses in SS433 in the range 2–10 keV in the model of eclipse of the thin relativistic jets (radiating in X-rays) by the optical star, there was obtained a very low mass ratio $q = \dfrac{M_x}{M_v} \simeq 0.15$. In this case, it was assumed that the optical star fills its Roche lobe.

Nevertheless, there are observational data indicating that the optical star in SS433 not only fills it inner critical Roche lobe but also overfills it. At the same time, the star outflows both through the inner Lagrange point L_1 and through the outer Lagrange point L_2. From the analysis of spectral data, arguments in favor of the outflow of the optical star's matter through the outer Lagrange point L_2 were suggested [42]. It indicates that the star most likely fills its outer Roche lobe whose size is significantly (by 15–20%) greater than the size of the inner critical Roche lobe. In papers by [43,44], the equatorial outflow of the matter in the plane perpendicular to the direction of the jets was discovered in S433. It also indicates that the star outflows through the L_2 point and that the star fills its outer Roche lobe. The conclusion about the filling of the outer Roche lobe by the optical star in SS433 and about the outflow through the L_2 point is supported by the optical and infrared spectroscopy of this object that revealed two-humped stationary hydrogen emission lines [45,46]. These lines indicate the presence of the circumbinary envelope which most likely is forming during the outflow of the optical star through the L_2 point.

Recently, several theoretical studies [47,48] showed that, due to the limited carrying capacity (for the outflowing material) of the surroundings of the L_1 point, the star (with the radiative envelope) during the mass transfer in the thermal time scale can be in a state of a significant overflow of its inner critical Roche lobe for a long time and the matter can outflow through the L_2 point.

So, taking into account that the radius of the occulting star in SS433 is significantly greater than the size of the inner critical Roche lobe, the estimate of the component mass ratio obtained by [39–41] should be considered as the lower limit of the mass ratio.

In the review by [7], the results of the interpretation of the eclipsing (orbital) and precessional variability in the hard X-ray range ($kT \approx 18$–60 keV) were summarized. It was shown that, since the X-ray spectrum hardness does not change with the phase of the orbital and precessional variability (the observed X-ray flux changes by about five times), the hard X-ray radiation in contradiction to the soft X-ray radiation ($kT \approx 2$–10 keV) is formed, not in narrow jets but in an extended quasi-isothermal hot corona of the accretion disc. According to [49], the temperature of the corona is ≈ 20 keV, its optical depth (Thompson scattering) is ≈ 0.2 and the mass loss rate in the jets is $\dot{M} \approx 3 \times 10^{-7} M_\odot$ year^{-1} (the mass loss rate in the wind from the disc is $\sim 10^{-4} M_\odot$ year^{-1}).

In the paper by [7], to analyze the X-ray eclipses and precessional variability of SS433 in the hard X-ray range ($kT = 18$–60 keV), the following model was used: the optical star overflows its inner critical Roche lobe including the case of the filling of the outer Roche lobe, and the star outflows through the L_2 point. The interpretation of observations was made in the model of precessing accretion disc with hot corona in its central parts. It became clear that q cannot be unambiguously found from the X-ray eclipsing light curve alone. Co-interpretation of both the eclipsing X-ray light curve and precessional light curve in the model of the filling of the outer Roche lobe let us obtain the estimate of the mass ratio $q = \dfrac{M_x}{M_v} > 0.4$–$0.8$. For the mass of the A7I optical star equal to $10 M_\odot$, it gives the estimate of the mass of the black hole $M_x \approx 4$–$8 M_\odot$.

So, the conclusion about the high mass ratio of components in the SS433 system and the presence of the black hole in this system are supported by numerous independent observational data.

6. Explanation of the Evolutionary Status of SS433 as a Semi-Detached Binary System

According to the predictions of the modern theory of the evolution of massive close binary systems [50], at the stage of secondary mass exchange, due to the very high rate of the matter flow from an optical star to a relativistic object, a common envelope should form. In this case, due to deceleration by dynamic friction, rapid decreasing of the separation of components occurs. As a result, depending on the initial angular momentum of the binary system, either a short-period WR + c system (of the Cyg X-3 type, $P_{orb} \simeq 4.8$ h) is formed or the Thorne–Zhitkov object [51], a completely convective supergiant with relativistic object in the center, is formed. According to [52], the life time for the Thorne–Zhitkov object can be very short due to the neutrino emission.

The SS433 system is at the stage of secondary mass exchange but its evolution, for some reason, took a different path. The system evolves as a semi-detached binary. The removal of mass and angular momentum from the system is not carried out into the common envelope but in the form of a radial outflow of a powerful wind ($\dot{M} \simeq 10^{-4} M_\odot$ year^{-1}, $v \simeq 10^3$ km s^{-1}) from the supercritical accretion disc. This feature of SS433 has long remained a mystery.

Recently, studies have appeared that have shown that the ratio of the masses of the components is decisive in the evolutionary fate of a massive X-ray binary system at the stage of secondary mass exchange. In a recent work [53], it was shown that when a massive donor star in an X-ray binary system fills or overfills its Roche lobe and outflows onto a relativistic object, then if the component mass ratio is $q = \dfrac{M_x}{M_v} \gtrsim 0.29$, the system can avoid the formation of a common envelope and remain semi-detached. Such a system evolves as a semi-detached system with a stable overflow of the Roche lobe with an optical star and with a stable flow of stellar matter on a relativistic object and with the formation of a supercritical accretion disc with a powerful outflow of stellar wind from it (the so-called isotropic re-emission mode or SS433-like mode). If the mass ratio q is small ($q < 0.29$), then a massive X-ray binary system at the stage of secondary mass exchange inevitably passes through the stage of evolution with a common envelope. The evolutionary scenario that described the formation of the object of SS433 type was considered in the paper by [54].

Thus, the central point in understanding the evolution of SS433 is knowing the real mass ratio of the system's components. Until recently, there was no complete clarity on this issue. Different methods gave estimates from $q \simeq 0.14$ to $q = 0.8$ (see reviews by [6,7] for all the necessary references). The new estimate $q > 0.8$ [26], obtained on the basis of our discovered increase in the orbital period of SS433, is very reliable and allows us to understand the reason why the SS433 system remains semi-detached. The SS433 system evolves as a semi-detached because the relativistic object here is a relatively large mass black hole ($M_x > 8 M_\odot$), which is typical of stellar mass black holes in X-ray binaries.

7. Conclusions

Object SS433 ("Enigma of the century") is the first example of a microquasar discovered in the Galaxy. The object is a natural laboratory where complex and extremely interesting physical processes take place that shed light on many problems: from relativistic astrophysics to the physics and evolution of stars and high energy astrophysics.

For example, recently, the gamma radiation of SS433 was discovered; it can be formed as the result of the interaction of the relativistic jets with the matter of the W50 nebula that surrounds SS433 [55]. In the work by [56], the variability of the gamma flux from SS433, with the period of the precession of the accretion disc, was found. The authors noted that such variability can be related with the formation of the gamma radiation close to the accretion disc of SS433. In this regard, let us note that, in papers by [57,58], in the central part of the supercritical accretion disc in the SS433 system, the hot ($kT \approx 20$ keV) extended corona radiating in the hard ($kT = 18$–60 keV) X-ray range was discovered. Recently, new observational data were obtained that indicated the similarity of physical processes in SS433 and in (still mysterious) objects such as ultra-luminous X-ray sources (ULXs) [59].

In this article, we have summarized the most fundamental problems associated with the object SS433 and showed how the latest observations and theoretical models allow us to understand the nature of this extremely interesting object.

Funding: The work was supported by the Russian Science Foundation grant 17-12-01241 and by the Scientific and Educational School of M. V. Lomonosov Moscow State University "Fundamental and applied space research". The author acknowledges support from the M. V. Lomonosov Moscow State University Program of Development.

Institutional Review Board Statement: Not applicable.

Informed Consent Statement: Not applicable.

Data Availability Statement: The data presented in this study are available on request from the corresponding author.

Conflicts of Interest: The author declares no conflict of interest.

References

1. Stephenson, C.B.; Sanduleak, N. New H-alpha emission stars in the Milky Way. *Astrophys. J. Suppl. Ser.* **1977**, *33*, 459–469. [CrossRef]
2. Margon, B.; Ford, H.C.; Grandi, S.A.; Stone, R.P.S. Enormous periodic Doppler shifts in SS433. *Astrophys. J.* **1979**, *233*, L63–L68. [CrossRef]
3. Clark, D.H.; Murdin, P. An unusual emission-line star/X-ray source radio star, possibly associated with an SNR. *Nature* **1978**, *276*, 44–45. [CrossRef]
4. Milgrom, M. On the interpretation of the large variations in the line positions in SS433. *A&A* **1979**, *76*, L3–L6.
5. Fabian, A.C.; Rees, M.J. SS 433: A double jet in action? *Mon. Not. R. Astron. Soc.* **1979**, *187*, 13P–16P. [CrossRef]
6. Fabrika, S. The jets and supercritical accretion disk in SS433. *Astrophys. Space Phys. Res.* **2004**, *12*, 1–152.
7. Cherepashchuk, A.; Postnov, K.; Molkov, S.; Antokhina, E.; Belinski, A. SS433: A massive X-ray binary in an advanced evolutionary stage. *New Astron. Rev.* **2020**, *89*, 101542. [CrossRef]
8. Crampton, D.; Cowley, A.P.; Hutchings, J.B. The probable binary nature of SS 433. *Astrophys. J.* **1980**, *235*, L131–L135. [CrossRef]
9. Cherepashchuk, A.M. SS 433 as an Eclipsing Binary. *Mon. Not. R. Astron. Soc.* **1981**, *194*, 761–769. [CrossRef]
10. Shakura, N.I.; Sunyaev, R.A. Reprint of 1973A&A....24..337S. Black holes in binary systems. Observational appearance. *A&A* **1973**, *500*, 33–51.
11. Katz, J.I. Thirty-five-day Periodicity in Her X-1. *Nat. Phys. Sci.* **1973**, *246*, 87–89. [CrossRef]
12. Roberts, W.J. A slaved disk model for Hercules X-1. *Astrophys. J.* **1974**, *187*, 575–584. [CrossRef]
13. Shakura, N.I. Disk Model of Gas Accretion on a Relativistic Star in a Close Binary System. *Soviet Ast.* **1973**, *16*, 756.
14. van den Heuvel, E.P.J.; Ostriker, J.P.; Petterson, J.A. An early-type binary model for SS 433. *A&A* **1980**, *81*, L7–L10.
15. Cherepashchuk, A.M. Longterm Variability of X-ray Binaries—The Aftermath of Supernovae. *Sov. Astron. Lett.* **1981**, *7*, 401–403.
16. Shapiro, P.R.; Milgrom, M.; Rees, M.J. Radiative acceleration of astrophysical jets—Line-locking in SS 433. In *Extragalactic Radio Sources*; Heeschen, D.S., Wade, C.M., Eds.; Cambridge University Press: Cambridge, UK, 1982; Volume 97, p. 209.
17. Shapiro, P.R.; Milgrom, M.; Rees, M.J. The Radiative Acceleration of Astrophysical Jets: Line Locking in SS 433. *Astrophys. J. Suppl. Ser.* **1986**, *60*, 393. [CrossRef]
18. Crampton, D.; Hutchings, J.B. The SS 433 binary system. *Astrophys. J.* **1981**, *251*, 604–610. [CrossRef]

19. Davydov, V.V.; Esipov, V.F.; Cherepashchuk, A.M. Spectroscopic monitoring of SS 433: A search for long-term variations of kinematic model parameters. *Astron. Rep.* **2008**, *52*, 487–506. [CrossRef]
20. Cherepashchuk, A.M.; Postnov, K.A.; Belinski, A.A. On masses of the components in SS433. *Mon. Not. R. Astron. Soc.* **2018**, *479*, 4844–4848. [CrossRef]
21. Zahn, J.P. Reprint of 1977A&A....57..383Z. Tidal friction in close binary stars. *A&A* **1977**, *500*, 121–132.
22. Zahn, J.P. Tidal evolution of close binary stars. I—Revisiting the theory of the equilibrium tide. *A&A* **1989**, *220*, 112–116.
23. Blundell, K.M.; Bowler, M.G.; Schmidtobreick, L. Fluctuations and symmetry in the speed and direction of the jets of SS 433 on different timescales. *A&A* **2007**, *474*, 903–910. [CrossRef]
24. Doolin, S.; Blundell, K.M. The Precession of SS433's Radio Ruff on Long Timescales. *Astrophys. J.* **2009**, *698*, L23–L26. [CrossRef]
25. Perez, M.S.; Blundell, K.M. Inflow and outflow from the accretion disc of the microquasar SS433: UKIRT spectroscopy. *Mon. Not. R. Astron. Soc.* **2009**, *397*, 849–856. [CrossRef]
26. Cherepashchuk, A.M.; Belinski, A.A.; Dodin, A.V.; Postnov, K.A. Discovery of orbital eccentricity and evidence for orbital period increase of SS433. *Mon. Not. R. Astron. Soc.* **2021**, *507*, L19–L23. [CrossRef]
27. Cherepashchuk, A.M.; Esipov, V.F.; Dodin, A.V.; Davydov, V.V.; Belinskii, A.A. Spectroscopic Monitoring of SS 433. Stability of Parameters of the Kinematic Model over 40 Years. *Astron. Rep.* **2018**, *62*, 747–763. [CrossRef]
28. Goranskij, V. Photometric Mass Estimate for the Compact Component of SS 433: And Yet It Is a Neutron Star. *Perem. Zvezdy* **2011**, *31*, 5.
29. Cherepashchuk, A.M.; Postnov, K.A.; Belinski, A.A. Mass ratio in SS433 revisited. *Mon. Not. R. Astron. Soc.* **2019**, *485*, 2638–2641. [CrossRef]
30. Gies, D.R.; Huang, W.; McSwain, M.V. The Spectrum of the Mass Donor Star in SS 433. *Astrophys. J.* **2002**, *578*, L67–L70. [CrossRef]
31. Hillwig, T.C.; Gies, D.R.; Huang, W.; McSwain, M.V.; Stark, M.A.; van der Meer, A.; Kaper, L. Identification of the Mass Donor Star's Spectrum in SS 433. *Astrophys. J.* **2004**, *615*, 422–431. [CrossRef]
32. Hillwig, T.C.; Gies, D.R. Spectroscopic Observations of the Mass Donor Star in SS 433. *Astrophys. J.* **2008**, *676*, L37. [CrossRef]
33. Kubota, K.; Ueda, Y.; Fabrika, S.; Medvedev, A.; Barsukova, E.A.; Sholukhova, O.; Goranskij, V.P. Subaru and Gemini Observations Of SS 433: New Constraint on the Mass of the Compact Object. *Astrophys. J.* **2010**, *709*, 1374–1386. [CrossRef]
34. Picchi, P.; Shore, S.N.; Harvey, E.J.; Berdyugin, A. An optical spectroscopic and polarimetric study of the microquasar binary system SS 433. *A&A* **2020**, *640*, A96. [CrossRef]
35. Blundell, K.M.; Bowler, M.G.; Schmidtobreick, L. SS 433: Observation of the Circumbinary Disk and Extraction of the System Mass. *Astrophys. J.* **2008**, *678*, L47. [CrossRef]
36. Bowler, M.G. Yet more on the circumbinary disk of SS 433. *A&A* **2013**, *556*, A149. [CrossRef]
37. Bowler, M.G. SS 433: Two robust determinations fix the mass ratio. *A&A* **2018**, *619*, L4. [CrossRef]
38. D'Odorico, S.; Oosterloo, T.; Zwitter, T.; Calvani, M. Evidence that the compact object in SS433 is a neutron star and not a black hole. *Nature* **1991**, *353*, 329–331. [CrossRef]
39. Kawai, N.; Matsuoka, M.; Pan, H.C.; Stewart, G.C. GINGA observations of the X-ray eclipse of SS 433. *Publ. Astron. Soc. Jpn.* **1989**, *41*, 491–507.
40. Brinkmann, W.; Kawai, N.; Matsuoka, M. SS 433—The puzzle continues. *A&A* **1989**, *218*, L13–L16.
41. Kotani, T.; Kawai, N.; Matsuoka, M.; Brinkmann, W. The ASCA Observation Campaign of SS433. In *The Hot Universe*; Koyama, K., Kitamoto, S., Itoh, M., Eds.; Cambridge University Press: Cambridge, UK, 1998; Volume 188, p. 358.
42. Fabrika, S.N. An extended disc around SS 433. *Mon. Not. R. Astron. Soc.* **1993**, *261*, 241–245. [CrossRef]
43. Blundell, K.M.; Mioduszewski, A.J.; Muxlow, T.W.B.; Podsiadlowski, P.; Rupen, M.P. Images of an Equatorial Outflow in SS 433. *Astrophys. J.* **2001**, *562*, L79–L82. [CrossRef]
44. Bowler, M.G. More on the circumbinary disk of SS 433. *A&A* **2011**, *531*, A107. [CrossRef]
45. Filippenko, A.V.; Romani, R.W.; Sargent, W.L.W.; Blandford, R.D. Possible Evidence for Disk Emission in SS 433. *Astron. J.* **1988**, *96*, 242. [CrossRef]
46. Perez M., S.; Blundell, K.M. SS433's circumbinary ring and accretion disc viewed through its attenuating disc wind. *Mon. Not. R. Astron. Soc.* **2010**, *408*, 2–8. [CrossRef]
47. Pavlovskii, K.; Ivanova, N. Mass transfer from giant donors. *Mon. Not. R. Astron. Soc.* **2015**, *449*, 4415–4427. [CrossRef]
48. Pavlovskii, K.; Ivanova, N.; Belczynski, K.; Van, K.X. Stability of mass transfer from massive giants: Double black hole binary formation and ultraluminous X-ray sources. *Mon. Not. R. Astron. Soc.* **2017**, *465*, 2092–2100. [CrossRef]
49. Krivosheyev, Y.M.; Bisnovatyi-Kogan, G.S.; Cherepashchuk, A.M.; Postnov, K.A. Monte Carlo simulations of the broad-band X-ray continuum of SS433. *Mon. Not. R. Astron. Soc.* **2009**, *394*, 1674–1684. [CrossRef]
50. Masevich, A.G.; Tutukov, A.V. *Ehvolyutsiya Zvezd: Teoriya i Nablyudeniya (Evolution of Stars: Theory and Observations)*; Nauka: Moscow, Russia, 1988.
51. Thorne, K.S.; Zytkow, A.N. Stars with degenerate neutron cores. I. Structure of equilibrium models. *Astrophys. J.* **1977**, *212*, 832–858. [CrossRef]
52. Bisnovatyi-Kogan, G.S.; Lamzin, S.A. Stars with Neutron Cores—The Possibility of the Existence of Objects with a Low Neutrino Luminosity. *Soviet Ast.* **1984**, *28*, 187–193.
53. van den Heuvel, E.P.J.; Portegies Zwart, S.F.; de Mink, S.E. Forming short-period Wolf-Rayet X-ray binaries and double black holes through stable mass transfer. *Mon. Not. R. Astron. Soc.* **2017**, *471*, 4256–4264. [CrossRef]

54. Bogomazov, A.I. A study of the evolution of the close binaries Cyg X-3, IC 10 X-1, NGC 300 X-1, SS 433, and M33 X-7 using the "scenario machine". *Astron. Rep.* **2014**, *58*, 126–138. [CrossRef]
55. Xing, Y.; Wang, Z.; Zhang, X.; Chen, Y.; Jithesh, V. Fermi Observation of the Jets of the Microquasar SS 433. *Astrophys. J.* **2019**, *872*, 25. [CrossRef]
56. Rasul, K.; Chadwick, P.M.; Graham, J.A.; Brown, A.M. Gamma-rays from SS433: Evidence for periodicity. *Mon. Not. R. Astron. Soc.* **2019**, *485*, 2970–2975. [CrossRef]
57. Cherepashchuk, A.M.; Sunyaev, R.A.; Postnov, K.A.; Antokhina, E.A.; Molkov, S.V. Peculiar nature of hard X-ray eclipse in SS433 from INTEGRAL observations. *Mon. Not. R. Astron. Soc.* **2009**, *397*, 479–487. [CrossRef]
58. Cherepashchuk, A.M.; Sunyaev, R.A.; Molkov, S.V.; Antokhina, E.A.; Postnov, K.A.; Bogomazov, A.I. INTEGRAL observations of SS433: System's parameters and nutation of supercritical accretion disc. *Mon. Not. R. Astron. Soc.* **2013**, *436*, 2004–2013. [CrossRef]
59. Fabrika, S.; Ueda, Y.; Vinokurov, A.; Sholukhova, O.; Shidatsu, M. Supercritical accretion disks in ultraluminous X-ray sources and SS 433. *Nat. Phys.* **2015**, *11*, 551–553. [CrossRef]

 *universe*

MDPI

Article
Multi-Component MHD Model of Hot Jupiter Envelopes

Andrey Zhilkin [†] **and Dmitri Bisikalo** *,[†]

Institute of Astronomy of the Russian Academy of Sciences, 48 Pyatnitskaya St., 119017 Moscow, Russia;
zhilkin@inasan.ru
* Correspondence: bisikalo@inasan.ru
† These authors contributed equally to this work.

Abstract: A numerical model description of a hot Jupiter extended envelope based on the approximation of multi-component magnetic hydrodynamics is presented. The main attention is focused on the problem of implementing the completed MHD stellar wind model. As a result, the numerical model becomes applicable for calculating the structure of the extended envelope of hot Jupiters not only in the super-Alfvén and sub-Alfvén regimes of the stellar wind flow around and in the trans-Alfvén regime. The multi-component MHD approximation allows the consideration of changes in the chemical composition of hydrogen–helium envelopes of hot Jupiters. The results of calculations show that, in the case of a super-Alfvén flow regime, all the previously discovered types of extended gas-dynamic envelopes are realized in the new numerical model. With an increase in magnitude of the wind magnetic field, the extended envelope tends to become more closed. Under the influence of a strong magnetic field of the stellar wind, the envelope matter does not move along the ballistic trajectory but along the magnetic field lines of the wind toward the host star. This corresponds to an additional (sub-Alfvénic) envelope type of hot Jupiters, which has specific observational features. In the transient (trans-Alfvén) mode, a bow shock wave has a fragmentary nature. In the fully sub-Alfvén regime, the bow shock wave is not formed, and the flow structure is shock-less.

Keywords: numerical simulation; magnetic hydrodynamics (MHD); hot Jupiters

Citation: Zhilkin, A.; Bisikalo, D. Multi-Component MHD Model of Hot Jupiter Envelopes. *Universe* **2021**, *7*, 422. https://doi.org/10.3390/universe7110422

Academic Editors: Nazar R. Ikhsanov, Galina L. Klimchitskaya and Vladimir M. Mostepanenko

Received: 8 October 2021
Accepted: 3 November 2021
Published: 5 November 2021

Publisher's Note: MDPI stays neutral with regard to jurisdictional claims in published maps and institutional affiliations.

1. Introduction

Hot Jupiters are giant exoplanets with masses on the order of Jupiter's mass, located in the immediate vicinity of a host star [1]. The first hot Jupiter was discovered in 1995 [2]. Due to the close location to the host star and the relatively large size, gas envelopes of hot Jupiters can overfill their Roche lobes, resulting in intense gas outflows both at the night side (near the Lagrange point L_2) and at the day side (near the inner Lagrange point L_1) of the planet [3,4]. The presence of such outflows is indirectly indicated by the excessive absorption of radiation in the near ultraviolet range observed in some hot Jupiters during their transit across the disk of their host star [5–11]. These conclusions are confirmed by direct numerical calculations in the framework of one-dimensional aeronomic models [1,12–15].

In this case, the matter of expanding the upper atmosphere of hot Jupiter is no longer collision-less, and a gas-dynamic approximation can be used to describe it. Such an exosphere is more correctly called the *extended envelope* of a hot Jupiter. These are located above the exobase, have a fairly large size, relatively high density, and are characterized by significant deviations from a spherical shape. The structure of an extended envelope and its physical properties are determined by the fact that several other forces act on each element of the flow in the envelope in addition to the planet gravity: the gravitational force of the star, the orbital centrifugal force, the orbital Coriolis force, as well as the forces determined by interaction with the stellar wind, the radiation of the star, and the magnetic field. Therefore, the study of the structure of gas envelopes of such objects is one of the most urgent problems of modern astrophysics [16].

In the three-dimensional approximation, the gas-dynamic structure of the extended envelope of hot Jupiters was studied in [17–22]. In these works, it is shown that, depending on the model parameters, structures of three main types can be formed during the interaction of stellar wind with the expanding envelope of a hot Jupiter [17]. If the atmosphere of a planet is completely located inside its Roche lobe, then a *closed envelope* is formed. If the flow of matter from the inner Lagrange point L_1 is stopped by the dynamic pressure of stellar winds, then *a quasi-closed envelope* is formed. Finally, if the dynamic pressure of the stellar wind is not sufficient to stop outflow from the Lagrange point L_1, an *open envelope* is formed. The type of forming envelope significantly determines the mass loss rate of the hot Jupiter [17]. On the other hand, in the works [23–27], noticeable dependence of the mass loss rate by the planet on the strength of stellar wind wasn't found, while the type of envelope changed from the open type to the closed one.

Hot Jupiters may have their own magnetic field, which can influence the process of their flow around by stellar wind. However, estimates of the intrinsic magnetic fields of hot Jupiters show that it is most likely quite weak. The characteristic value of the magnetic moment of hot Jupiters, apparently, is 0.1–0.2 μ_J, where $\mu_J = 1.53 \times 10^{30}$ G·cm^3 is the Jupiter magnetic moment. This value is in a good agreement with both observational [28–31] and theoretical [32] estimates.

The relatively low value of the dipole moment is explained by the inefficiency of the dynamo process of magnetic field generation in the bowels of these planets. This is due to the fact that hot Jupiters are located close to the host star; therefore, due to strong tidal disturbances, the proper rotation of a typical hot Jupiter should pass into a state of synchronization with its orbital motion over a period of about several million years [33]. In the state of synchronous rotation, the efficiency of dynamo generation of magnetic field drops sharply.

The process of dynamo-generation of magnetic field occurs in the bowels of the planet and is largely determined by its internal structure (see, for example, [34,35]). However, it should be noted that the magnetic field of hot Jupiters can be generated not only in the bowels but also in the upper layers of atmosphere. The estimates made in [21] show that the upper atmosphere of hot Jupiters consists of almost completely ionized gas. This is due to the processes of thermal ionization and hard radiation of the host star.

Therefore, the upper part of hot Jupiter atmosphere can be called the ionospheric envelope. It is shown in [36] that the proper magnetic field of hot Jupiter should influence the formation of large-scale (zonal) currents in its atmosphere. Detailed three-dimensional calculations [37,38] demonstrate a complex picture of the distribution of winds in the upper atmosphere in which magnetic fields play an important role.

In particular, electromagnetic forces can shift the hot spot forming at point facing the host star to the west. This effect can also be manifested on the light curves of hot Jupiters. For example, a comparison of the observed light curves with those calculated for the planet HAT-P-7b allows estimation of the characteristic magnetic field in the atmosphere of 6 G [39]. Most likely, this estimate is greatly overvalued. Recall that, at the level of the Jupiter cloud layer, the magnitude of the magnetic field is 4–5 G.

It is important to note one more circumstance. Due to their close location to the host star, hot Jupiters can have a quite strong magnetic field induced by the stellar wind magnetic field. As the calculations presented in [40] show, the corresponding magnetic moment of such a field can be from 10% to 20% of the Jupiter magnetic moment. Summing up all these comments, we can conclude that the question of magnitude and configuration of magnetic field in hot Jupiters is still open.

Analysis of the influence of stellar wind magnetic field on the process of its flow around the atmosphere of hot Jupiter [41] shows that in the case of hot Jupiters, this effect can be extremely important. This is due to the fact that almost all hot Jupiters are located in the so-called sub-Alfvén zone of stellar wind of the host star, where the wind speed is less than the Alfvénic one. Taking into account the orbital motion of the planet, the flow velocity turns out to be close to the Alfvénic one.

Therefore, depending on the specific situation (the major semi-axis of the planet orbit, the spectral class of host star, the proper rotation period of star, the features of stellar wind), both super-Alfvén and sub-Alfvén flow regimes can be realized. Note that in the super-Alfvén regime, the magnetosphere of hot Jupiter will contain all the main elements (the bow shock wave, the magnetopause, the magnetospheric tail at the night side, and others) present in the magnetospheres of planets of the Solar system [42,43]. In the case of sub-Alfvén flow regime, the bow shock wave will be absent in the magnetosphere structure [44].

The problem of the magnetic field influence on the dynamics of extended envelopes of hot Jupiters is of constant great interest to the scientific community. In recent years, several attempts have been made to account for the magnetic field in one-dimensional [45–48], in two-dimensional [49] and in three-dimensional [47,50,51] numerical aeronomic models of hot Jupiter atmospheres. However, in these studies, only the immediate vicinity of the planet was considered, and estimates of the mass loss rate were performed without considering the presence of extended envelopes.

An exception is the work [51], in which the authors performed three-dimensional numerical modeling in a wide spatial domain and obtained MHD solutions for exoplanets with open and quasi-closed envelopes. In our recent work, we investigated the influence of both the planet own magnetic field [52,53] and the wind field [41,54–57] on the envelope dynamics of hot Jupiters. The results of these investigations are presented in the review [16].

In this paper, we consider the further development of our numerical MHD codes. The main attention in the paper is focused on obtaining a numerical model that is self-consistent with the stellar wind and includes additional physical effects due to the complex chemical composition of the hydrogen–helium envelopes of hot Jupiters. Taking into account the self-consistent model with the wind allows, in particular, to more correctly determine the position of the Alfvén point.

In our previous numerical models, this parameter was either estimated approximately from the condition of constancy in the computational domain of the radial wind speed, or was set separately. Another important development direction is the creation of a numerical model of multi-fluid (multi-component) magnetic hydrodynamics for describing the structure of hot Jupiter extended envelope. In the paper, we did not consider specific processes that lead to local changes in the concentration of components (chemical reactions, ionization, etc.), as well as the corresponding heating–cooling processes, since the main goal was to include the stellar wind model.

However, in the future, this will allow for not only consideration of the chemical composition of the hydrogen–helium envelopes of hot Jupiters but also to trace the distributions and dynamics of components of particular interest (e.g., biomarkers). The multi-fluid MHD model described in this paper was developed on the basis of already existing single-fluid MHD model for describing the structure of a hot Jupiter extended envelope [16].

The paper is organized as follows. Section 2 describes the MHD model of stellar wind we used. Section 3 describes the three-dimensional numerical model of a multi-component MHD. Section 4 describes the numerical model of the hot Jupiter extended envelope based on multi-component MHD. Section 5 presents the results of calculations. The main conclusions of the study are summarized in Section 6. Finally, some computation method details are described in the Appendix A.

2. Stellar Wind Model

2.1. Basic Equations

To describe the structure (including the magnetic field) of stellar wind in the vicinity of hot exoplanets in our numerical model, we will rely on the well-studied properties of the solar wind. As shown in numerous ground- and space-based investigations (see, for example, the review of [58]), the picture of magnetic field of the solar wind is fairly complex. Schematically, this structure is illustrated in Figure 1 (see, e.g., our paper [41]). In the corona region, the magnetic field is mainly determined by the Sun intrinsic magnetism, and therefore it is essentially non-radial.

At the boundary of the corona, which is located at a distance of several radii of the Sun, magnetic field becomes purely radial with high accuracy. Beyond this region is the heliospheric area, the magnetic field that is significantly determined by the properties of the solar wind. In this region, the magnetic field lines with a distance from the center gradually twist in the form of a spiral due to rotation of the Sun, and therefore (especially at large distances) the magnetic field of the wind can be described with good accuracy using a simple Parker model [59].

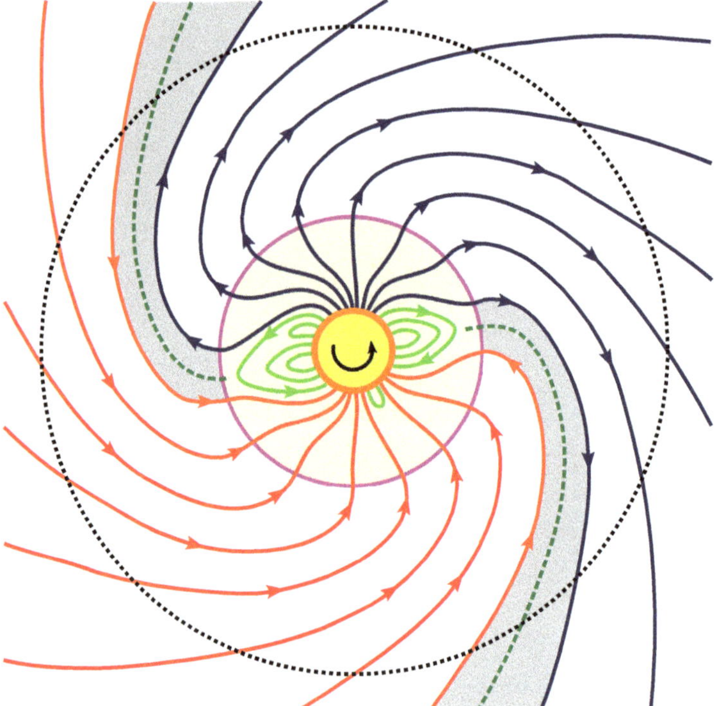

Figure 1. Schematic representation of the solar wind structure in the ecliptic plane. The Sun corresponds to a small colored circle in the center. The arrow shows the direction of rotation of the Sun. The boundary of the middle circle indicates the region of the corona, at the edge of which the magnetic field becomes purely radial. The shaded gray areas correspond to the zones of the heliospheric current sheet (shown by dotted lines running from the corona to the periphery), which separates the solar wind magnetic field with different directions of magnetic field lines (from the Sun or to the Sun). The orbit of a hot exoplanet is shown by a dotted circle, which is located in the heliospheric region.

However, the observed magnetic field in the solar wind is not axisymmetric but has a manifested sector structure. This is due to the fact that at different points of the spherical surface of the corona, the field may have different polarity (the direction of field lines relative to the direction of normal vector), for example, due to the inclination of magnetic axis of the Sun to its rotation axis.

As a result, two clearly distinguished sectors with different directions of the magnetic field are formed in the solar wind in the ecliptic plane. In one sector, the magnetic field lines are directed towards the Sun, and in the opposite sector, away from the Sun. These two sectors are separated by the heliospheric current sheet, which rotates together with the

Sun and therefore the Earth, during its orbit around the Sun, crosses it many times per year, passing from the solar wind sector with one magnetic field polarity to the neighboring sector with the opposite magnetic field polarity.

In our calculations, we neglect the possible sector structure of the wind magnetic field, as well as the presence of a heliospheric current sheet in it, focusing on the influence of its global parameters. We reasonably assume that the orbit of a hot exoplanet is located in the heliospheric region beyond the boundary of the corona. In Figure 1, it is shown as a dotted circle. To describe the structure of the wind (including its magnetic field) in the heliospheric region, an axisymmetric (or even a spherically-symmetric) model [60] can be used as the first approximation.

We will consider the wind model in an inertial reference frame in spherical coordinates (r, θ, φ). We assume that the center of the spherical coordinate system coincides with the center of the star. At the same time, we can ignore the dependence of the wind parameters on the angle θ, since we are interested in the flow structure only near the orbit plane of a hot exoplanet. Therefore, for simplicity, we will assume that all values depend only on the radial coordinate r.

The stationary structure of the wind under such conditions is determined by the continuity equation

$$\frac{1}{r^2}\frac{d}{dr}\left(r^2\rho v_r\right) = 0,\tag{1}$$

the equation of motion for the radial v_r and azimuthal v_φ components of the velocity vector v

$$v_r\frac{dv_r}{dr} - \frac{v_\varphi^2}{r} = -\frac{1}{\rho}\frac{dP}{dr} - \frac{GM_s}{r^2} - \frac{B_\varphi}{4\pi\rho r}\frac{d}{dr}\left(rB_\varphi\right),\tag{2}$$

$$v_r\frac{dv_\varphi}{dr} + \frac{v_r v_\varphi}{r} = \frac{B_r}{4\pi\rho r}\frac{d}{dr}\left(rB_\varphi\right),\tag{3}$$

the equation of induction

$$\frac{1}{r}\frac{d}{dr}\left(rv_r B_\varphi - rv_\varphi B_r\right) = 0\tag{4}$$

and the Maxwell equation ($\nabla \cdot \boldsymbol{B} = 0$)

$$\frac{1}{r^2}\frac{d}{dr}\left(r^2 B_r\right) = 0.\tag{5}$$

Here, ρ represents the density, P is the pressure, G is the gravitational constant, and M_s is the mass of the central star. Density, pressure, and temperature satisfy the equation of state for an ideal polytropic gas,

$$P = K_\kappa \rho^\kappa = \frac{2k_B}{m_p}\rho T,\tag{6}$$

where K_κ is the constant, κ is the polytropic index, k_B is the Boltzmann constant, and m_p is the proton mass. The average molecular weight of the wind matter is considered to be equal to $1/2$, which corresponds to a fully ionized hydrogen plasma consisting only of electrons and protons.

From the Maxwell Equation (5), we find

$$r^2 B_r = B_s R_s^2,\tag{7}$$

where R_s is the radius of the star, and B_s is the magnitude of field at the star surface. From the continuity Equation (1), we can obtain an integral of motion corresponding to the law of mass conservation,

$$4\pi r^2 \rho v_r = \dot{M}_s,\tag{8}$$

where the integration constant $\dot{M}_s$ determines the mass loss rate by star due to the outflow of matter in the form of a stellar wind. The Equation (2) can be rewritten as:

$$\frac{d}{dr}\left(\frac{v_r^2}{2} + \frac{c_s^2}{\kappa - 1} - \frac{GM_s}{r} + \frac{B_\varphi^2}{4\pi\rho}\right) - \frac{v_\varphi^2}{r} - rB_\varphi\frac{d}{dr}\left(\frac{B_\varphi}{4\pi\rho r}\right) = 0,$$

(9)

where the square of the speed of sound

$$c_s^2 = \frac{\kappa P}{\rho} = \kappa K_\kappa \rho^{\kappa-1}.$$

(10)

We can multiply the Equation (3) by v_φ and divide by v_r. Then, it can be represented as:

$$\frac{d}{dr}\left(\frac{v_\varphi^2}{2} - \frac{B_r B_\varphi^2 v_\varphi}{4\pi\rho v_r}\right) + \frac{v_\varphi^2}{r} + rB_\varphi\frac{d}{dr}\left(\frac{B_r v_\varphi}{4\pi\rho r v_r}\right) = 0.$$

(11)

Summing the Equations (9) and (11), we find:

$$\frac{d}{dr}\left(\frac{v_r^2}{2} + \frac{v_\varphi^2}{2} + \frac{c_s^2}{\kappa - 1} - \frac{GM_s}{r} + \frac{B_\varphi^2}{4\pi\rho} - \frac{B_r B_\varphi^2 v_\varphi}{4\pi\rho v_r}\right)$$
$$+ rB_\varphi\frac{d}{dr}\left(\frac{B_r v_\varphi}{4\pi\rho r v_r} - \frac{B_\varphi}{4\pi\rho r}\right) = 0.$$

(12)

The last term on the left side of this equation turns to zero, because

$$\frac{d}{dr}\left(\frac{B_r v_\varphi}{4\pi\rho r v_r} - \frac{B_\varphi}{4\pi\rho r}\right) = \frac{1}{\dot{M}_s}\frac{d}{dr}\left(rB_r v_\varphi - rB_\varphi v_r\right) = 0.$$

(13)

Here, we used the law of conservation of mass (8) and the equation of induction (4). This circumstance allows us to derive the integral of motion corresponding to the law of energy conservation,

$$\frac{v_r^2}{2} + \frac{v_\varphi^2}{2} + \frac{c_s^2}{\kappa - 1} - \frac{GM_s}{r} + \frac{B_\varphi^2}{4\pi\rho} - \frac{B_r B_\varphi v_\varphi}{4\pi\rho v_r} = Q_s,$$

(14)

where the constant Q_s determines the density of energy flux in the stellar wind, the value of which is $\dot{M}_s Q_s$.

Note that, from (7) and (8) follows

$$\frac{B_r}{4\pi\rho v_r} = \frac{B_s R_s^2}{\dot{M}_s} = \text{const.}$$

(15)

This circumstance allows from the Equations (3) and (4) to find two another integrals of motion:

$$rv_\varphi - \frac{B_r}{4\pi\rho v_r}rB_\varphi = L,$$

(16)

$$rv_r B_\varphi - rv_\varphi B_r = F.$$

(17)

The first integral determines the law of conservation of angular momentum. The second integral is related to the vertical component of the electric field E_θ, since it is obvious that $F = cE_\theta r$, where c is the speed of light. For the convenience of further calculations, we introduce the following notation:

$$a_r = \frac{B_r}{\sqrt{4\pi\rho}}, \quad a_\varphi = \frac{B_\varphi}{\sqrt{4\pi\rho}}.$$

(18)

Then, from Equations (16) and (17), it can be obtained that:

$$\left(v_r^2 - a_r^2\right)v_\varphi = \frac{1}{r}\left(v_r^2 L + \frac{a_r F}{\sqrt{4\pi\rho}}\right),\tag{19}$$

$$\left(v_r^2 - a_r^2\right)a_\varphi = \frac{v_r}{r}\left(a_r L + \frac{F}{\sqrt{4\pi\rho}}\right).\tag{20}$$

The value of the constant F in the last integral of motion (17) can be found from the boundary conditions at some point $r = r_0$. However, in the reference frame rotating with the Sun, the value of this constant should be equal to zero. This is due to the fact that, in this frame of reference, the vector of the magnetic field $\boldsymbol{B'}$ must be collinear with the velocity vector $\boldsymbol{v'}$ of the ideally conducting wind plasma: $\boldsymbol{v'} \times \boldsymbol{B'} = 0$. For non-relativistic motion, the magnetic field $\boldsymbol{B'} = \boldsymbol{B}$, and the velocity

$$v_r' = v_r, \quad v_\varphi' = v_\varphi - \Omega_s r,\tag{21}$$

where Ω_s is the angular velocity of the proper rotation of the star. Therefore,

$$F = -\Omega_s r^2 B_r = -\Omega_s R_s^2 B_s.\tag{22}$$

Substituting the expression (22) into the Equations (19) and (20), we find:

$$\left(v_r^2 - a_r^2\right)v_\varphi = \frac{1}{r}\left(v_r^2 L - a_r^2 \Omega_s r^2\right),\tag{23}$$

$$\left(v_r^2 - a_r^2\right)a_\varphi = \frac{v_r a_r}{r}\left(L - \Omega_s r^2\right).\tag{24}$$

In the obtained expressions, there is a singularity at some point $r = r_A$, where the radial wind velocity v_r becomes equal to the Alfvén one

$$u_A = |a_r| = \frac{|B_r|}{\sqrt{4\pi\rho}},\tag{25}$$

that corresponding to the radial component of magnetic field B_r. Let us call this special point as a *Alfvén* point. Near the surface of the star, the radial wind velocity v_r should be less than the Alfvén one u_A. At large distances, the radial velocity v_r, on the contrary, exceeds the Alfvén one u_A. At the Alfvén point $r = r_A$, there is a transition from the sub-Alfvén flow regime to the super-Alfvén one. Therefore, the region $r < r_A$ can be called the *sub-Alfvén* zone of the stellar wind, and the region $r > r_A$, respectively, the *super-Alfvén* zone.

The azimuthal components of the velocity v_φ and the magnetic field B_φ in the expressions (23) and (24) must remain continuous at the Alfvén point. Therefore, in order to satisfy this condition, it is necessary to set the value of the integration constant

$$L = \Omega_s r_A^2.\tag{26}$$

As a result, we find the final solution

$$v_\varphi = \frac{\Omega_s}{r}\frac{v_r^2 r_A^2 - a_r^2 r^2}{v_r^2 - a_r^2},\tag{27}$$

$$a_\varphi = \frac{\Omega_s}{r}v_r a_r \frac{r_A^2 - r^2}{v_r^2 - a_r^2}.\tag{28}$$

Let us introduce the Alfvén Mach number for the radial components of the velocity and the magnetic field,

$$\lambda = \frac{\sqrt{4\pi\rho}\,v_r}{B_r}.$$
(29)

It is not difficult to make sure that

$$\lambda^2 = \frac{v_r r^2}{v_A r_A^2} = \frac{\rho_A}{\rho},$$
(30)

where v_A and ρ_A are the values of the radial velocity and wind density at the Alfvén point. For numerical modeling, a convenient record of expressions for the azimuthal components of the velocity and magnetic field are

$$v_\varphi = \Omega_s r \frac{1 - \lambda^2 r_A^2/r^2}{1 - \lambda^2},$$
(31)

$$B_\varphi = B_r \frac{\Omega_s r}{v_r} \lambda^2 \frac{1 - r_A^2/r^2}{1 - \lambda^2}.$$
(32)

Note that in the sub-Alfvén zone of the stellar wind $\lambda < 1$, while in the super-Alfvén zone $\lambda > 1$. At the Alfvén point, $\lambda = 1$. The obtained relations, considering the integrals of mass (8) and energy (14), allow us to algebraically find the distributions of all magnetohydrodynamic quantities describing the wind structure.

2.2. Method of Solution

From the Equations (1)–(5) using simple manipulations, it can be derived that

$$\left(v_r^2 - u_S^2\right)\left(v_r^2 - u_F^2\right)\frac{r}{v_r}\frac{dv_r}{dr} = \left(v_r^2 - a_r^2\right)\left(2c_s^2 + v_\varphi^2 - \frac{GM_s}{r}\right) + 2v_r v_\varphi a_r a_\varphi,$$
(33)

where

$$u_{S,F}^2 = \frac{1}{2}\left[c_s^2 + a^2 \mp \sqrt{(c_s^2 + a^2)^2 - 4c_s^2 u_A^2}\right],$$
(34)

$$a^2 = a_r^2 + a_\varphi^2.$$
(35)

The quantities u_S and u_F determine the values of slow and fast magnetosonic velocities, respectively. It can be seen from the Equation (33) that there are two special points in the solution: *slow magnetosonic* $r = r_S$, in which the speed $v_r = u_S$, and *fast magnetosonic* $r = r_F$, in which the speed $v_r = u_F$. At these points, the coefficient for the derivative dv_r/dr turns to zero. In order for the solution to remain smooth, the right side of the Equation (33) at these points must also vanish. However, as we saw above, the Alfvén point $r = r_A$, in which the wind speed $v_r = u_A$, is also a solution critical point. This circumstance can be expressed directly by substituting the relations (27) and (33) into the Equation (28). As a result, we find:

$$\left(v_r^2 - u_A^2\right)^2\left(v_r^2 - u_S^2\right)\left(v_r^2 - u_F^2\right)\frac{r}{v_r}\frac{dv_r}{dr} = \left(v_r^2 - u_A^2\right)^3\left(2c_s^2 - \frac{GM_s}{r}\right)$$
$$+ \frac{\Omega_s^2}{r^2}\left(v_r^2 r_A^2 - u_A^2 r^2\right)\left(v_r^4 r_A^2 + u_A^4 r^2 - 3v_r^2 u_A^2 r^2 + v_r^2 u_A^2 r_A^2\right).$$
(36)

Further, we have

$$\left(v_r^2 - u_S^2\right)\left(v_r^2 - u_F^2\right) = v_r^4 - (c_s^2 + u_A^2)v_r^2 + c_s^2 u_A^2 - a_\varphi^2 v_r^2$$
$$= \left(v_r^2 - c_s^2\right)\left(v_r^2 - u_A^2\right) - a_\varphi^2 v_r^2.$$
(37)

Therefore, considering (28)

$$\left(v_r^2 - u_A^2\right)^2 \left(v_r^2 - u_S^2\right)\left(v_r^2 - u_F^2\right) = \left(v_r^2 - c_s^2\right)\left(v_r^2 - u_A^2\right)^3 - \frac{\Omega_s^2}{r^2} v_r^4 u_A^2 (r_A^2 - r^2)^2. \tag{38}$$

Now, the equation for the radial velocity (33) can be written as:

$$\left[\left(v_r^2 - c_s^2\right)\left(v_r^2 - u_A^2\right)^3 - \frac{\Omega_s^2}{r^2} v_r^4 u_A^2 (r_A^2 - r^2)^2\right]\frac{r}{v_r}\frac{dv_r}{dr}$$

$$= \left(v_r^2 - u_A^2\right)^3 \left(2c_s^2 - \frac{GM_s}{r}\right) \tag{39}$$

$$+\frac{\Omega_s^2}{r^2}\left(v_r^2 r_A^2 - u_A^2 r^2\right)\left(v_r^4 r_A^2 + u_A^4 r^2 - 3v_r^2 u_A^2 r^2 + v_r^2 u_A^2 r_A^2\right).$$

It is known from observations that in the solar wind, the radial velocity monotonically increases with the distance from the Sun. Therefore, we are interested in a solution with an accelerating wind, when the radial velocity gradient has a positive value at any distance from the star, $dv_r/dr > 0$. This means that in the Equation (33), the coefficient for the derivative in the left part must change sign simultaneously with the right part.

Let us study the asymptotic behavior of the solution of interest to us at small distances in the limit at $r \to 0$. Suppose that the radial velocity changes in this case according to the power law $v_r = Ar^\sigma$, where A is a certain coefficient. Consider the case when the radial velocity v_r is much smaller than any of the characteristic velocities u_S, u_F and u_A. Then, the coefficient for the derivative on the left side of the Equation is (33)

$$\left(v_r^2 - u_S^2\right)\left(v_r^2 - u_F^2\right) = u_S^2 u_F^2 = c_s^2 u_A^2. \tag{40}$$

From the relations (27) and (28) at our limit, we find

$$v_\varphi = \Omega_s r, \quad a_\varphi = -a_r\frac{\Omega_s r}{v_A}. \tag{41}$$

It follows that the right-hand side of the Equation (33) is approximately equal to

$$u_A^2 \frac{GM_s}{r} - 2v_r u_A^2 \frac{\Omega_s^2 r^2}{v_A} \approx u_A^2 \frac{GM_s}{r}. \tag{42}$$

As a result, neglecting non-essential terms, we come to the equation:

$$c_s^2 \frac{r}{v_r}\frac{dv_r}{dr} = \frac{GM_s}{r}. \tag{43}$$

Taking into account the law of mass conservation (8) and the expression for the speed of sound (10), we find the values of constants

$$\sigma = \frac{3 - 2\kappa}{\kappa - 1}, \quad A = \frac{\dot{M}_s}{4\pi}\left[\frac{3 - 2\kappa}{\kappa - 1}\frac{\kappa K_\kappa}{GM_s}\right]^{\frac{1}{\kappa - 1}}. \tag{44}$$

Since in such a solution the index of σ must be positive, then the index of polytrope $\kappa < 3/2$. Thus, the solution with an accelerating wind (a positive value of the radial velocity gradient, $dv_r/dr > 0$) is realized only in the case of polytropic index $\kappa < 3/2$. This value is less than the adiabatic index $\gamma = 5/3$. Therefore, effective sources of heating must be present in the stellar wind.

In fact, the entropy of an ideal gas $s = c_V \ln(P/\rho^\gamma)$, where c_V is the specific heat capacity at a constant volume. Therefore, the full derivative

$$\frac{ds}{dt} = c_V\left(\frac{d}{dt}\ln P - \gamma\frac{d}{dt}\ln\rho\right). \tag{45}$$

Taking into account the polytropic equation of state (6), we can write

$$T\frac{ds}{dt} = -c_V T(\gamma - \kappa)\frac{1}{\rho}\frac{d\rho}{dt}. \tag{46}$$

The expression on the right-hand side of this equation determines the heating function Γ. Since the specific internal energy of an ideal gas $\varepsilon = c_V T$, then using the continuity Equation (1), we obtain:

$$\Gamma = (\gamma - \kappa)\varepsilon\frac{1}{r^2}\frac{d}{dr}(r^2 v_r). \tag{47}$$

We can see that in the case of an accelerating wind ($dv_r/dr > 0$) and $\kappa < \gamma$ this value is positive. Effective heating in the solar wind is determined by the fundamental driving mechanism. An overview of possible wind driving mechanisms for stars of various types can be found in [61].

As boundary conditions in our wind model, we will use the values of density ρ_0, radial velocity v_0 and temperature T_0 at some point r_0. Using these values, we can calculate the speed of sound

$$c_0 = \sqrt{\frac{2k_B T_0}{m_p}} \tag{48}$$

and the integration constant

$$\dot{M}_s = 4\pi r_0^2 \rho_0 v_0. \tag{49}$$

The last expression follows from the law of mass conservation (8). The value of polytropic index κ, which lies in the range $1 < \kappa < 3/2$, is considered as unknown. Its specific value must be such that the boundary conditions are satisfied.

The law of mass conservation (8) also implies the relation

$$\dot{M}_s = 4\pi r_A^2 \rho_A v_A. \tag{50}$$

We find from here

$$\rho_A = \frac{1}{4\pi}\left(\frac{\dot{M}_s}{r_0^2 B_0}\right)^2, \tag{51}$$

where by

$$B_0 = B_s\left(\frac{R_s}{r_0}\right)^2 \tag{52}$$

the value of the radial magnetic field at the point r_0 is indicated. We introduce dimensionless quantities $x_A = r_A/r_0$ and $y_A = v_A/v_0$. It is not difficult to verify that they satisfy the relation

$$y_A = \frac{\beta}{x_A^2}, \tag{53}$$

where is the dimensionless parameter

$$\beta = \frac{\rho_0}{\rho_A} = \frac{B_0}{v_0\sqrt{4\pi\rho_A}}. \tag{54}$$

To solve the equations describing the wind structure, it is convenient to rewrite them in a dimensionless form. To do this, we denote the dimensionless variables $\xi = r/r_A$ and

$\eta = v_r/v_A$. In this case, the Alfvén Mach number (29) turns out to be equal to $\lambda = \xi\sqrt{\eta}$. In addition, we denote

$$z = \left(\frac{\rho}{\rho_0}\right)^{\frac{\kappa-1}{2}} = \left(\beta\lambda^2\right)^{\frac{1-\kappa}{2}}. \tag{55}$$

In dimensionless variables, the Equation (33) takes the form

$$\frac{\xi}{\eta}\frac{d\eta}{d\xi} = \frac{Y}{X}, \tag{56}$$

where

$$X = (\lambda^2 - 1)^3\left(y_A^2\eta^2 - \alpha z^2\right)\xi^2 - \omega\lambda^4 x_A^2(\xi^2 - 1)^2, \tag{57}$$

$$Y = (\lambda^2 - 1)^3\left(2\alpha\xi z^2 - \frac{\mu}{x_A}\right)\xi + \omega x_A^2(\lambda^2 - \xi^2)\left(\lambda^4 - 3\lambda^2\xi^2 + \lambda^2 + \xi^2\right), \tag{58}$$

$$\alpha = \frac{c_0^2}{v_0^2}, \quad \mu = \frac{GM_s}{r_0 v_0^2}, \quad \omega = \frac{\Omega_s^2 r_0^2}{v_0^2}. \tag{59}$$

As was noted above, the slow magnetosonic ($\xi = \xi_S$, $\eta = \eta_S$) and fast magnetosonic ($\xi = \xi_F$, $\eta = \eta_F$) points are critical solution points. However, the Alfvén point ($\xi = 1$, $\eta = 1$) is also critical, since in it the values X and Y also simultaneously turn to zero.

The expression determining the conservation of energy (14) in dimensionless variables can be written as follows:

$$Z = \frac{y_A^2\eta^2}{2} + \frac{\alpha z^2}{\kappa - 1} - \frac{\mu}{x_A\xi} + \frac{\omega x_A^2}{2\xi^2(\lambda^2 - 1)^2}\left[\lambda^4 + (2\lambda^2 - 1)(\xi^2 - 2)\xi^2\right] - q = 0, \tag{60}$$

where it is denoted

$$q = \frac{Q_s}{v_0^2}. \tag{61}$$

At the point r_0 we have $\xi_0 = 1/x_A$, $\eta_0 = 1/y_A$, $\lambda_0 = 1/\sqrt{\beta}$, $z_0 = 1$. Therefore, at this point, Equation (60) takes the form:

$$Z_0 = \frac{1}{2} + \frac{\alpha}{\kappa - 1} - \mu + \frac{\omega}{2(\lambda_0^2 - 1)^2}\left[\lambda_0^4 x_A^4 + (2\lambda_0^2 - 1)(1 - 2x_A^2)\right] - q = 0. \tag{62}$$

We will use ξ_S, η_S, ξ_F, η_F, x_A, q and κ as unknown quantities in solving the problem. In total, we have 7 unknowns. The system includes four equations $X = 0$, $Y = 0$ at the points ξ_S and ξ_F, two Equation (60) written at the points ξ_S and ξ_F, as well as the Equation (62). This system of non-linear algebraic equations can be solved numerically by iterations. The algorithm for constructing the solution is as follows. First, we numerically solve the system of equations for the wind parameters ξ_S, η_S, ξ_F, η_F, x_A, q and κ. After these parameters are determined, for each value of ξ, we numerically solve the Equation (60) and find the dependence $\eta(\xi)$ corresponding to the curve passing through all three critical points.

2.3. Calculation Example

Let us consider the results of a demonstration calculation of the stellar wind structure. The model parameters were the values of density $n_0 = 1400$ cm^{-3}, velocity $v_0 = 130$ km/s, and temperature $T_0 = 7.3 \times 10^5$ K at a distance of $r_0 = 10R_\odot$ from the star [62]. For example, let us take a typical hot Jupiter HD 209458b, which is used in our numerical simulations described below. The host star is characterized by the following parameters: spectral class G0, mass $M_s = 1.1M_\odot$, and radius $R_s = 1.2R_\odot$. The period of star proper rotation $P_{\text{rot}} = 14.4$ days, which corresponds to the angular velocity $\Omega_s = 5.05 \times 10^{-6}$ s^{-1} or the linear velocity at the equator $v_{\text{rot}} = 4.2$ km/s.

The major semi-axis of the planet orbit $A = 10.2 R_\odot$, which is close to the selected boundary point r_0 and corresponds to the period of rotation around the star $P_{\text{orb}} = 84.6$ h. The magnetic field at the stellar surface was set to $B_s = 0.5$ G, which approximately corresponds to the value of average magnetic field at the surface of the Sun in a quiet state $B_\odot = 1$ G, if we compare the corresponding magnetic fluxes, $B_s R_s^2 \approx B_\odot R_\odot^2$.

It should be noted that the parameter B_s corresponds to the average value of magnetic field on the star surface only formally. The solution under consideration is valid only in the heliospheric area (see Figure 1). In the corona region, the star intrinsic magnetic field plays a decisive role. Therefore, strictly speaking, the value of B_s is obtained from the average value of the radial magnetic field B_c at the corona boundary $r = R_c$, recalculated by considering the conservation of the magnetic flux by the star radius, $B_s R_s^2 = B_c R_c^2$.

On the other hand, it is known that the magnetic fields of solar-type stars can lie in the range from about 0.1 to several Gauss [63,64]. In addition, the host stars for hot Jupiters can be not only of the solar type, since their spectral classes lie in the wide range from class F to class M. In addition to the radial component, the azimuthal component of the magnetic field is also present in the stellar wind, which is due to the star proper rotation. The angular velocity of the star proper rotation Ω_s, in turn, also depends on the spectral class [64]. These remarks significantly expand the set of possible models of the stellar wind in vicinity of hot Jupiters.

Hence, the following parameter values are obtained: mass loss rate $\dot{M}_s = 1.85 \times 10^{11}$ g/s $= 2.94 \times 10^{-15} M_\odot$/year, density at the Alfvén point $\rho_A = 2.25 \times 10^{-22}$ g/cm^3, dimensionless parameters $\alpha = 0.713$, $\beta = 10.424$, $\mu = 1.241$, $\omega = 0.0725$. Note that the magnitude of the mass loss rate of the star $\dot{M}_s$ in our stellar wind model is only one of its parameters. It does not coincide with the real mass loss rate, since we solved the problem only in the plane of planet orbit. The real wind does not have spherical symmetry, and therefore these values can differ several times.

The results of calculations of the wind structure are shown in Figure 2. A family of integral curves corresponding to different types of solutions is shown. Some of these curves pass through the critical points, the positions of which are indicated by circles. In this case, the Alfvén critical point in the plane of the variables ξ, η has coordinates $(1, 1)$. As a solution corresponding to the stellar wind, it is necessary to use an integral curve passing through all three critical points. The corresponding curve in Figure 2 is shown as a bold line.

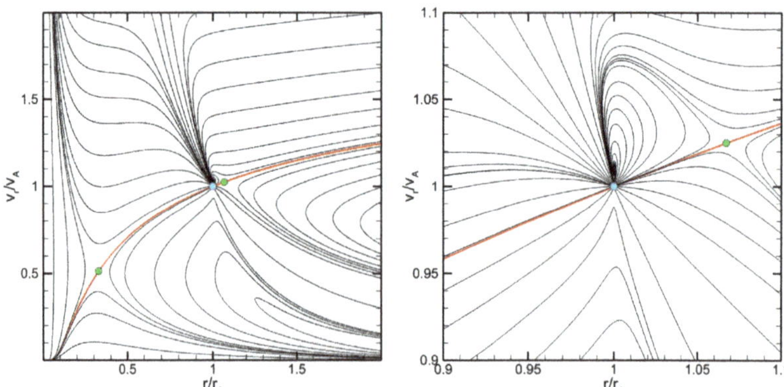

Figure 2. Integral curves corresponding to various types of solutions for the magnetohydrodynamic stellar wind. Critical points are shown as circles. The Alfvén critical point has coordinates $(1, 1)$. The bold line corresponds to the solution passing through all three critical points. The right diagram shows the vicinity of the Alfvén point.

The numerical solution of a system of non-linear algebraic equations describing the stellar wind structure gives the following coordinates of magnetosonic singular points: $\xi_S = 0.328$, $\eta_S = 0.514$, $\xi_F = 1.067$, $\eta_F = 1.025$. On the left panel of Figure 2, it can be seen that a slow magnetosonic point is a critical point of the "saddle" type. The fast magnetosonic point is located near the Alfvén point, the vicinity of which is shown on an enlarged scale on the right panel of Figure 2.

It can be seen that a fast magnetosonic point is also a critical point of the "saddle" type. The Alfvén critical point has a more complex topology, since it has a higher singularity order. For the Alfvén point, the value $x_A = 2.511$ is obtained, which corresponds to the radius $r_A = 25.114 R_\odot$. The polytropic index turned out to be $\kappa = 1.055$, and the parameter $q = 12.774$. The polytropic index is close to unity. Therefore, the stellar wind in vicinity of hot Jupiter can be considered almost iso-thermal. This result is in good agreement with the observations.

It is known that, at short distances from the Sun ($r < 15 R_\odot$), the effective adiabatic index $\gamma = 1.1$ [65,66]. At large distances $r > 25 R_\odot$, the effective adiabatic index can be estimated by the value $\gamma = 1.46$ [67]. Since the orbits of hot Jupiters are located at close distances from the host star in the region of wind acceleration $r < 20 R_\odot$, the wind structure in vicinity of the planet is found to be almost iso-thermal with good accuracy.

The calculation results are shown in Figures 3 and 4. Figure 3 shows the distributions of the radial velocity v_r (solid line), the speed of sound c_s, the Alfvén velocity u_A, as well as the slow u_S and fast u_F magnetosonic velocities depending on the radius r. The fast magnetosonic point is located close to the Alfvén point, slightly exceeding it. The slow magnetosonic point coincided with the sound point at which the radial wind speed $v_r = c_s$. The latter circumstance is due to the fact that in this region, because of the small contribution of the azimuthal component of the magnetic field, the slow magnetosonic velocity $u_S \approx \min(u_A, c_s) = c_s$.

Figure 4 shows the radial profiles of the particle number density $n(r)$, temperature $T(r)$, azimuthal velocity $v_\varphi(r)$, as well as the radial $B_r(r)$ (dotted line) and azimuthal $B_\varphi(r)$ (solid line) components of the magnetic field. The dependence $v_\varphi(r)$ is non-monotonic. The maximum values of the azimuthal velocity are reached approximately in the area of the hot Jupiter's location $13 R_\odot$. However, the characteristic values of 15 km/s are small compared to the radial velocity of 130 km/s.

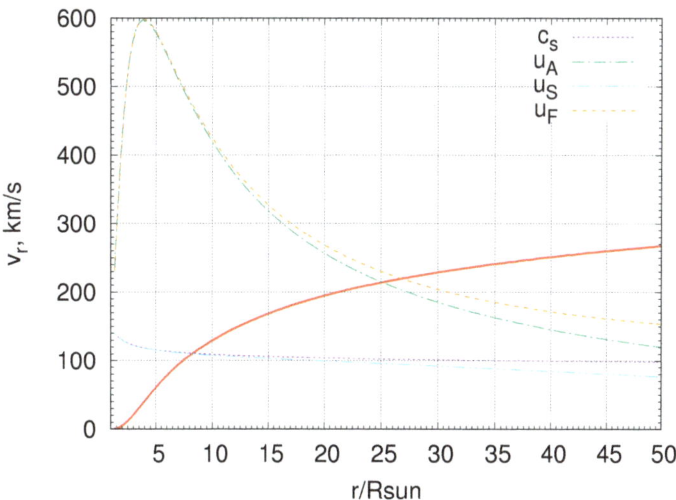

Figure 3. Distributions of the radial velocity v_r (solid line), the speed of sound c_s, the Alfvén velocity u_A, as well as the slow u_S and fast u_F magnetosonic velocities depending on the radius r.

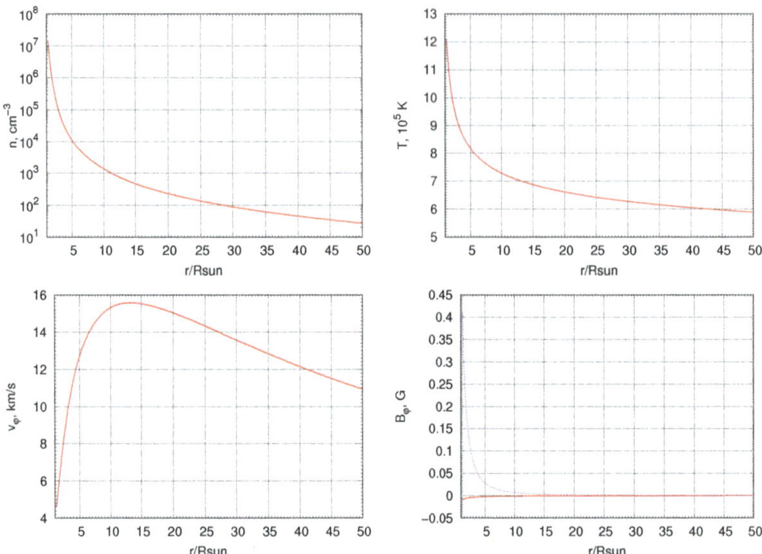

Figure 4. Particle number density profiles $n(r)$ (**top left**), temperature profiles $T(r)$ (**top right**), azimuthal velocity $v_\varphi(r)$ (**bottom left**), as well as the radial $B_r(r)$ (dotted line) and azimuthal $B_\varphi(r)$ (solid line) components of the magnetic field (**bottom right**).

3. Multi-Component Magnetic Hydrodynamics

Let us consider an approximation of multi-component (multi-fluid) magnetic hydrodynamics, which uses equations for mass quantities, the induction equation, as well as the continuity equations for individual plasma components. The individual components (electrons, ions, and neutrals of various kinds) of plasma will be marked with the index s. Denote by v the average mass velocity of matter. Then, the continuity equations for each component of the kind s can be written as:

$$\frac{\partial \rho_s}{\partial t} + \nabla \cdot (\rho_s v) = D_s + S_s. \tag{63}$$

where ρ_s is the density of the component s, the value

$$D_s = \nabla \cdot (\rho_s w_s) \tag{64}$$

determines diffusion, $w_s = v - v_s$ is the diffusion velocities. The last term in the right hand side of (63) takes into account the source function, which describes changes in the number of particles of the kind s due to chemical reactions, as well as the processes of dissociation, ionization, and recombination. For the total density

$$\rho = \sum_s \rho_s \tag{65}$$

due to the conditions

$$\sum_s \rho_s w_s = 0, \quad \sum_s S_s = 0, \tag{66}$$

we have the usual continuity equation,

$$\frac{\partial \rho}{\partial t} + \nabla \cdot (\rho v) = 0. \tag{67}$$

Let us introduce the values ζ_s that describe the mass fraction of the component s and satisfy the relations:

$$\rho_s = \zeta_s \rho, \quad \sum_s \zeta_s = 1. \tag{68}$$

Then, from the Equation (63), we can find

$$\frac{\partial}{\partial t}(\rho \zeta_s) + \nabla \cdot (\rho \zeta_s \boldsymbol{v}) = D_s + S_s. \tag{69}$$

Note that one of the values ζ_s can be excluded, since it can always be expressed in terms of the others using conditions (68). It is convenient to consider electrons as such a component. We will use a separate index for it, and, for the other components, we assume that s runs through the values from 1 to N. Therefore, all values ζ_s can be considered independent. The mass fraction of electrons in this case is equal to

$$\zeta_e = 1 - \sum_s \zeta_s. \tag{70}$$

Since the mass of electrons is much smaller than the mass of ions and neutrals, ζ_e turns out to be a value close to zero.

The equations of motion and energy for mass quantities can be written in the following form:

$$\rho\left[\frac{\partial \boldsymbol{v}}{\partial t} + (\boldsymbol{v} \cdot \nabla)\boldsymbol{v}\right] + \nabla P + \frac{\boldsymbol{B} \times \nabla \times \boldsymbol{B}}{4\pi} = \rho \boldsymbol{f} - \boldsymbol{D}_v, \tag{71}$$

$$\rho\left[\frac{\partial \varepsilon}{\partial t} + (\boldsymbol{v} \cdot \nabla)\varepsilon\right] + P \nabla \cdot \boldsymbol{v} = D_\varepsilon + \rho Q, \tag{72}$$

where P is the pressure, $\boldsymbol{B}$ is the magnetic field, $\boldsymbol{f}$ is the specific external force, the values

$$\boldsymbol{D}_v = \sum_s \nabla \cdot (\rho_s \boldsymbol{w}_s \boldsymbol{w}_s), \tag{73}$$

$$D_\varepsilon = \sum_s [\nabla \cdot (\rho_s \varepsilon_s \boldsymbol{w}_s) + P_s \nabla \cdot \boldsymbol{w}_s] \tag{74}$$

are determined by the diffusion velocities $\boldsymbol{w}_s$, and the value of Q describes the heating–cooling sources. In the expression (74), the notations P_s and ε_s are used for the partial pressures and specific internal energies of the components. To these equations, we should add the induction equation describing the evolution of the magnetic field,

$$\frac{\partial \boldsymbol{B}}{\partial t} - \nabla \times (\boldsymbol{v} \times \boldsymbol{B}) = -\nabla \times (\eta \nabla \times \boldsymbol{B}) + \nabla \times (\boldsymbol{v}_{\mathrm{D}} \times \boldsymbol{B}). \tag{75}$$

Here, the first term in the right hand side describes the ohmic diffusion of the magnetic field, where η is the corresponding magnetic viscosity. The second term determines the effect of ambipolar diffusion occurring in an incompletely ionized plasma. The vector $\boldsymbol{v}_{\mathrm{D}}$ represents the speed of ambipolar diffusion. Our numerical code takes into account the effects caused by both magnetic viscosity and ambipolar diffusion. However, in this paper we do not consider them, and we will assume that the values $\eta = 0$, $\boldsymbol{v}_{\mathrm{D}} = 0$.

The density ρ, pressure P, internal energy ε, and temperature T satisfy the equation of state for an ideal gas

$$P = \frac{k_{\mathrm{B}}}{\mu m_{\mathrm{p}}} \rho T, \quad \varepsilon = \frac{k_{\mathrm{B}} T}{\mu m_{\mathrm{p}}(\gamma - 1)}, \tag{76}$$

where k_B is the Boltzmann constant, m_p is the proton mass, γ is the adiabatic index. The average molecular weight μ is determined by the expression $\rho = \mu m_p n$, where n is the total number density. Let us write down the condition of quasi-neutrality of plasma

$$-e n_e + \sum_s e_s n_s = 0, \tag{77}$$

where e is the elementary charge, n_e is the electron number density, e_s is the charge of particles of the kind s, and n_s is the number density of the component s. Hence, the number density of electrons

$$n_e = \sum_s Z_s n_s, \tag{78}$$

where $Z_s = e_s/e$ is the charge number of particles of the kind s (for neutrals it is zero). Taking into account this expression, we find

$$\frac{1}{\mu m_p} = \sum_s (1 + Z_s) \frac{\zeta_s}{m_s}, \tag{79}$$

where m_s is the mass of particles of the kind s.

The numerical method we use to solve the equations of multi-component magnetic hydrodynamics is described in the Appendix A.

4. Model for Envelope of Hot Jupiter

4.1. Model Description

The structure and dynamics of plasma flow in the vicinity of hot Jupiter will be described in the approximation of multi-component (multi-fluid) magnetic hydrodynamics, which was discussed in the previous Section 3. In this approximation, it is possible, in particular, to take into account the complex chemical composition of the extended envelope of hot Jupiter. It is convenient to explicitly distinguish the background magnetic field [16,41,52,68,69], when the total magnetic field B is represented as a superposition of the background magnetic field H and the magnetic field b induced by electrical currents in plasma itself, $B = H + b$. This approach allows us to significantly minimize the numerical errors that occur during arithmetic operations with large numbers, provides greater stability of the scheme, and improves the quality of calculations in rarefied regions with a strong magnetic field—for example, in the magnetosphere of hot Jupiter.

In our formulation of the problem, the background field is created by sources distributed inside the star (or, more precisely, inside the corona), as well as in the bowels of the planet. Therefore, these sources are absent in the computational domain, and hence the background field should satisfy the potentiality condition, $\nabla \times H = 0$. This allows it to partially exclude from the equations of magnetic hydrodynamics [70,71].

In general case, the background magnetic field is not stationary, $\partial H/\partial t \neq 0$. However, our model assumes that this property is possessed by the magnetic field of the planet. The proper rotation of hot Jupiter, due to strong tidal interactions from a closely located host star, turns out to be synchronized with the orbital rotation. As a result, the period of planet rotation around its proper axis will be equal to its orbital period, and, consequently, a hot Jupiter will always face the same side to its host star.

Therefore, in a rotating frame of reference associated with the orbital motion of the planet, the orientation of its own magnetic field will not change over time. On the other hand, the magnitude of field (for example, the magnetic moment) changes on much larger time scales compared to the orbital period and, in our calculations, this change can be ignored. The same comments can be made about the induced magnetic field of the planet.

As theoretical estimates show (see, e.g., [1,17]), a hydrodynamic approach is applicable to describe the flow of matter near a typical hot Jupiter. This is due to the fact that the extended envelope of hot Jupiter has a sufficiently high density and therefore, its matter is not collision-less. In addition, due to the processes of thermal ionization and hard radiation of the host star, the upper atmosphere and the extended envelope of hot Jupiters consists of almost fully ionized gas [21].

Taking into account the complex chemical composition of the extended envelope, this makes it possible to use the multi-component magnetic hydrodynamics approximation. On the other hand, the analysis of the characteristic collision frequencies [13,46,47,72] of the components in the hydrogen–helium envelope of hot Jupiter allows us to neglect the diffusion effects [48] with fairly good accuracy and assume that the average velocities of all components are equal to the average mass velocity of matter, $v_s = v$. We can also assume that all the components are in thermodynamic equilibrium, and therefore their temperatures are equal to the temperature of matter, $T_s = T$.

Considering the above circumstances, the equations of multi-component magnetic hydrodynamics can be written as

$$\frac{\partial \rho}{\partial t} + \nabla \cdot (\rho v) = 0, \tag{80}$$

$$\frac{\partial v}{\partial t} + (v \cdot \nabla)v = -\frac{\nabla P}{\rho} - \frac{b \times \nabla \times b}{4\pi\rho} - \frac{H \times \nabla \times b}{4\pi\rho} + f, \tag{81}$$

$$\frac{\partial b}{\partial t} = \nabla \times (v \times b + v \times H) - \frac{\partial H}{\partial t}, \tag{82}$$

$$\frac{\partial \varepsilon}{\partial t} + (v \cdot \nabla)\varepsilon + \frac{P}{\rho}\nabla \cdot v = Q, \tag{83}$$

$$\frac{\partial}{\partial t}(\rho \zeta_s) + \nabla \cdot (\rho \zeta_s v) = S_s, \quad s = 1, \ldots, N. \tag{84}$$

To close this system of equations, the equation of state for an ideal gas (76) with the adiabatic index $\gamma = 5/3$ is used. In this paper, we focus on a more correct accounting of the stellar wind. Therefore, in the simulations described below, we assumed that the source functions S_s due to chemical reactions, processes of ionization, recombination and dissociation of molecules, as well as the corresponding contributions to the heating function Q are equal to zero. This implies that various plasma components that we considered as passive admixtures transported together with the matter. In addition, for the same reason, in this work, we did not take into account the effects of magnetic viscosity and ambipolar diffusion.

We assume that the planet orbit is circular. Therefore, in a non-inertial reference frame rotating together with a binary system with an angular velocity Ω, the locations of the star and the planet centers do not change. In this case, the specific external force is determined by the expression

$$f = -\nabla\Phi - 2(\Omega \times v). \tag{85}$$

Here, the first term on the right hand side describes the force due to the gradient of the Roche potential

$$\Phi = -\frac{GM_s}{|r - r_s|} - \frac{GM_p}{|r - r_p|} - \frac{1}{2}[\Omega \times (r - r_c)]^2, \tag{86}$$

where M_s is the mass of the star, M_p is the mass of the planet, r_s is the radius vector of the star center, r_p is the radius vector of the planet center, r_c is the radius vector of the center of mass of the system. The second term in the right hand side of (85) describes the Coriolis force.

In this numerical model, the stellar wind is taken into account according to the same scheme as for the purely gas-dynamic case [17]. However, this does not use a constant value of the radial wind velocity but the profile $v_r(r)$ obtained from the wind model (see Section 2). Profiles for the remaining values are calculated using this profile: $\rho(r)$, $v_\varphi(r)$, $B_\varphi(r)$. This allows finding the wind parameters at any point $r = (x, y, z)$ of the computational domain. At the same time, in a non-inertial reference frame, the magnetic field of the wind does not change, and the wind velocity is recalculated using the formula

$$v_w = u_w - \Omega \times (r - r_c),$$

(87)

where u_w is the wind velocity in the inertial reference frame. These distributions are used to set initial conditions in the region occupied by the stellar wind, as well as to implement boundary conditions. The wind model used assumes a polytropic equation of state (6). The Equations (80)–(84) use the model of adiabatic magnetic hydrodynamics with the adiabatic index γ. This means that, from the point of view of the adiabatic model, there are effective heating processes in the stellar wind. In order to reconcile these two models, the corresponding function (47) [25] is added to the energy Equation (83)

$$Q_w = (\gamma - \kappa)\varepsilon_w \nabla \cdot v_w,$$

(88)

where ε_w is the specific internal energy of the wind matter. This takes into account the expression (87) wind velocity divergence

$$\nabla \cdot v_w = \nabla \cdot u_w = \frac{1}{r^2}\frac{d}{dr}\left(r^2 v_r\right).$$

(89)

Since the stellar wind accelerates (the velocity v_r increases with the distance r from the star), the divergence of its velocity turns out to be positive. Consequently, the function $Q_w > 0$.

4.2. Upper Atmosphere

At the initial moment of time, a spherically symmetric iso-thermal atmosphere was set around the planet, the density distribution in which was determined from the hydrostatic equilibrium condition:

$$\rho = \rho_a \exp\left[-\frac{GM_p}{R_p A_{gas} T_a}\left(1 - \frac{R_p}{|r - r_p|}\right)\right].$$

(90)

Here, ρ_a is the density at the photometric radius of hot Jupiter R_p, T_a is the atmospheric temperature, $A_{gas} = k_B/(\mu m_p)$ is the gas constant, μ is the average molecular weight. A similar expression can be written for the pressure P. For the hot Jupiter HD 209458b dimensionless parameter,

$$\eta = \frac{GM_p}{R_p A_{gas} T_a} = 10.3\,\mu\left(\frac{T_a}{10^4\,\text{K}}\right)^{-1}.$$

(91)

The initial thickness of the atmosphere was determined from the pressure equilibrium condition with the matter of the stellar wind. The total wind pressure

$$P_{tot} = P_w + \rho_w v_w^2 + \frac{B_w^2}{8\pi}.$$

(92)

Therefore, from the equality of pressure at the point of a frontal impact, we find the initial radius of the atmosphere

$$R_a = \left(1 - \frac{1}{\eta}\ln\frac{P_a}{P_{tot}}\right)^{-1} R_p.$$

(93)

The value R_a is determined to a greater extent by the atmosphere parameters ρ_a and T_a and to a lesser extent by the magnetic field of the wind B_w. In particular, the ratio between the initial radius of the atmosphere R_a and the size of the Roche lobe of the planet determines the type of extended envelope of the hot Jupiter (closed or open).

We assume that the hydrogen–helium atmosphere of hot Jupiter has a homogeneous chemical composition. In any allocated volume of the atmosphere, the parameter $\chi = [\text{He}/\text{H}]$, equal to the ratio of the helium nuclei number to the hydrogen one number, remains constant. The following significant components of the atmosphere are taken into account [16,73]: molecular hydrogen H_2, atomic hydrogen H, ionized hydrogen H^+, atomic helium He, once ionized helium He^+. The mass fraction of hydrogen X and helium Y are determined by the expressions:

$$X = \frac{\rho(H_2) + \rho(H) + \rho(H^+)}{\rho} = \xi(H_2) + \xi(H) + \xi(H^+), \tag{94}$$

$$Y = \frac{\rho(He) + \rho(He^+)}{\rho} = \xi(He) + \xi(He^+). \tag{95}$$

From here, we can find: $Y/X = 4\chi$. Taking into account the equality $X + Y = 1$, we find

$$X = \frac{1}{1 + 4\chi}, \quad Y = \frac{4\chi}{1 + 4\chi}. \tag{96}$$

Let us consider the degree of hydrogen ionization x_1, the degree of dissociation of molecular hydrogen x_2, and the degree of helium ionization x_3,

$$x_1 = \frac{\rho(H^+)}{\rho(H^+) + \rho(H)}, \quad x_2 = \frac{\rho(H)}{\rho(H) + \rho(H_2)}, \quad x_3 = \frac{\rho(He^+)}{\rho(He^+) + \rho(He)}. \tag{97}$$

Then, for the mass fractions of the components, it can be written:

$$\xi(H_2) = \frac{(1 - x_1)(1 - x_2)}{1 - x_1 + x_2 x_2} X, \tag{98}$$

$$\xi(H) = \frac{(1 - x_1) x_2}{1 - x_1 + x_2 x_2} X, \tag{99}$$

$$\xi(H^+) = \frac{x_1 x_2}{1 - x_1 + x_2 x_2} X, \tag{100}$$

$$\xi(He) = (1 - x_3) Y, \tag{101}$$

$$\xi(He^+) = x_3 Y. \tag{102}$$

Using these definitions, from the expression (79) for the average molecular weight, we find

$$\frac{1}{\mu} = \frac{1}{2} X \left(1 + \frac{x_2 + 2 x_1 x_2}{1 + x_1 - x_1 x_2} \right) + \frac{1}{4} Y (1 + x_3). \tag{103}$$

The total degree of ionization of atmospheric matter

$$\xi = \xi(H^+) + \xi(He^+) = \frac{x_1 x_2}{1 - x_1 + x_1 x_2} X + x_3 Y. \tag{104}$$

In the simulations of the flow structure in the vicinity of the hot Jupiter HD 209458b given below, we set the following parameters of the chemical composition of the atmosphere: $\chi = [\text{He}/\text{H}] = 0.05$ [25], $x_1 = x_2 = x_3 = 0.9$. From here, we find the mass fraction of hydrogen $X = 0.83$ and helium $Y = 0.17$. These parameters correspond to the mass fractions of the components $\xi(H_2) = 0.009$, $\xi(H) = 0.08$, $\xi(H^+) = 0.74$, $\xi(He) = 0.02$, $\xi(He^+) = 0.15$. At the same time, the average molecular weight of the atmospheric matter is $\mu = 0.69$, and the total degree of ionization is $\xi = 0.89$.

The boundary conditions (density ρ_a, temperature T_a and velocity v_a) at the photometric radius were set based on the results of calculations carried out within the framework of one-dimensional aeronomic models considering supra-thermal particles [16,73]. Therefore, in a certain sense, our numerical model is hybrid. It follows from aeronomic calculations that a planetary wind is formed in the upper atmosphere of hot Jupiter under the influence of star hard radiation, which determines the mass loss rate $\dot{M}_a \approx 10^9$–10^{10} g/s. Taking into account the expression $\dot{M}_a = 4\pi R_p^2 \rho_a v_a$ from here, we can find the speed of atmospheric outflow v_a at the photometric radius.

4.3. Magnetic Field

The background magnetic field can be set as a superposition of several separate fields:

$$\boldsymbol{H} = \boldsymbol{H}_s + \boldsymbol{H}_w + \boldsymbol{H}_p + \boldsymbol{H}_a. \tag{105}$$

Here, $\boldsymbol{H}_s$ describes the proper field of the star, $\boldsymbol{H}_w$ is determined by the wind field, $\boldsymbol{H}_p$ is the proper field of the planet, and $\boldsymbol{H}_a$ is determined by the field of atmosphere. Let us characterize the contribution of each term.

As noted in Section 2, the proper magnetic field of the star $\boldsymbol{H}_s$ plays a significant role in the corona region. Therefore, if the planet orbit is located in the heliospheric region, then this field can be neglected. The main role in this zone is played by the wind field. The magnetic field of the host star should be taken into account when the planet orbit is located near the outer boundary of the corona or even inside it. Currently, hot Jupiters of this type are observed and in some papers (see, for example, [51]) the field of the star was taken into account explicitly. In our work, we consider the hot Jupiter HD 209458b, whose orbit is located in the heliospheric region. This allows us to neglect the proper field of the host star in the calculations.

The wind magnetic field, which is determined from the model described in Section 2, does not make sense to include (105) entirely in the background field. The fact is that the total magnetic field of the wind $\boldsymbol{B}_w$ does not satisfy the condition of potentiality, $\nabla \times \boldsymbol{B}_w \neq 0$, since the rotor of the magnetic field just determines the electromagnetic force affecting the dynamics of the wind plasma. However, the radial magnetic field of the wind B_R can be included in the background field. On the one hand, this field, as it is easy to see, satisfies the condition of potentiality. On the other hand, in the vicinity of hot Jupiter, located close to the host star, the radial component of the magnetic field dominates in the wind plasma, since the ratio B_φ / b_r is small in absolute value (see Figure 4). Taking into account the assumed spherical symmetry of the stellar wind, we obtain (see Section 2)

$$\boldsymbol{H}_w = \frac{B_s R_s^2}{|\boldsymbol{r} - \boldsymbol{r}_s|^2} \boldsymbol{n}_s, \tag{106}$$

where the unit vector $\boldsymbol{n}_s = (\boldsymbol{r} - \boldsymbol{r}_s)/|\boldsymbol{r} - \boldsymbol{r}_s|$ determines the direction from the center of star $\boldsymbol{r}_s$ to the observation point $\boldsymbol{r}$.

In our numerical model, we assume that the intrinsic magnetic field of hot Jupiter is purely dipole,

$$\boldsymbol{H}_p = \frac{\mu_p}{|\boldsymbol{r} - \boldsymbol{r}_p|^3} \left[3(\boldsymbol{d}_p \cdot \boldsymbol{n}_p)\boldsymbol{n}_p - \boldsymbol{d}_p \right], \tag{107}$$

where μ_p is the magnetic moment of the planet, $\boldsymbol{n}_p = (\boldsymbol{r} - \boldsymbol{r}_p)/|\boldsymbol{r} - \boldsymbol{r}_p|$, $\boldsymbol{d}_p$ is a unit vector directed along the magnetic axis and determining the vector of magnetic moment $\boldsymbol{\mu}_p = \mu_p \boldsymbol{d}_p$. In our calculations, we assumed the value of magnetic moment of the hot Jupiter HD 209458b to be $\mu_p = 0.1 \mu_J$. The orientation of magnetic dipole was determined by the angles θ and ϕ, which were used as model parameters. The components of the unit vector $\boldsymbol{d}_p$ directed along the magnetic axis in the Cartesian coordinate system are described by the expressions:

$$\boldsymbol{d}_p = (\sin\theta\cos\phi, \quad \sin\theta\sin\phi, \quad \cos\theta). \tag{108}$$

At the same time, we assumed that the proper rotation of the planet is synchronized with the orbital one, and the axes of proper and orbital rotation are collinear.

For hot Jupiters, the magnetic field induced in the upper atmosphere and the extended envelope seems to play an important role. Since plasma is almost completely ionized in this region, any movements in it lead to the appearance of electric currents and, consequently, to the generation of the magnetic field. This field, in particular, can arise due to zonal currents in the upper atmosphere caused by uneven heating from the radiation of the host star [39]. On the other hand, due to its close location to the host star, a quite strong magnetic field can arise in the upper atmosphere of hot Jupiter, induced by currents shielding the magnetic field of the stellar wind [40]. The configuration of induced magnetic field is such that inside the atmosphere it completely compensates for the magnetic field of the wind, and outside it is dipole. Therefore, in the region beyond the initial atmosphere, we can write

$$H_a = \frac{\mu_a}{|r - r_p|^3} \left[3(d_a \cdot n_p)n_p - d_a \right], \tag{109}$$

where the corresponding magnetic moment $\mu_a = R_a^3 B_w / 2$, and the unit vector $d_a = -B_w / B_w$ is directed against the vector B_w. The magnitude of induced magnetic moment μ_a depends on the initial radius of the atmosphere R_a and the wind field B_w in the orbit of the planet. If we take an average field of the star $B_s = 0.5$ Gs, then, for the characteristic dimensions of the atmosphere $R_a = (4\text{--}5)R_p$, we can obtain the values of magnetic moment $\mu_a = (0.05\text{--}0.2)\mu_J$. In other words, in order of magnitude, the induced magnetic moment μ_a turns out to be equal to the intrinsic magnetic moment μ_p of the hot Jupiter. Note that the induced magnetic field H_a is completely determined by the wind field B_w. Therefore, the structure of induced field (in particular, the vector of induced magnetic moment $\mu_a = \mu_a d_a$) will track the direction to the star when the planet moves along its orbit.

As noted above, in a non-inertial reference frame rotating together with a binary system consisting of a star and a planet, the magnetic fields H_p and H_a are stationary, $\partial H_p / \partial t = 0$, $\partial H_a / \partial t = 0$. The proper field of a star, on the contrary, is non-stationary, $\partial H_s / \partial t \neq 0$, since it rotates together with the star at the angular velocity $\Omega_s - \Omega$. The radial component of the wind magnetic field H_w is also rigidly associated with the star. In this case, it can be assumed that the field lines velocity of the radial magnetic field in a non-inertial reference frame is equal to

$$v_s = -\Omega \times (r - r_s), \tag{110}$$

since the azimuthal component of the field, due to the proper rotation of the star Ω_s, has already been taken into account in the vector b. The change in the magnetic field H_w in time is determined by the equation

$$\frac{\partial H_w}{\partial t} = \nabla \times (v_s \times H_w). \tag{111}$$

Substituting the expressions (106) and (110) here, it is not difficult to verify by direct calculations that the right hand side of this equation vanishes due to the spherical symmetry of the field H_w. Consequently, the left hand side should also be equal to zero, $\partial H_w / \partial t = 0$. Thus, in our model, the total background magnetic field (105) turns out to be stationary, $\partial H / \partial t = 0$.

4.4. Numerical Method

To numerically solve the equations of multi-component magnetic hydrodynamics (80)–(84), we use a combination of difference schemes of Roe (see Section 3) and Lax–Friedrichs [74,75]. The solution algorithm consists of several successive stages resulting from the application of the splitting method by physical processes. Suppose that we know the distribution of all values on the computational mesh at the moment of time t.

Then, to obtain the values at the next time moment $t + \Delta t$, we decompose the complete system of Equations (80)–(84) into two subsystems.

The first subsystem corresponds to the equations of ideal multi-component magnetic hydrodynamics with intrinsic magnetic field $\boldsymbol{b}$ of plasma without considering the background magnetic field $\boldsymbol{H}$:

$$\frac{\partial \rho}{\partial t} + \nabla \cdot (\rho \boldsymbol{v}) = 0, \tag{112}$$

$$\frac{\partial \boldsymbol{v}}{\partial t} + (\boldsymbol{v} \cdot \nabla)\boldsymbol{v} = -\frac{\nabla P}{\rho} - \frac{\boldsymbol{b} \times \nabla \times \boldsymbol{b}}{4\pi\rho} + \boldsymbol{f}, \tag{113}$$

$$\frac{\partial \boldsymbol{b}}{\partial t} = \nabla \times (\boldsymbol{v} \times \boldsymbol{b}), \tag{114}$$

$$\frac{\partial \varepsilon}{\partial t} + (\boldsymbol{v} \cdot \nabla)\varepsilon + \frac{P}{\rho}\nabla \cdot \boldsymbol{v} = Q, \tag{115}$$

$$\frac{\partial}{\partial t}(\rho \zeta_s) + \nabla \cdot (\rho \zeta_s \boldsymbol{v}) = 0, \quad s = 1, \ldots, N. \tag{116}$$

In our numerical model, a method based on Roe's scheme was used to solve this system (see Appendix A).

The second subsystem corresponds to considering the influence of the background field:

$$\frac{\partial \boldsymbol{v}}{\partial t} = -\frac{\boldsymbol{H} \times \nabla \times \boldsymbol{b}}{4\pi\rho}, \tag{117}$$

$$\frac{\partial \boldsymbol{b}}{\partial t} = \nabla \times (\boldsymbol{v} \times \boldsymbol{H}). \tag{118}$$

The first equation in this subsystem describes the influence of the electromagnetic force caused by the background field, and the second equation does the generation of a magnetic field. At the same time, it is assumed that at this stage of the algorithm, the density ρ and the specific internal energy ε do not change. To solve the second subsystem, the Lax–Friedrichs scheme was used with increasing TVD (total variation diminishing) corrections [68].

To clear the divergence of the magnetic field $\boldsymbol{b}$, we used the method of the generalized Lagrange multiplier [76]. The choice of this method is due to the fact that the flow in vicinity of hot Jupiter is essentially non-stationary, especially in the flow on the night side forming the magnetospheric tail.

5. Results of Simulations

5.1. Model Parameters

As an object of study, we use a typical hot Jupiter HD 209458b, which has the mass $M_p = 0.71 M_J$ and the photometric radius $R_p = 1.38 R_J$, where M_J and R_J is the mass and radius of Jupiter. The host star is characterized by the following parameters: spectral class G0, mass $M_s = 1.1 M_\odot$, radius $R_s = 1.2 R_\odot$. The period of proper rotation of the star $P_{rot} = 14.4$ days, which corresponds to the angular velocity $\Omega_s = 5.05 \times 10^{-6}$ s^{-1} or linear velocity at the equator $v_{rot} = 4.2$ km/s. The major semi-axis of the planet orbit $A = 10.2 R_\odot$, which corresponds to the period of revolution around the star $P_{orb} = 84.6$ h.

In the simulations, the temperature of atmosphere T_a is varied, while the particle number density at the photometric radius was set equal to a fixed value $n_a = 10^{10}$ cm^{-3}. The atmospheric outflow rate $\dot{M}_a = 10^9$ g/s corresponds to the velocity of the planetary wind at the photometric radius $v_a = 78$ cm/s. The values of these parameters correspond to the results obtained in aeronomic models for atmospheres of hot Jupiters [12,14–16,48,73]. We assumed that the magnitude of the magnetic moment μ_p of the planet is 0.1 of the magnetic moment of Jupiter, and the orientation of magnetic dipole axis (108) is determined by the angle values $\theta = 90°$ and $\varphi = 60°$.

At the same time, as already mentioned above, we believed that the proper rotation of the planet is synchronized with the orbital one, and the axes of proper and orbital rotations are collinear. The average magnetic field at stellar surface B_s in various models was set to 0.01 G (weak field), 0.2 G (medium field), and 0.5 G (strong field). These values are in the physically acceptable range (see, e.g., [63]).

Figure 5 shows the distributions of the radial wind velocity v_r (left panel) and the heating function Q_w, which is determined by the Equation (88) (right panel), for all three cases of values B_s. The characteristic value of the heating function in the vicinity of the planet $Q_w = 5 \times 10^9$ erg·g^{-1}·s^{-1}. In this case, the maximum values of the heating function are reached at distances of approximately $4R_\odot$ from the star center.

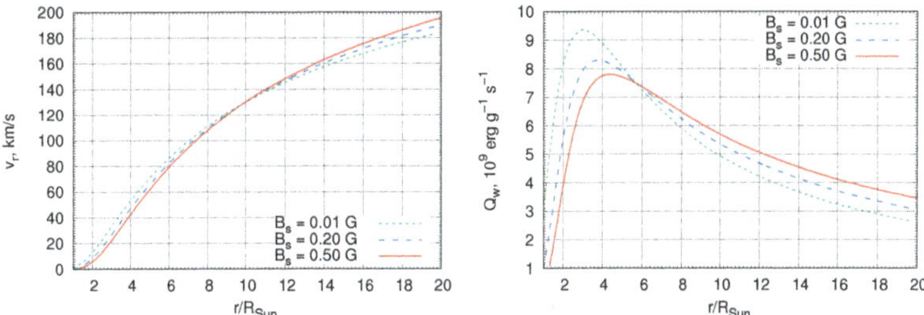

Figure 5. Distributions of the radial wind velocity v_r (**left**) and the heating function Q_w (88) (**right**) for the magnetic field values at the stellar surface B_s used in simulations.

For numerical simulations, we used the Cartesian coordinate system (x, y, z) in a non-inertial reference frame rotating together with the binary system "star–planet" around the center of masses. The origin of the coordinate system was chosen at the planet center $r_p = (0,0,0)$. The x axis was located along the line connecting the centers of the star and the planet, while the center of the star was at the point $r_s = (-A,0,0)$. The y axis was chosen so that its direction coincided with the direction of the orbital motion of the planet. Finally, considering the selected orientations of the axes x and y, the third axis z will coincide in the direction with the vector of the orbital angular velocity Ω.

In order to check the correctness of considering the complete wind model, we performed separate numerical calculations of the flow structure in the region where the planet is absent. To do this, it was enough to choose a computational domain symmetrically located relative to the center of the star (i.e., in the vicinity of the point $x = -2A$). Calculation results for cases of weak ($B_s = 0.01$ G) and strong ($B_s = 0.5$ G) magnetic field of the wind are shown in Figures 6 and 7, respectively.

These figures show the distributions of density (color and iso-lines), velocity (arrows) and magnetic field (solid lines) at the initial moment of time (left panels) and after one orbital period (right panels). The density values are given in units of magnitude $\rho_w = 2.3 \times 10^{-21}$ g/cm^3, corresponding to the wind density at a distance of $10R_\odot$ from the center of the star. The dotted line shows the boundary of the Roche lobe. The star is located to the right of the computational domain.

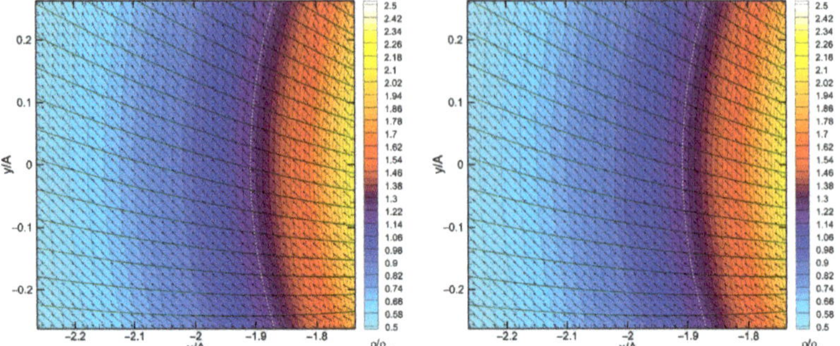

Figure 6. Distributions of density (color), velocity (arrows) and magnetic field (lines) in the stellar wind for the case when the magnetic field at the surface of the star $B_s = 0.01$ G at the initial time (**left**) and after one orbital period (**right**). The dotted line shows the boundary of the Roche lobe.

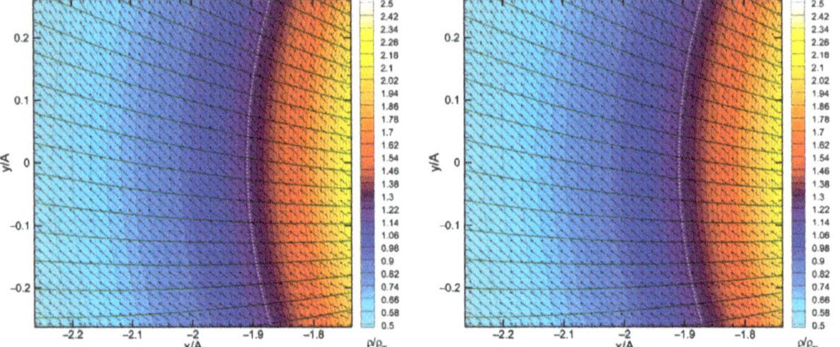

Figure 7. The same as in Figure 6, but for the case, when the magnetic field at the stellar surface $B_s = 0.5$ G.

The analysis of these figures allows us to conclude that during the time of order of the orbital period, in the numerical model, the distributions of wind parameters for both the case of weak and strong fields practically do not change. Small perturbations introduced into the solution by boundary conditions are not essential for our purposes. Therefore, we can assume that the accounting of the complete wind model is carried out correctly.

Figure 8 shows the initial distributions of density (color) and magnetic field (arrow lines) in the vicinity of the planet for the case, when the temperature of the atmosphere $T_a = 6000$ K. The left panel corresponds to the magnitude of the magnetic field at the surface of the star $B_s = 0.01$ G (weak field), and the right panel does to the case for the strong field $B_s = 0.5$ G. The dotted line again shows the boundary of the Roche lobe and the white circle corresponds to the photometric radius of the planet. The star is located on the left side. The radius of the atmosphere in the case of the strong field (right panel) is smaller compared to the case of the weak field (left panel). This is due to the fact that an increase in the field B_s leads to an increase in the total wind pressure (92) and, consequently, to a decrease in the initial radius of the atmosphere (93).

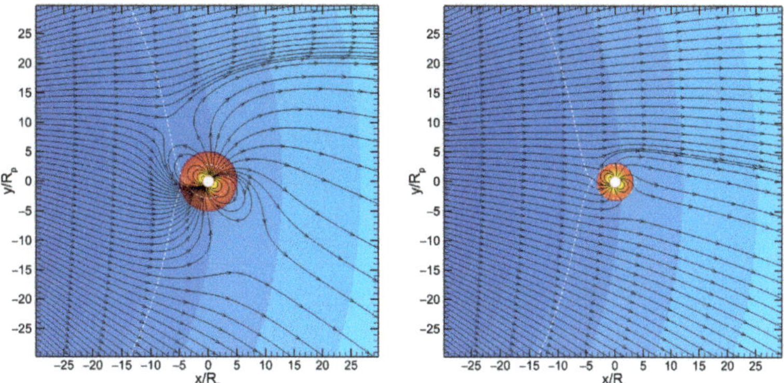

Figure 8. The initial distributions of density (color) and magnetic field (lines with arrows) for the case, when the temperature of the atmosphere $T_a = 6000$ K, and the magnetic field at the stellar surface B_s is 0.01 G (left) and 0.5 G (right). The dotted line shows the boundary of the Roche lobe. The white circle corresponds to the photometric radius of the planet.

It can be noticed that the magnetic field is clearly divided into four zones. In the first zone (above and below the planet in the figures), the magnetic field is characterized by open force lines of the star, which begin at its surface and go to infinity. In the second zone (to the right of the planet), the magnetic field is determined by the open force lines of the planet. In the third zone (to the left of the planet), magnetic lines are common for the star and the planet. They start at the surface of the star and finish at the surface of the planet, forming a kind of magnetic "bridge" connecting the star with the planet and performing orbital rotation with them together. Finally, the last zone consists of the closed lines of the planet forming the inner part of the magnetosphere.

There are two neutral points in the orbital plane, in which, due to the superposition of individual fields, the total induction vector $B = 0$ and therefore the direction of the magnetic field becomes indeterminate. In space, the set of these neutral points forms a certain line, the shape of which, in particular, is determined by the orientation parameters of the magnetic axis of the planet. In case of a strong wind field (Figure 8, right panel) neutral points and closed dipole lines are located in a more compact region around the planet. When the stellar wind flows around the planet, a more complex flow pattern is formed and the structure of the magnetic field is significantly distorted. However, in general, the described topology (separation into magnetic zones and the presence of neutral points) is preserved.

Below, we present the results of two blocks of numerical simulations. In the first block, we set a weak field of the wind corresponding to the super-Alfvén regime of the stellar wind flowing around hot Jupiter, and varied the temperature of the atmosphere. This led to the formation of various types of super-Alfvén envelopes. In the second block, with fixed atmospheric parameters, we varied the magnitude of the magnetic field of the wind in order to trace how the envelope structure changes in this case. The parameters of the models, as well as the flow characteristics, are presented in Table 1.

Table 1. Parameters and characteristics of the models: T_a is the temperature of the upper atmosphere, B_s is the magnetic field of the star, and $\dot{M}_p$ is the mass loss rate of the planet.

Model	T_a, K	B_s, G	Flow Regime	Envelope Type	$\dot{M}_p$, 10^9 g/s
1	5000	0.01	Super-Alfvénic	Closed	1
2	5500	0.01	Super-Alfvénic	Quasi-closed	1
3	6000	0.01	Super-Alfvénic	Quasi-Opened	4
4	6500	0.01	Super-Alfvénic	Opened	10
5	6500	0.2	Trans-Alfvénic	Quasi-Opened	9
6	6500	0.5	Sub-Alfvénic	Quasi-Opened	7

5.2. Super-Alfvén Flow Regime

This section presents the results of numerical simulation of the flow structure in the vicinity of hot Jupiter for the case, when the magnetic field at the surface of the star $B_s = 0.01$ G. The Alfvén Mach number (29) in the orbit of the planet is $\lambda = 15.5$. This means that the planet is located in the super-Alfvén zone of the stellar wind. The total Alfvén Mach number, considering the velocity of the orbital motion of the planet, $\lambda = 23.2$. Therefore, in this case, the flow around of hot Jupiter by the stellar wind should occur in the super-Alfvén regime.

Calculations were performed for four models differing in atmospheric temperature values T_a (see Table 1). Namely, the temperature of atmosphere T_a was set equal to 5000 K (model 1), 5500 K (model 2), 6000 K (model 3), 6500 K (model 4). The simulation was carried out in the computational domain $-30 \leq x/R_p \leq 30$, $-30 \leq y/R_p \leq 30$, $-15 \leq z/R_p \leq 15$ with the number of cells $192 \times 192 \times 96$.

The simulation results are demonstrated by Figure 9. The density distributions (color, iso-lines), velocities (arrows) and magnetic field (lines) in the orbital plane are represented on various panels of this figure. The density is expressed in units of wind density in the vicinity of the planet ρ_w. The boundary of the Roche lobe is shown by a dotted line. The planet is located in the center of the computational domain and is represented by a light circle, the radius of which corresponds to the photometric one. All the solutions obtained correspond to the time moment of the order of one third of the orbital period from the beginning of the computation. During this period of time, a stable quasi-stationary flow pattern is formed.

In all four calculation variants, as a result of the stellar wind flowing around, a wide (on the order of several radii of the planet) hydrogen–helium turbulent plume is formed at the night side of the planet. The interaction of the stellar wind with the envelope of hot Jupiter leads to the appearance of bow shock wave. The position and shape of the bow shock in this numerical model are slightly different from those we received in our previous models [16,17,41,52,54–57,73]. This is due to a more correct consideration of wind parameters. In the old model, wind velocity and temperature were considered as constants, but in the new model they change in space.

In addition, we previously accelerated the wind by disabling the gravitational forces of the star and the planet in the area occupied by the wind plasma. As a result, the flow velocity increased and the shock wave pressed closer to the planet. In the new numerical model, the flow velocity is calculated from the wind structure and as a result, the shock wave moves further away from the planet. Another effect is due to the fact that the sound point is located inside the computational domain at a distance of about 20 radii of the planet towards the star (see Figure 3). Therefore, the lower left edge of the shock wave, reaching this point, breaks off. This is clearly visible on the upper panels of Figure 9.

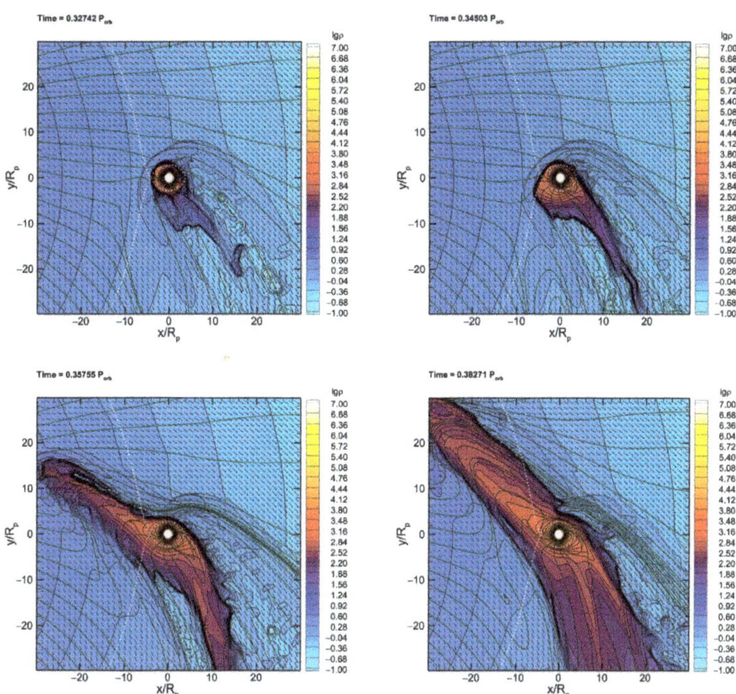

Figure 9. Distributions of density (color, iso-lines), velocity (arrows) and magnetic field (lines) in the orbital plane at a time moment approximately equal to a one third of the orbital period for models 1 ($T_a = 5000$ K, top left), 2 ($T_a = 5500$ K, top right), 3 ($T_a = 6000$ K, bottom left) and 4 ($T_a = 6500$ K, bottom right). The density is expressed in units of ρ_w. The dotted line shows the border of the Roche lobe. The white circle corresponds to the photometric radius of the planet.

In model 1 ($T_a = 5000$ K, upper left panel Figure 9) a compact envelope of hot Jupiter is formed, except a relatively weak plume located inside the Roche lobe. The bow shock has an almost spherical shape. According to the classification proposed in [17], such a flow configuration corresponds to the *closed* envelope of hot Jupiter. In model 2 ($T_a = 5500$ K, upper right panel Figure 9) a compact envelope of hot Jupiter is also forming. However, in this case, there is a small cusp in the surface of contact discontinuity (envelope boundary) directed towards the inner Lagrange point L_1. Therefore, the outflow of matter from the envelope occurs not only due to a turbulent plume forming at the night side, but also due to a weak outflow from the day side of the planet. The shape of bow shock is also close to spherical. An extended envelope of this type can be called *quasi-closed*. The mass loss rate for these models $\dot{M}_p \approx 10^9$ g/s and determines by the outflow from the night side of the planet (see the last column in Table 1).

A rather complex flow pattern is observed in model 3 ($T_a = 6000$ K, lower left panel Figure 9) and model 4 ($T_a = 6500$ K, lower right panel Figure 9). In these models, two powerful flows are formed from the vicinity of Lagrange points L_1 and L_2. The first stream, as in the previous two models, begins at the night side and forms a wide turbulent plume behind the planet. The second stream is formed at the day side, directed towards the star and, therefore, moves against the wind due to the stellar gravity.

A stream of hydrogen–helium matter from the inner Lagrange point significantly distorts the shape of the bow shock, while pushing it further away from the planet. We can say that the shock wave consists of two separate parts, one of which arises around the atmosphere of the planet, and the other arises around the stream from the inner Lagrange point L_1. The flow of matter in the stream is directed against the wind and therefore the Kelvin–Helmholtz instability develops along its surface.

In model 3, the flow has a lower intensity and it is largely deflected and blocked by the stellar wind. Such an extended envelope can be called *quasi-open*. The corresponding mass loss rate $\dot{M}_\mathrm{p} \approx 4 \times 10^9$ g/s and determines mainly by the outflow from inner Lagrange point L_1. Finally, in model 4, the intensity of the flow is so high that it does not stop under impact of the stellar wind and continues to move away towards the star. This leads to a significant loss of matter from the hot Jupiter envelope. We find that in this model the mass loss rate $\dot{M}_\mathrm{p} \approx 10^{10}$ g/s. An extended envelope of this type can be called *open*.

Thus, in the new numerical model, all the previously discovered [17] types of extended envelopes remain. However, their boundaries in the parameter space have shifted to the region of lower densities and temperatures. This is mainly due to the fact that, in this model, we take into account the chemical composition of the atmosphere.

5.3. Sub-Alfvén Flow Regime

In this section, we present the results of the second block of numerical simulations. The temperature of atmosphere was set equal to $T_\mathrm{a} = 6500$ K, which corresponds in the case of a weak field ($B_\mathrm{s} = 0.01$ G) to an open extended envelope under conditions of super-Alfvén flow by the stellar wind (model 4 from the previous Section 5.2). In the second block, calculations were performed for the cases $B_\mathrm{s} = 0.2$ G (model 5) and $B_\mathrm{s} = 0.5$ G (model 6). The parameters of the computational domain and the mesh were used the same as in the previous models 1–4. The results of the simulations are shown in Figure 10, which shows the flow structure for these two variants at a time moment of about one third of the orbital period from the beginning of computation.

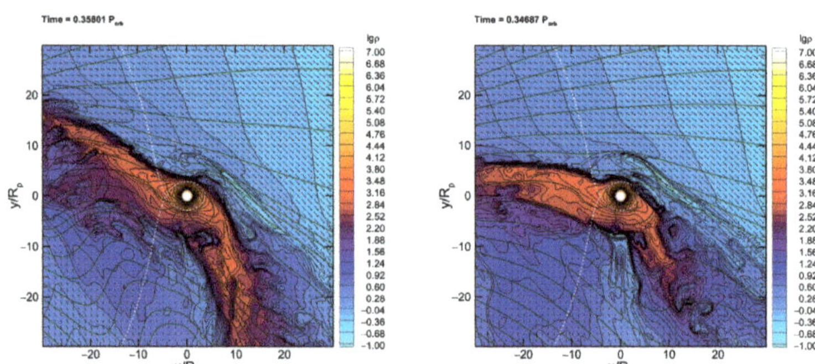

Figure 10. Distributions of density (color, iso-lines), velocity (arrows), and magnetic field (lines) in the orbital plane at a time moment approximately equal to one third of the orbital period for Model 5 ($B_\mathrm{s} = 0.2$ G, **left**) and Model 6 ($B_\mathrm{s} = 0.5$ G, **right**). The density is expressed in units of ρ_w. The dotted line shows the border of the Roche lobe. The white circle corresponds to the photometric radius of the planet.

In both models, the planet is located in the sub-Alfvén wind zone, but between the points r_S (slow magnetosonic) and r_A (Alfvén). Consequently, in the orbit of the planet, the wind velocity v_w exceeds the slow magnetosonic velocity u_S, but becomes less than the Alfvén velocity u_A. In model 5, the Alfvén Mach number at the planet orbit is $\lambda = 0.77$ and therefore the planet is located in the sub-Alfvén zone of the stellar wind. The total Alfvén Mach number, considering the velocity of the orbital motion of the planet $\lambda = 1.16$.

We can say that this situation corresponds to the *trans-Alfvén regime* of the stellar wind flowing around the planet, since hot Jupiter is located in the sub-Alfvén wind zone, but the flow occurs in the super-Alfvén regime. In model 6, the Alfvén Mach number on the orbit of the planet is $\lambda = 0.31$, and the total Alfvén Mach number is $\lambda = 0.46$. Therefore, in this case, the conditions for the implementation of the sub-Alfvén regime of the hot Jupiter flow around by the stellar wind are completely satisfied.

In the super-Alfvén flow around regime, an extended open-type envelope was formed for these atmospheric parameters (see the lower right panel in Figure 9). Under the conditions of a strong magnetic field of the wind, the envelope structure has undergone significant changes. The intensity of the matter flow from the inner Lagrange point L_1 at the day side weakened and the envelope became closer to the quasi-open type. We obtained that the mass loss rate $\dot{M}_p \approx 9 \times 10^9$ g/s for the model 5 and $\dot{M}_p \approx 7 \times 10^9$ g/s for the model 6. It follows from these results that, at fixed parameters of the atmosphere (density ρ_a and temperature T_a), the rate of mass loss $\dot{M}_p$ for hot Jupiter decreases with an increase in the magnetic field of the wind (the parameter B_s in our numerical model).

In addition, the flow direction changed, since, in a strong field, plasma tends to move mainly along magnetic force lines. This is especially noticeable in the case of model 6 (right panel in Figure 10). Since the wind field is almost radial in the vicinity of the planet, the envelope matter moves directly to the star and, continuing to move along such a trajectory, falls immediately onto the star.

Thus, in the case of a strong magnetic wind field (sub-Alfvén flow around regime), we have some new type of extended envelope, complementing the previous classification [17]. In order to distinguish these envelopes types, we can call them *super-Alfvén* and *sub-Alfvén*, respectively. For example, in this case, sub-Alfvén extended envelopes of open or quasi-open type are formed. It is obvious that the observational manifestations of such envelopes should have important differences in comparison with ones formed in the super-Alfvén flow around regime [56].

It is also interesting to note that a magnetic barrier is formed in front of the planet in the direction of its orbital motion. It manifests itself as an elongated area of condensation of magnetic field lines (see Figure 10). Within this region, the magnetic field induction increases.

The process of the stellar wind flowing around the planet is basically shock-less. In model 5 (the left panel in Figure 10) a weak shock wave is formed around the planet, but towards the end of the stream it disappears, since the conditions of the sub-Alfvén flow regime are already being realized in this region. In model 6 (the right panel in Figure 10) in the entire computational domain, the flow occurs in the sub-Alfvén mode. Therefore, shock waves are not formed either around the atmosphere of the planet, or around the flow of matter flowing from the inner Lagrange point L_1.

In model 6 (the right panel in Figure 10), a non-physical spherical structure appeared around the planet. Its appearance is due to the fact that to describe the induced magnetic field of the atmosphere, we used the formula (109), which follows from the exact solution for the problem of the magnetic field of an ideally conducting sphere placed in a homogeneous external magnetic field. In our case, the magnetic field of the wind is not uniform. Therefore, this solution does not allow to accurately compensate the external field in the entire volume of the atmosphere and parasitic currents will be induced all the time near the surface of the initial contact discontinuity. This is noticeable in models with a strong wind field. In future works, we will attempt to overcome this problem.

6. Conclusions

To investigate the process of the stellar wind matter flowing around hot Jupiters, considering both the planet own magnetic field and the wind magnetic field, we developed a three-dimensional numerical model based on the approximation of multi-component magnetic hydrodynamics. Our numerical model is based on the Roe–Einfeldt–Osher difference scheme of high resolution for the equations of multi-component MHD.

The total magnetic field is represented as a superposition of the external magnetic field and the magnetic field induced by electric currents in the plasma itself. The superposition of the planet proper magnetic field, the induced magnetic field of the atmosphere, and the radial component of the wind magnetic field was used as the external field. In the numerical algorithm, the factors associated with the presence of an external magnetic field were taken into account at a separate stage using an appropriate Godunov-type difference scheme.

The main attention was focused on the inclusion of a complete MHD model of the stellar wind. This, in particular, makes it possible to more correctly calculate the location of the Alfvén point. As a result, the numerical model is applicable for calculating the structure of the extended envelope of hot Jupiters not only in the super-Alfvén [55] and sub-Alfvén [56] regimes of stellar wind flow around but also in the trans-Alfvén regime. The multi-component MHD approximation in the future will be used by us to account for changes in the chemical composition of the hydrogen–helium envelopes of hot Jupiters. In this paper, we did not consider these processes, assuming all the individual components as passive admixtures moving together with the matter.

As an example of the object to study, we considered a typical hot Jupiter HD 209458b. The numerical simulations presented in the paper can be divided into two blocks. The first block includes calculations with a weak wind field (super-Alfvén flow regime), in which we changed the parameters of atmosphere (temperature). This made it possible to simulate different types of super-Alfvén envelopes: closed, quasi-closed, quasi-open and open. In the second block, with fixed atmospheric parameters (open envelope), we changed the magnetic field of the wind and analyzed how the envelope structure changes.

In the case of the super-Alfvén flow regime, all previously discovered types of extended envelopes are also implemented in the new numerical model. However, their boundaries in the parameter space have shifted to the region of lower densities and temperatures. This is due to the fact that in this model we take into account the chemical composition of the atmosphere. Position and shape of the bow shock in the new numerical model differ from those we obtained in the previous models. This can be explained by a more correct consideration of wind parameters. In the old model, wind velocity and temperatures were considered constant throughout the computational domain, and in the new model they change in space according to the analytical solution.

In addition, in the old model, the gravitational forces of the star and planet were turned off to accelerate the wind. As a result, the flow velocity naturally increased and the shock wave pressed closer to the planet. In the new model, the flow velocity is obtained directly from the wind model (considering all forces) and as a result, the shock wave moves further away from the planet.

With the increase in the magnitude of wind magnetic field, the total wind pressure enlarges. As a result, the stream of matter from the inner Lagrange point L_1 is stopped by the stellar wind earlier. Therefore, for example, an open envelope tends to become quasi-open with the growth of field. In addition, in the strong magnetic field of the wind, the direction of movement of stream changes. The stream plasma will attempt to move along the magnetic force lines. As a result, an additional type of envelopes is realized— sub-Alfvén ones, which have their own specific observational features.

In the trans-Alfvén regime, the bow shock wave has a fragmentary nature. In the completed sub-Alfvén flow around regime, the bow shock wave is not formed at all. It should also be noted that with an increase in the magnetic field of the wind, the induced field of the atmosphere growths. The corresponding magnetic moment becomes greater than the planet intrinsic magnetic moment. As a result, the magnetic pole shifts. This may affect the overall configuration of the magnetosphere and, in particular, the position of the dead zones and the auroral zone.

Author Contributions: A.Z. and D.B. developed the analytical and numerical model, obtained the results and wrote the manuscript. All authors have read and agreed to the published version of the manuscript.

Funding: The authors are grateful to the Government of the Russian Federation and the Ministry of Higher Education and Science of the Russian Federation for the support (grant 075-15-2020-780 (no. 13.1902.21.0039)).

Institutional Review Board Statement: Not applicable.

Informed Consent Statement: Not applicable.

Data Availability Statement: Not applicable.

Acknowledgments: The study was carried out using computing facilities of the Interdepartmental Supercomputer Center of the Russian Academy of Sciences, www.jscc.ru, accessed on 20 October 2021.

Conflicts of Interest: The authors declare no conflict of interest.

Appendix A. Difference Scheme for the Equations of Multi-Component Magnetic Hydrodynamics

Appendix A.1. Roe Matrix

In this section, we describe the adaptation of the Roe scheme [77] to the case of multi-component magnetic hydrodynamics equations. We will discard all source terms, since they can be accounted separately. The hyperbolic part of the system consists of the equations of magnetic hydrodynamics

$$\frac{\partial \rho}{\partial t} + \nabla \cdot (\rho v) = 0, \tag{A1}$$

$$\frac{\partial}{\partial t}(\rho v) + \nabla \cdot (\rho v v + P_T + BB) = 0, \tag{A2}$$

$$\frac{\partial B}{\partial t} - \nabla \times (v \times B) = 0, \tag{A3}$$

$$\frac{\partial E_T}{\partial t} + \nabla \cdot [v(E_T + P_T) - B(v \cdot B)] = 0 \tag{A4}$$

and equations for the mass fractions of components

$$\frac{\partial}{\partial t}(\rho \zeta_s) + \nabla \cdot (\rho \zeta_s v) = 0, \quad s = 1, \dots, N. \tag{A5}$$

Here, all the equations are written in a conservative form and notations for total pressure and total energy density are used

$$P_T = P + \frac{B^2}{2}, \quad E_T = \rho \varepsilon + \rho \frac{v^2}{2} + \frac{B^2}{2}. \tag{A6}$$

For the convenience of numerical simulation, a system of units is used in these equations, in which the multiplier 4π does not occur in the expression for the electromagnetic force.

For the numerical solving of three-dimensional equations of multi-component magnetic hydrodynamics (A1)–(A5), the splitting technique by spatial directions can be used. As a result, the solving of a three-dimensional problem is reduced to solving a series of one-dimensional problems. In the case of using Godunov-type schemes [78], numerical fluxes in each spatial direction are calculated based on the corresponding one-dimensional Riemann problem on the decay of an arbitrary discontinuity.

Consider the case of a plane flow corresponding to the coordinate direction x. Since in the plane flow the magnetic field component $B_x = \text{const}$, the equations of one-dimensional multi-component magnetic hydrodynamics in a conservative form can be written as

$$\frac{\partial \mathbf{u}}{\partial t} + \frac{\partial \mathbf{F}}{\partial x} = 0, \tag{A7}$$

where the vectors of conservative variables and fluxes are defined by expressions:

$$
\mathbf{u} = \begin{pmatrix} \rho \\ \rho v_x \\ \rho v_y \\ \rho v_z \\ B_y \\ B_z \\ E_T \\ \rho \zeta_s \end{pmatrix}, \quad
\mathbf{F} = \begin{pmatrix} \rho v_x \\ \rho v_x^2 + P_T \\ \rho v_x v_y - B_x B_y \\ \rho v_x v_z - B_x B_z \\ v_x B_y - v_y B_x \\ v_x B_z - v_z B_x \\ \rho h v_x - B_x (\boldsymbol{v} \cdot \boldsymbol{B}), \\ \rho v_x \zeta_s \end{pmatrix}.
\tag{A8}
$$

Here, it is assumed that the index s runs through values from 1 to N and a notation is used for the total enthalpy density h, determined by the relation

$$
\rho h = E_T + P_T.
\tag{A9}
$$

Note that the equation for the B_x component is excluded from this system. In fact, this value is a flow parameter.

The Roe scheme refers to Godunov-type schemes and is based on an approximate solving of the Riemann problem on the decay of an arbitrary discontinuity. In this method, instead of solving the Riemann problem for the original system of non-linear Equation (A7), a linearized problem is solved

$$
\frac{\partial \mathbf{u}}{\partial t} + \hat{A}(\mathbf{u}_L, \mathbf{u}_R) \cdot \frac{\partial \mathbf{u}}{\partial x} = 0
\tag{A10}
$$

with initial conditions: $u(x,0) = u_L$ for $x < 0$ and $u(x,0) = u_R$ for $x > 0$.

In order for the solutions of the original (A7) and the linearized (A10) problems to be consistent, the matrix $\hat{A}(u_L, u_R)$ must satisfy the following three conditions.

(1) Hyperbolicity. Matrix $\hat{A}(u_L, u_R)$ must be hyperbolic. Otherwise, the Riemann problem for a system of linearized Equation (A10) loses its meaning.
(2) Consistency. Matrix $\hat{A}(u_L, u_R)$ should make smooth transition to the hyperbolicity matrix $\hat{A}(u) = \partial F / \partial u$ in the limit at $u_L \to u_R = u$.
(3) Conservation. Matrix $\hat{A}(u_L, u_R)$ must satisfy the condition of conservation relative to discontinuities:

$$
\hat{A}(\mathbf{u}_L, \mathbf{u}_R) \cdot \Delta \mathbf{u} = \Delta \mathbf{F},
\tag{A11}
$$

where is denoted $\delta u = u_R - u_L$, $\delta F = F_R - F_L$. In this case, solving of the linearized discontinuity decay problem (A10) will satisfy the same integral conservation laws as the solving of the original non-linear problem (A7).

As is known, the solution of the Riemann problem for a linear hyperbolic system of Equation (A10) is a set of strong discontinuities whose velocities are equal to the eigenvalues λ_α of the Roe matrix $\hat{A}(u_L, u_R)$, where α is the index of the characteristic. Corresponding jumps of values at discontinuities

$$
[\mathbf{u}]_\alpha = \mathbf{r}_\alpha \Delta S^\alpha,
\tag{A12}
$$

where square brackets denote differencies between the right and the left values when crossing the discontinuity, $\Delta S^\alpha = l^\alpha \cdot \delta u$ is the characteristic amplitudes, and r_α, l^α is the right and left eigenvectors of Roe matrix. At each discontinuity, the corresponding Hugoniot conditions must be satisfied:

$$
\lambda_\alpha [\mathbf{u}]_\alpha = [\mathbf{F}]_\alpha.
\tag{A13}
$$

The Roe matrix for the equations of magnetic hydrodynamics (A1)–(A4), as well as all the characteristic parameters necessary for constructing the scheme for the special case of the adiabatic index $\gamma = 2$, are described in [79]. In [80], the corresponding Roe matrix with the adiabatic index $1 < \gamma \leq 2$ was constructed for the first time. A detailed description of this scheme is given in our monograph [68]. We will not dwell on this here, considering all these properties to be known. Recall, however, that this matrix of dimension 7×7 has the following set of eigenvalues:

$$\lambda_{\pm F} = v_x \pm u_F, \quad \lambda_{\pm S} = v_x \pm u_S, \quad \lambda_{\pm A} = v_x \pm u_A, \quad \lambda_E = v_x, \tag{A14}$$

where the indices F, S, A, and E correspond to fast, slow, Alfvén, and entropy characteristics. The values u_F and u_S describe fast and slow magnetosonic velocities, and u_A is the Alfvén velocity. Following the paper of [80], we introduce intermediate values (Roe's averages) for density, velocity, magnetic field and total enthalpy:

$$\rho = \sqrt{\rho_L \rho_R}, \quad v = \frac{\sqrt{\rho_L} v_L + \sqrt{\rho_R} v_R}{\sqrt{\rho_L} + \sqrt{\rho_R}}, \tag{A15}$$

$$B = \frac{\sqrt{\rho_R} B_L + \sqrt{\rho_L} B_R}{\sqrt{\rho_L} + \sqrt{\rho_R}}, \quad h = \frac{\sqrt{\rho_L} h_L + \sqrt{\rho_R} h_R}{\sqrt{\rho_L} + \sqrt{\rho_R}}. \tag{A16}$$

The Roe matrix for the equations of one-dimensional multi-component magnetic hydrodynamics (A7), (A8) has the following structure

$$\hat{A} = \left(\begin{array}{ccccc|ccc} & & & & & 0 & \cdots & 0 \\ & & A_{ik} & & & \vdots & \ddots & \vdots \\ & & & & & 0 & \cdots & 0 \\ \hline -\xi_1 v_x & \xi_1 & 0 & \cdots & 0 & v_x & \cdots & 0 \\ \vdots & \vdots & \vdots & \ddots & \vdots & \vdots & \ddots & \vdots \\ -\xi_N v_x & \xi_N & 0 & \cdots & 0 & 0 & \cdots & v_x \end{array} \right). \tag{A17}$$

As it can be seen, this matrix consists of four blocks. In the upper left block of dimension 7×7 there are elements of the A_{ik} Roe matrix for magnetic hydrodynamics. The indices i and k run through values from 1 to 7. All elements of the upper right block of dimension $N \times 7$ are equal to zero. The lower left block has dimension $7 \times N$. In the line with the number s of this block, the first element is equal to $-\xi_s v_x$, the second element is equal to ξ_s, and the remaining elements are zero. The lower right block of dimension $N \times N$ is diagonal, and all its diagonal elements are equal to v_x.

Let us write out the Roe's conditions (A11) for the matrix (A17). For the matrix components corresponding to the equations of magnetic hydrodynamics, they do not give anything new:

$$\Delta F^i = \sum_{k=1}^{7} A_{ik} \Delta u^k. \tag{A18}$$

Consequently, the structure of magnetohydrodynamic part of the matrix A_{ik} is not affected and remains the same. For the selected component $\xi_s = \xi$ we have the relation:

$$\Delta(\rho \xi v_x) = -\xi v_x \Delta \rho + \xi \Delta(\rho v_x) + v_x \Delta(\rho \xi). \tag{A19}$$

From here, we find the Roe's average for the value ξ,

$$\xi = \frac{\xi_L \sqrt{\rho_L} + \xi_R \sqrt{\rho_R}}{\sqrt{\rho_L} + \sqrt{\rho_R}}. \tag{A20}$$

This expression is valid for any ξ_s component.

Taking into account the block structure of the Roe matrix $\hat{A}_*$ it can be written

$$\det(\hat{A} - \lambda \hat{I}) = \det(A_{ik} - \lambda \delta_{ik})(v_x - \lambda)^N, \tag{A21}$$

where δ_{ik} is a unit matrix of dimension 7×7, and $\hat{I}$ is a unit matrix of full size. It follows that the eigenvalues of the Roe matrix $\hat{A}$ are the eigenvalues (A14) of the matrix A_{ik}, as well as N-multiple eigenvalues

$$\lambda_s = v_x, \quad s = 1, \ldots, N, \tag{A22}$$

corresponding to the Equation (A5) for the mass fractions of ξ_s.

Appendix A.2. Eigenvectors

The right eigenvectors of the Roe matrix $\hat{A}$ are denoted as follows:

$$\mathbf{r} = \left(r^1, \ldots, r^7, \tilde{r}^1, \ldots, \tilde{r}^N \right)^T, \tag{A23}$$

where T denotes transposition, the first 7 components of the vector correspond to the magnetohydrodynamic part, the remaining N components, marked with tildes, correspond to the Equation (A5) for the mass fractions ξ_s. Let us write out the equations for the right eigenvectors

$$\hat{A} \cdot \mathbf{r} = \lambda \mathbf{r}. \tag{A24}$$

We have:

$$\sum_{k=1}^{7} A_{ik} r^k = \lambda r^i, \quad i = 1, \ldots, 7, \tag{A25}$$

$$-\xi_s v_x r^1 + \xi_s r^2 + v_x \tilde{r}^s = \lambda \tilde{r}^s, \quad s = 1, \ldots, N. \tag{A26}$$

Let us first consider the MHD characteristics when λ are determined by the eigenvalues of (A14) of the matrix A_{ik}. It follows from the first Equation (A25) that in this case the components of r^i coincide with the corresponding components of the right eigenvectors in magnetic hydrodynamics. The remaining components of the right eigenvectors can be found from the additional Equation (A26). If $\lambda \neq v_x$, then for each component $\tilde{r}^s$ of the right vector, it can be written:

$$\tilde{r}^s = \frac{\xi_s}{\lambda - v_x} \left(r^2 - v_x r^1 \right). \tag{A27}$$

Hence, for fast and slow characteristics we find $\tilde{r}^s = \alpha_f \xi_s$ and $\tilde{r}^s = \alpha_s \xi_s$, respectively, and for Alfvén $\tilde{r}^s = 0$. The coefficients α_f and α_s determine the normalization of Roe matrix eigenvectors in magnetic hydrodynamics [79,80] (see also [68]). For the entropy characteristic $\lambda = v_x$, $r^1 = 1$, $r^2 = v_x$ and, consequently, all additional Equation (A26) are satisfied automatically. This means that in this case we can choose $\tilde{r}^s$ in an arbitrary way. Without limiting generality, they can be put equal to zero. As a result, we obtain the following right eigenvectors

$$\mathbf{r}_{\pm F} = \left(r^1_{\pm F}, \ldots, r^7_{\pm F}, \alpha_F \xi_1, \ldots, \alpha_F \xi_N \right)^T,$$

$$\mathbf{r}_{\pm S} = \left(r^1_{\pm S}, \ldots, r^7_{\pm S}, \alpha_S \xi_1, \ldots, \alpha_S \xi_N \right)^T,$$

$$\mathbf{r}_{\pm A} = \left(r^1_{\pm A}, \ldots, r^7_{\pm A}, 0, \ldots, 0 \right)^T,$$

$$\mathbf{r}_E = \left(r^1_E, \ldots, r^7_E, 0, \ldots, 0 \right)^T. \tag{A28}$$

For additional characteristics corresponding to the eigenvalues of λ_s, we come to the following equations:

$$\sum_{k=1}^{7} A_{ik} r^k = v_x r^i, \quad i = 1, \ldots, 7, \tag{A29}$$

$$r^2 - v_x r^1 = 0, \quad s = 1, \ldots, N. \tag{A30}$$

The equations for the r^i components coincide with the corresponding equations for the entropy characteristic. Therefore, there should be $r^i = r^i_E$. The relation (A30) is also satisfied identically. The values of $\tilde{r}^s$ remain arbitrary. They must be selected in such a way as to obtain a linearly independent set of vectors. To do this, it is enough to set $\tilde{r}^s = 1$ for the characteristic corresponding to the index s, and to set the remaining components equal to zero. As a result, we obtain the following set of additional right eigenvectors

$$\mathbf{r}_s = \left(r^1_E, \dots, r^7_E, 0, \dots, 1, \dots, 0\right)^T. \tag{A31}$$

Here, the unit is located in place of the component numbered $7 + s$. Since the vectors $\mathbf{r}_s$ and $\mathbf{r}_E$ are linearly independent, and the eigenvalues for these characteristics are the same ($\lambda = v_x$), then, without violating linear independence, the vector $\mathbf{r}_s$ can be replaced by the difference $\mathbf{r}_s - \mathbf{r}_E$. In this case, we have a set of unit vectors

$$\mathbf{r}_s = (0, \dots, 0, 0, \dots, 1, \dots, 0)^T. \tag{A32}$$

Let us now consider the left eigenvectors of the Roe matrix $\hat{A}$:

$$\mathbf{l} = \left(l_1, \dots, l_7, \tilde{l}_1, \dots, \tilde{l}_N\right), \tag{A33}$$

where the first seven components of the vector again correspond to the magnetohydrodynamic part, and the remaining N components, marked with tildes, correspond to the Equation (A5) for the mass fractions ζ_s. Describing relations for left eigenvectors

$$\mathbf{l} \cdot \hat{A} = \lambda \mathbf{l}, \tag{A34}$$

we come to the following equations:

$$\sum_{k=1}^{7} l_k A_{k1} - \sum_{r=1}^{N} \tilde{l}_r \zeta_r v_x = \lambda l_1, \tag{A35}$$

$$\sum_{k=1}^{7} l_k A_{k2} + \sum_{r=1}^{N} \tilde{l}_r \zeta_r = \lambda l_2, \tag{A36}$$

$$\sum_{k=1}^{7} l_k A_{kn} = \lambda l_n, \quad n = 3, \dots, 7, \tag{A37}$$

$$v_x \tilde{l}_s = \lambda \tilde{l}_s, \quad s = 1, \dots, N. \tag{A38}$$

For fast, slow, and Alfvén characteristics, the eigenvalue is $\lambda \neq v_x$. Therefore, for them all the additional components are $\tilde{l}_s = 0$. For the entropy characteristic, the values of $\tilde{l}_s$ are arbitrary. It can also be chosen them $\tilde{l}_s = 0$, so as not to change anything in the magnetohydrodynamic part of the left eigenvector. As a result, for all MHD characteristics, the left eigenvectors will have the following form:

$$\begin{aligned}
\mathbf{l}^{\pm F} &= \left(l_1^{\pm F}, \dots, l_7^{\pm F}, 0, \dots, 0\right), \\
\mathbf{l}^{\pm S} &= \left(l_1^{\pm S}, \dots, l_7^{\pm S}, 0, \dots, 0\right), \\
\mathbf{l}^{\pm A} &= \left(l_1^{\pm A}, \dots, l_7^{\pm A}, 0, \dots, 0\right), \\
\mathbf{l}^{E} &= \left(l_1^{E}, \dots, l_7^{E}, 0, \dots, 0\right).
\end{aligned} \tag{A39}$$

For additional characteristics λ_s, the last Equation (A38) is satisfied automatically. The remaining equations are reduced to the form:

$$\sum_{k=1}^{7} l_k A_{k1} - \sum_{r=1}^{N} \tilde{l}_r \zeta_r v_x = v_x l_1, \tag{A40}$$

$$\sum_{k=1}^{7} l_k A_{k2} + \sum_{r=1}^{N} \tilde{l}_r \zeta_r = v_x l_2, \tag{A41}$$

$$\sum_{k=1}^{7} l_k A_{kn} = v_x l_n, \quad n = 3,\dots,7. \tag{A42}$$

We can see that in the case of $\tilde{l}_r = 0$, $l_i = l_i^E$ follows from these equations. However, then we find a linearly dependent set of vectors. To obtain a linearly independent set of vectors, for an additional characteristic with the index s, we choose $\tilde{l}_s = 1$, and set the remaining additional components equal to zero. For the rest of the components, we will look for a solution in a more general form:

$$l_i = \chi l_i^E + x_i, \tag{A43}$$

where χ is the normalizing factor, and x_i is the unknown quantities. Since

$$\sum_{k=1}^{7} l_k^E A_{ki} = v_x l_i^E, \quad i = 1,\dots,7, \tag{A44}$$

then, for x_i, we come to the equations:

$$\sum_{k=1}^{7} x_k A_{k1} - \zeta_s v_x = v_x x_1, \tag{A45}$$

$$\sum_{k=1}^{7} x_k A_{k2} + \zeta_s = v_x x_2, \tag{A46}$$

$$\sum_{k=1}^{7} x_k A_{kn} = v_x x_n, \quad n = 3,\dots,7. \tag{A47}$$

Without limiting generality, we can assume that only the value x_1 is non-zero. Recall [80] that the first row of the matrix A_{ik} contains a single non-zero element $A_{12} = 1$. Therefore, the Equation (A47) is satisfied automatically, and the Equations (A45) and (A46) coincide and give the solution $x_1 = -\zeta_s$. As a result, the left eigenvectors for additional characteristics will have the following form:

$$\mathbf{l}^s = \left(\chi l_1^E - \zeta_s, \chi l_2^E, \dots, \chi l_7^E, 0, \dots, 1, \dots, 0 \right). \tag{A48}$$

Here, the unit is located again in place of the component with the index $7 + s$.

The coefficient χ should be found from the condition of orthonormality of eigenvectors

$$\mathbf{l}^\alpha \cdot \mathbf{r}_\beta = \delta_\beta^\alpha. \tag{A49}$$

For MHD characteristics, when α and β correspond to $\pm F$, $\pm S$, $\pm A$ or E, all components of $\tilde{l}_s = 0$. Therefore

$$\mathbf{l}^\alpha \cdot \mathbf{r}_\beta = \sum_{k=1}^{7} l_k^\alpha r_\beta^k = \delta_\beta^\alpha \tag{A50}$$

due to the orthonormality of the MHD vectors. If α corresponds to MHD characteristics, and β is the additional characteristics, then it's obvious,

$$\mathbf{l}^\alpha \cdot \mathbf{r}_s = 0. \tag{A51}$$

If we consider vectors for additional characteristics $\alpha = r$, $\beta = s$, then we also find the necessary relation,

$$\mathbf{l}^r \cdot \mathbf{r}_s = \delta_s^r. \tag{A52}$$

Finally, let us consider the case, when $\alpha = s$, and β corresponds to the MHD characteristics. We have:

$$\mathbf{l}^s \cdot \mathbf{r}_\beta = \chi \sum_{k=1}^{7} l_k^E r_\beta^k - \xi_s r_\beta^1 + \tilde{r}_\beta^s. \tag{A53}$$

If $\alpha = \pm F$, then the first term in the right part vanishes due to the orthonormality of the MHD vectors. Then, $r_{\pm F}^1 = \alpha_f$, $\tilde{r}_{\pm F}^s = \alpha_f \xi_s$ and, consequently, the entire right part turns out to be zero. The situation is similar in cases where α corresponds to $\pm S$ and $\pm A$. For the entropy characteristic $\alpha = E$ we have:

$$\mathbf{l}^s \cdot \mathbf{r}_E = \chi \sum_{k=1}^{7} l_k^E r_E^k - \xi_s r_E^1 = 0. \tag{A54}$$

Hence, using the normalization condition of entropy MHD vectors, as well as the equality $r_E^1 = 1$, we find $\chi = \xi_s$. Thus, we finally find

$$\mathbf{l}^s = \left(\xi_s l_1^E - \xi_s, \xi_s l_2^E, \dots, \xi_s l_7^E, 0, \dots, 1, \dots, 0 \right). \tag{A55}$$

Here, as before, the unit is located again in place of the component with the index $7 + s$.

Using the expressions obtained for the left eigenvectors, it is easy to calculate the characteristic amplitudes of $\Delta S^\alpha = \mathbf{l}^\alpha \cdot \Delta \mathbf{u}$. Since for all MHD characteristics the additional components of the left eigenvectors $\tilde{l}_s = 0$, the corresponding expressions for characteristic amplitudes do not change. For additional characteristics of λ_s we have:

$$\Delta S^s = \xi_s \sum_{k=1}^{7} l_k^E \Delta u^k - \xi_s \Delta \rho + \Delta(\xi_s \rho). \tag{A56}$$

Taking into account (A20), this expression can be simplified,

$$\Delta S^s = \xi_s \Delta S^E + \rho \Delta \xi_s, \tag{A57}$$

where ρ is the Roe's average for density (A15).

Appendix A.3. Test Calculations

To numerically solve the equations of multi-component magnetic hydrodynamics (A1)–(A5), we use the finite-difference Roe scheme [77], some details of which are described above. To improve the accuracy, we apply the Osher increasing correction [81]. The resulting difference scheme belongs to the class of TVD (total variation diminishing) schemes [82] and for the case of single-fluid magnetic hydrodynamics is described in detail in [83]. The scheme has the first order of approximation in time and the third order in space. Note that the magnetohydrodynamic version of the Roe scheme in our scheme is presented in such a way that in the absence of a magnetic field ($\mathbf{B} = 0$) this scheme exactly passes into the Roe-Einfeldt-Osher scheme we used in purely gas-dynamic simulations [17].

The main disadvantage of the Roe method should, apparently, be considered that the linear system (A10), qualitatively repeating the solution of the original problem (A7), does not reproduce centered rarefaction waves. Instead, the solution consists of a jumps system propagating at speeds corresponding to the eigenvalues of the Roe matrix $\hat{A}(\mathbf{u}_L, \mathbf{u}_R)$, and some of these jumps may not satisfy the condition of evolutionarity. However, for most cases, the Roe method works well, even if the exact solution of the original problem involves rarefaction waves.

The exception is solutions with trans-sonic rarefaction waves, for the correct accounting of which special evolutionary corrections are used. In the case of magnetic hydrodynamics equations (single-fluid or multi-fluid), such corrections should be used for fast and slow magnetosonic characteristics, in which rarefaction waves with corresponding properties may sometimes occur. For fast magnetosonic characteristics, our difference scheme uses the Einfeldt [84] correction. In the case of slow magnetosonic characteristics, the initial Einfeldt correction does not agree with a purely gas-dynamic version of the difference Roe scheme. Therefore, for such characteristics, we use a modified entropy correction [83].

In multi-dimensional problems, when using Godunov's methods for solving equations of magnetic hydrodynamics, a separate procedure is required to clear the divergence of the magnetic field, since the difference scheme operates with values averaged over the volume of a cell and, therefore, does not conserve the magnetic flux. The method we use to clear the magnetic field divergence is described in Section 4.4. In the test calculations, the results of which are given in this section, divergence cleaning was not applied.

In the first test calculation, numerical simulation of the decay of an arbitrary MHD discontinuity was carried out. At the initial moment of time, the resting matter (velocity $v = 0$) was considered in a homogeneous magnetic field $B_x = 1/4$, $B_y = \sqrt{3}/4$, $B_z = 0$. An arbitrary discontinuity was located at the point $x = 0$.

In the region to the left of the discontinuity $x < 0$, the following values were set: density $\rho = 1$, pressure $P = 1$. In the region to the right of the discontinuity $x > 0$, the parameters were set: density $\rho = 0.125$, pressure $P = 0.1$. The adiabatic index $\gamma = 5/3$. A calculated grid containing 512 cells was used. A two-component mixture consisting of hydrogen ions H^+ and helium He^+ was considered. We assume that at the initial moment of time, the matter to the left of the discontinuity consists only of helium ions, and the matter to the right of the discontinuity consists only of hydrogen ions.

The results of the numerical solution of this problem obtained at the time $t = 0.15$ are shown in Figure A1. The decay of the initial arbitrary discontinuity leads to the formation of two rarefaction waves (fast 1 and slow 2) propagating into the region of helium matter. The contact discontinuity 3 propagates to the right. Two shock waves are formed in the region of hydrogen plasma (slow 4 and fast 5). The panel on the bottom right shows the resulting distribution of the mass fractions of helium ions ξ_1 and hydrogen ions ξ_2. The contact boundary between the matters shifts to the right, is clearly localized in space and does not contain any non-physical oscillations. Analysis of the figure shows that the scheme approximates all types of running waves quite well.

As a second demonstration example, let us consider the results of a test calculation of the problem on a volume-distributed explosion in a medium with a homogeneous magnetic field. The initial conditions correspond to the situation when, in the entire volume of an infinitely long cylinder with a radius of $R = 0.2$ with a density of $\rho = 1$, the pressure instantly increases by 10 times compared to the pressure in the external environment. As a result of such an explosion, the matter of the cloud begins to expand into the external environment.

The density and pressure in the external environment were set as follows: $\rho_{ext} = 0.125$, $P_{ext} = 0.1$. The magnetic field at the initial time in the entire volume of the computational domain was homogeneous: $B_x = 1/4$, $B_x = \sqrt{3}/4$, $B_x = 0$. The cloud was filled with helium ions He^+, and the external environment was filled with hydrogen ions H^+. Due to the formulation of the problem in each plane $z = \text{const}$, the flow pattern will be the same.

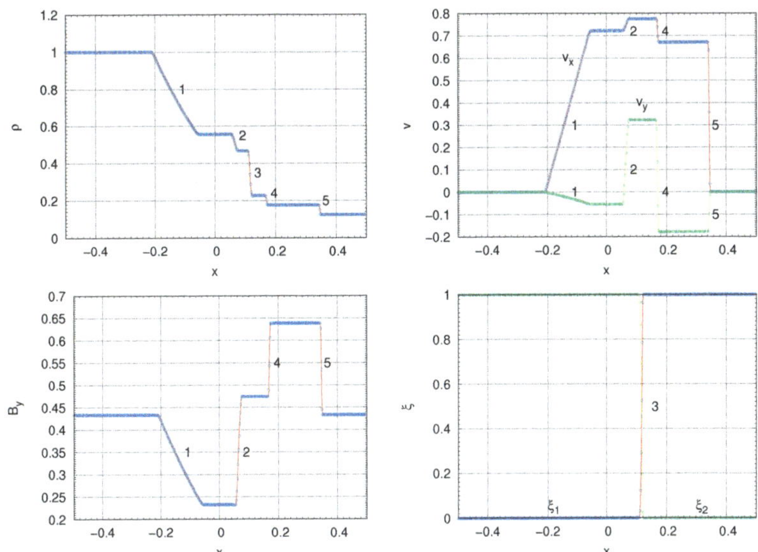

Figure A1. The results of the test calculation of the Riemann problem on the decay of an arbitrary discontinuity (see the description of the initial conditions in the text). Density distributions are shown (**top left**), velocity components v_x and v_y (**top right**), magnetic field components b_y (**bottom left**), as well as mass fractions ξ_1 and ξ_2 (**bottom right**) at time $t = 0.15$. The figures correspond to the digits: 1—fast rarefaction wave, 2—slow rarefaction wave, 3—contact gap, 4—slow shock wave, and 5—fast shock wave.

Therefore, the problem is actually two-dimensional. Calculations were carried out in the computational domain $-0.5 \leq x \leq 0.5$, $-0.5 \leq y \leq 0.5$ on a mesh with the number of cells 512×512. It should be noted that according to its formulation, the previous test problem corresponds to the decay of an arbitrary discontinuity on the right edge of the cloud along the x coordinate.

The results of the calculation of the second test problem at time moment $t = 0.1$ are shown in Figure A2. The left panel shows the distribution of density (color) and velocity (arrows). The boundary of the cloud that separates helium and hydrogen corresponds to a solid line. The right panel shows the distribution of density (iso-lines) and magnetic field (lines with arrows). As it can be seen from the figure, the flow structure is determined by a complex system of strong MHD discontinuities propagating outward (fast and slow shock waves, contact discontinuity), as well as fast and slow rarefaction MHD waves propagating to the center.

The presence of the magnetic field leads to the fact that the flow pattern is anisotropic. In particular, the contact boundary along the magnetic field propagates faster than in the direction across the field. As a result, the initially symmetrical helium cloud becomes elongated along the magnetic force lines. The analysis of the figure allows us to conclude that the computational qualities of the difference scheme we used, noted above for a one-dimensional problem, are preserved in the multi-dimensional case.

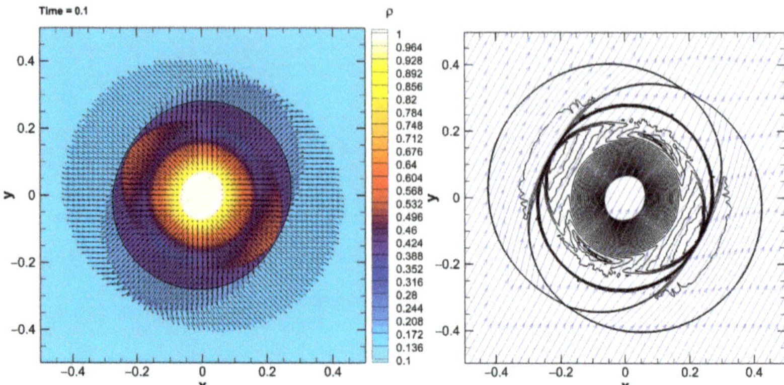

Figure A2. The results of the test calculation of the problem on a volume-distributed explosion in a homogeneous magnetic field. The density (color) and velocity distributions (arrows) are shown on the left. The solid line corresponds to the border of the cloud. The density distributions (iso-lines) and magnetic field distributions (lines with arrows) are shown on the right. The distributions of the quantities are presented at time moment $t = 0.1$.

References

1. Murray-Clay, R.A.; Chiang, E.I.; Murray, N. Atmospheric escape from hot Jupiters. *Astrophys. J.* **2009**, *693*, 23–42. [CrossRef]
2. Mayor, M.; Queloz, D. A Jupiter-mass companion to a solar-type star. *Nature* **1995**, *378*, 355–359. [CrossRef]
3. Lai, D.; Helling, C.; van den Heuvel, E.P.J. Mass transfer, transiting stream, and magnetopause in close-in exoplanetary systems with applications to WASP-12. *Astrophys. J.* **2010**, *721*, 923–928. [CrossRef]
4. Li, S.-L.; Miller, N.; Lin, D.N.C.; Fortney, J.J. WASP-12b as a prolate, inflated and disrupting planet from tidal dissipation. *Nature* **2010**, *463*, 1054–1056. [CrossRef] [PubMed]
5. Vidal-Madjar, A.; Lecavelier des Etangs, A.; Desert, J.-M.; Ballester, G.E.; Ferlet, R.; Hébrard, G.; Mayor, M. An extended upper atmosphere around the extrasolar planet HD209458b. *Nature* **2003**, *422*, 143–146. [CrossRef]
6. Vidal-Madjar, A.; Lecavelier des Etangs, A.; Desert, J.-M.; Ballester, G.E.; Ferlet, R.; Hébrard, G.; Mayor, M. Exoplanet HD 209458b (Osiris): Evaporation strengthened. *Astrophys. J.* **2008**, *676*, L57. [CrossRef]
7. Ben-Jaffel, L. Exoplanet HD 209458b: Inflated hydrogen atmosphere but no sign of evaporation. *Astrophys. J.* **2007**, *671*, L61–L64. [CrossRef]
8. Vidal-Madjar, A.; Desert, J.-M.; Lecavelier des Etangs, A.; Hébrard, G.; Ballester, G.E.; Ehrenreich, D.; Ferlet, R.; McConnell, J.C.; Mayor, M.; Parkinson, C.D. Detection of oxygen and carbon in the hydrodynamically escaping atmosphere of the extrasolar planet HD 209458b. *Astrophys. J.* **2004**, *604*, L69–L72. [CrossRef]
9. Ben-Jaffel, L.; Sona Hosseini, S. On the existence of energetic atoms in the upper atmosphere of exoplanet HD209458b. *Astrophys. J.* **2010**, *709*, 1284–1296. [CrossRef]
10. Linsky, J.L.; Yang, H.; France, K.; Froning, C.S.; Green, J.C.; Stocke, J.T.; Osterman, S.N. Observations of mass loss from the transiting exoplanet HD 209458b. *Astrophys. J.* **2010**, *717*, 1291–1299. [CrossRef]
11. Lecavelier des Etangs, A.; Bourrier, V.; Wheatley, P.J.; Dupuy, H.; Ehrenreich, D.; Vidal-Madjar, A.; Hébrard, G.; Ballester, G.E.; Désert, J.-M.; Ferlet, R.; et al. Temporal variations in the evaporating atmosphere of the exoplanet HD 189733b. *Astron. Astrophys.* **2012**, *543*, id.L4. [CrossRef]
12. Yelle, R.V. Aeronomy of extra-solar giant planets at small orbital distances. *Icarus* **2004**, *170*, 167–179. [CrossRef]
13. Garcia Munoz, A. Physical and chemical aeronomy of HD 209458b. *Planet. Space Sci.* **2007**, *55*, 1426–1455. [CrossRef]
14. Koskinen, T.T.; Harris, M.J.; Yelle, R.V.; Lavvas, P. The escape of heavy atoms from the ionosphere of HD209458b. I. A photochemical-dynamical model of the thermosphere. *Icarus* **2013**, *226*, 1678–1694. [CrossRef]
15. Ionov, D.E.; Shematovich, V.I.; Pavlyuchenkov, Y.N. Influence of photoelectrons on the structure and dynamics of the upper atmosphere of a hot Jupiter. *Astron. Rep.* **2017**, *61*, 387–392. [CrossRef]
16. Bisikalo, D.V.; Shematovich, V.I.; Kaygorodov, P.V.; Zhilkin, A.G. Gas envelopes of exoplanets—Hot Jupiters. *Phys. Uspekhi* **2021**, *64*, 747–800. [CrossRef]
17. Bisikalo, D.V.; Kaigorodov, P.V.; Ionov, D.E.; Shematovich, V.I. Types of gaseous envelopes of "hot Jupiter" exoplanets. *Astron. Rep.* **2021**, *57*, 715–725. [CrossRef]
18. Cherenkov, A.A.; Bisikalo, D.V.; Kaigorodov, P.V. Mass-loss rates of "hot-Jupiter" exoplanets with various types of gaseous envelopes. *Astron. Rep.* **2014**, *58*, 679–687. [CrossRef]
19. Bisikalo, D.V.; Cherenkov, A.A. The influence of coronal mass ejections on the gas dynamics of the atmosphere of a "hot Jupiter" exoplanet. *Astron. Rep.* **2016**, *60*, 183–192. [CrossRef]

20. Cherenkov, A.; Bisikalo, D.; Fossati, L.; Möstl, C. The influence of coronal mass ejections on the mass-loss rates of hot-Jupiters. *Astrophys. J.* **2017**, *846*, 31. [CrossRef]
21. Cherenkov, A.A.; Bisikalo, D.V.; Kosovichev, A.G. Influence of stellar radiation pressure on flow structure in the envelope of hot-Jupiter HD 209458b. *Mon. Not. R. Astron. Soc.* **2018**, *475*, 605–613. [CrossRef]
22. Bisikalo, D.V.; Cherenkov, A.A.; Shematovich, V.I.; Fossati, L.; Möstl, C. The influence of a stellar flare on the dynamical state of the atmosphere of the exoplanet HD 209458b. *Astron. Rep.* **2018**, *62*, 648–653. [CrossRef]
23. Shaikhislamov, I.F.; Khodachenko, M.L.; Lammer, H.; Kislyakova, K.G.; Fossati, L.; Johnstone, C.P.; Prokopov, P.A.; Berezutsky, A.G.; Zakharov, Y.P.; Posukh, V.G. Two regimes of interaction of a hot Jupiter's escaping atmosphere with the stellar wind and heneration of energized atomic hydrogen corona. *Astrophys. J.* **2016**, *832*, 173. [CrossRef]
24. Shaikhislamov, I.F.; Khodachenko, M.L.; Lammer, H.; Berezutsky, A.G.; Miroshnichenko, I.B.; Rumenskikh, M.S. 3D aeronomy modelling of close-in exoplanets. *Mon. Not. R. Astron. Soc.* **2018**, *481*, 5315–5323. [CrossRef]
25. Shaikhislamov, I.F.; Khodachenko, M.L.; Lammer, H.; Berezutsky, A.G.; Miroshnichenko, I.B.; Rumenskikh, M.S. Three-dimensional modelling of absorption by various species for hot Jupiter HD 209458b. *Mon. Not. R. Astron. Soc.* **2020**, *491*, 3435–3447.
26. Khodachenko, M.L.; Shaikhislamov, I.F.; Lammer, H.; Kislyakova, K.G.; Fossati1, L.; Johnstone, C.P.; Arkhypov, O.V.; Berezutsky, A.G.; Miroshnichenko, I.B.; Posukh, V.G. Lyα absorption at transits of HD 209458b: A comparative study of various mechanisms under different conditions. *Astrophys. J.* **2017**, *847*, 126. [CrossRef]
27. Khodachenko, M.L.; Shaikhislamov, I.F.; Lammer, H.; Berezutsky, A.G.; Miroshnichenko, I.B.; Rumenskikh, M.S.; Kislyakova, K.G.; Dwivedi, N.K. Global 3D hydrodynamic modeling of in-transit Lyα absorption of GJ 436b. *Astrophys. J.* **2019**, *885*, 67. [CrossRef]
28. Grießmeier, J.-M.; Stadelmann, A.; Penz, T.; Lammer, H.; Selsis, F.; Ribas, I.; Guinan, E.F.; Motschmann, U.; Biernat, H.K.; Weiss, W.W. The effect of tidal locking on the magnetospheric and atmospheric evolution of "Hot Jupiters". *Astron. Astrophys.* **2004**, *425*, 753-762. [CrossRef]
29. Sanchez-Lavega, A. The magnetic field in giant extrasolar planets. *Astrophys. J.* **2004**, *609*, L87–L90. [CrossRef]
30. Vidotto, A.A.; Jardine, M.; Helling, C. Prospects for detection of exoplanet magnetic fields through bow-shock observations during transits. *Mon. Not. R. Astron. Soc.* **2011**, *411*, L46–L50. [CrossRef]
31. Kislyakova, K.G.; Holmström, M.; Lammer, H.; Odertand, P.; Khodachenkol, M.L. Magnetic moment and plasma environment of HD 209458b as determined from Lyα observations. *Science* **2014**, *346*, 981–984. [CrossRef]
32. Stevenson, D.J. Planetary magnetic fields. *Rep. Prog. Phys.* **1983**, *46*, 555–620. [CrossRef]
33. Showman, A.P.; Guillot, T. Atmospheric circulation and tides of "51 Pegasus b-like" planets. *Astron. Astrophys.* **2002**, *385*, 166–180. [CrossRef]
34. Jones, C.A. Planetary magnetic fields and fluid dynamos. *Annu. Rev. Fluid Mech.* **2011**, *43*, 583–614. [CrossRef]
35. Jones, C.A. A dynamo model of Jupiter's magnetic field. *Icarus* **2014**, *241*, 148–159. [CrossRef]
36. Batygin, K.; Stanley, S.; Stevenson, D.J. Magnetically controlled circulation on hot extrasolar planets. *Astrophys. J.* **2013**, *776*, 53. [CrossRef]
37. Rogers, T.M.; Showman, A.P. Magnetohydrodynamic simulations of the atmosphere of HD 209458b. *Astrophys. J.* **2014**, *782*, L4. [CrossRef]
38. Rogers, T.M.; Komacek, T.D. Magnetic effects in hot Jupiter atmospheres. *Astrophys. J.* **2014**, *794*, 132. [CrossRef]
39. Rogers, T.M. Constraints on the magnetic field strength of HAT-P-7b and other hot giant exoplanets. *Nat. Astron.* **2017**, *1*, 0131. [CrossRef]
40. Erkaev, N.V.; Odert, P.; Lammer, H.; Kislyakova, K.G.; Fossati, L.; Mezentsev, A.V.; Johnstone, C.P.; Kubyshkina, D.I.; Shaikhislamov, I.F.; Khodachenko, M.L. Effect of stellar wind induced magnetic fields on planetary obstacles of non-magnetized hot Jupiters. *Mon. Not. R. Astron. Soc.* **2017**, *470*, 4330–4336. [CrossRef]
41. Zhilkin, A.G.; Bisikalo, D.V. On possible types of magnetospheres of hot Jupiters. *Astron. Rep.* **2019**, *63*, 550–564. [CrossRef]
42. Belen'kaya, E.S. Magnetospheres of planets with an intrinsic magnetic field. *Phys. Uspekhi* **2009**, *52*, 765–788. [CrossRef]
43. Russell, C.T. Planetary magnetospheres. *Rep. Prog. Phys.* **1993**, *56*, 687–732. [CrossRef]
44. Ip, W.-H.; Kopp, A.; Hu, J.H. On the star-magnetosphere interaction of close-in exoplanets. *Astrophys. J.* **2004**, *602*, L53–L56. [CrossRef]
45. Koskinen, T.T.; Cho, J.Y.-K.; Achilleos, N.; Aylward, A.D. Ionization of extrasolar giant planet atmospheres. *Astrophys. J.* **2010**, *722*, 178–187. [CrossRef]
46. Koskinen, T.T.; Yelle, R.V.; Lavvas, P.; Lewis, N.K. Characterizing the thermosphere of HD 209458b with UV transit observations. *Astrophys. J.* **2010**, *723*, 116–128. [CrossRef]
47. Trammell, G.B.; Arras, P.; Li, Z.-Y. Hot Jupiter magnetospheres. *Astrophys. J.* **2011**, *728*, 152. [CrossRef]
48. Shaikhislamov, I.F.; Khodachenko, M.L.; Sasunov, Y.L.; Lammer, H.; Kislyakova, K.G.; Erkaev, N.V. Atmosphere expansion and mass loss of close-orbit giant exoplanets heated by stellar XUV. I. Modeling of hydrodynamic escape of upper atmospheric material. *Astrophys. J.* **2014**, *795*, 132. [CrossRef]
49. Khodachenko, M.L.; Shaikhislamov, I.F.; Lammer, H.; Prokopov, P.A. Atmosphere expansion and mass loss of close-orbit giant exoplanets heated by stellar XUV. II. Effects of planetary nagnetic field; structuring of inner magnetosphere. *Astrophys. J.* **2015**, *813*, 50. [CrossRef]
50. Trammell, G.B.; Li, Z.-Y.; Arras, P. Magnetohydrodynamic simulations of hot Jupiter upper atmospheres. *Astrophys. J.* **2014**, *788*, 161. [CrossRef]

51. Matsakos, T.; Uribe, A.; Königl, A. Classification of magnetized star-planet interactions: Bow shocks, tails, and inspiraling flows. *Astron. Astrophys.* **2015**, *578*, A6. [CrossRef]
52. Arakcheev, A.S.; Zhilkin, A.G.; Kaigorodov, P.V.; Bisikalo, D.V.; Kosovichev, A.G. Reduction of mass loss by the hot Jupiter WASP-12b due to its magnetic field. *Astron. Rep.* **2017**, *61*, 932–941. [CrossRef]
53. Bisikalo, D.V.; Arakcheev, A.S.; Kaigorodov, P.V. Pulsations in the atmospheres of hot Jupiters possessing magnetic fields. *Astron. Rep.* **2017**, *61*, 925–931. [CrossRef]
54. Zhilkin, A.G.; Bisikalo, D.V.; Kaygorodov, P.V. Coronal mass ejection effect on envelopes of hot Jupiters. *Astron. Rep.* **2020**, *64*, 159–167. [CrossRef]
55. Zhilkin, A.G.; Bisikalo, D.V.; Kaygorodov, P.V. The orientation influence of a hot Jupiter's intrinsic dipole magnetic field on the flow structure in its extended envelope. *Astron. Rep.* **2020**, *64*, 259–271. [CrossRef]
56. Zhilkin, A.G.; Bisikalo, D.V. Possible new envelope types of hot Jupiters. *Astron. Rep.* **2020**, *64*, 563–577. [CrossRef]
57. Zhilkin, A.G.; Bisikalo, D.V.; Kolymagina, E.A. MHD model of the interaction of a coronal mass ejection with the hot Jupiter HD 209458b. *Astron. Rep.* **2021**, *65*, 676–692. [CrossRef]
58. Owens, M.J.; Forsyth, R.J. The heliospheric magnetic field. *Living Rev. Sol. Phys.* **2013**, *10*, 5. [CrossRef]
59. Parker, E.N. Dynamics of the interplanetary gas and magnetic fields. *Astrophys. J.* **1958**, *128*, 664–676. [CrossRef]
60. Weber, E.J.; Davis, L., Jr. The angular momentum of the solar wind. *Astrophys. J.* **1967**, *148*, 217–227. [CrossRef]
61. Lamers, H.J.; Cassinelli, J.P.; Cassinelli, J. *Introduction to Stellar Winds*; Cambridge University Press: Camridge, UK, 1999.
62. Withbroe, G.L. The temperature structure, mass, and energy flow in the corona and inner solar wind. *Astrophys. J.* **1988**, *325*, 442–467. [CrossRef]
63. Fabbian, D.; Simoniello, R.; Collet, R.; Criscuoli, S.; Korhonen, H.; Krivova, N.A.; Oláh, K.; Jouve, L.; Solanki, S.K.; Alvarado-Gómez, J.D.; et al. The variability of magnetic activity in solar-type stars. *Astron. Nachrichten* **2017**, *338*, 753–772. [CrossRef]
64. Lammer, H.; Gudel, M.; Kulikov, Y.; Ribas, I.; Zaqarashvili, T.V.; Khodachenko, M.L.; Kislyakova, K.G.; Gröller, H.; Odert, P.; Leitzinger, M.; et al. Variability of solar/stellar activity and magnetic field and its influence on planetary atmosphere evolution. *Earth Planets Space* **2021**, *64*, 179–199. [CrossRef]
65. Steinolfson, R.S.; Hundhausen, F.J. Density and white light brightness in looplike coronal mass ejections: Temporal evolution. *J. Geophys. Res.* **1988**, *93*, 14269–14276. [CrossRef]
66. Roussev, I.I.; Gombosi, T.I.; Sokolov, I.V.; Velli, M.; Manchester, W., IV; DeZeeuw, D.L.; Liewer, P.; Tóth, G.; Luhmann, J. A three-dimensional model of the solar wind incorporating solar magnetogram observations. *Astrophys. J.* **2003**, *595*, L57–L61. [CrossRef]
67. Totten, T.L.; Freeman, J.W.; Arya, S. An empirical determination of the polytropic index for the free-streaming solar wind using Helios 1 data. *J. Geophys. Res.* **1995**, *100*, 13–18. [CrossRef]
68. Bisikalo, D.V.; Zhilkin, A.G.; Boyarchuk, A.A. *Gas Dynamics of Close Binary Stars*; Fizmatlit: Moscow, Russia, 2013. (In Russian)
69. Zhilkin, A.G.; Bisikalo, D.V.; Boyarchuk, A.A. Flow structure in magnetic close binary stars. *Phys. Uspekhi* **2021**, *55*, 115–136. [CrossRef]
70. Tanaka, T. Finite volume TVD scheme on an unstructured grid system for three-dimensional MHD simulation of inhomogeneous systems including strong background potential fields. *J. Comput. Phys.* **1994**, *111*, 381–389. [CrossRef]
71. Powell, K.G.; Roe, P.L.; Linde, T.J.; Gombosi, T.I.; De Zeeuw, D.L. A solution-adaptive upwind scheme for ideal magnetohydrodynamics. *J. Comput. Phys.* **1999**, *154*, 284–309. [CrossRef]
72. Guo, J.H. Escaping particle fluxes in the atmospheres of close-in exoplanets. I. Model of hydrogen. *Astrophys. J.* **2011**, *733*, 98. [CrossRef]
73. Bisikalo, D.V.; Shematovich, V.I.; Kaigorodov, P.V.; Zhilkin, A.G. *Gaseous Envelopes of Exoplanets—Hot Jupiters*; Nauka: Moscow, Russia, 2020. (In Russian)
74. Lax, P.D. Weak solutions of nonlinear hyperbolic equations and their numerical computation. *Commun. Pure Appl. Math.* **1954**, *7*, 159–193. [CrossRef]
75. Friedrihs, K.O. Symmetric hyperbolic linear differential equations. *Commun. Pure Appl. Math.* **1954**, *7*, 345–392. [CrossRef]
76. Dedner, A.; Kemm, F.; Kroner, D.; Munz, C.-D.; Schnitzera, T.; Wesenberg, M. Hyperbolic divergence cleaning for the MHD equations. *J. Comput. Phys.* **2002**, *175*, 645–673. [CrossRef]
77. Roe, P.L. The use of the Riemann problem in finite difference schemes. *Lect. Notes Phys.* **1981**, *141*, 354–359.
78. Godunov, S.K. A difference scheme for numerical solution of discontinuous solution of hydrodynamic equations. *Math. Sbornik* **1959**, *47*, 271–306. Translated US Joint Publ. Res. Service, JPRS 7225 Nov. 29, 1960.
79. Brio, M.; Wu, C.C. An upwind differencing scheme for the equations of ideal magnetohydrodynamics. *J. Comput. Phys.* **1988**, *75*, 400–422. [CrossRef]
80. Cargo, P.; Gallice, G. Roe matrices for ideal MHD and systematic construction of Roe matrices for systems of conservation laws. *J. Comput. Phys.* **1997**, *136*, 446–466. [CrossRef]
81. Chakravarthy, S.R.; Osher, S. A new class of high accuracy TVD schemes for hyperbolic conservationlaws. In Proceedings of the 23rd Aerospace Sciences Meeting, Reno, NV, USA, 14–17 January 1985. [CrossRef]
82. Harten, A.; Hyman, J. Self-adjusting grid methods for one-dimensional hyperbolic conservation laws. *J. Comput. Phys.* **1983**, *50*, 235–269. [CrossRef]

83. Zhilkin, A.G.; Sobolev, A.V.; Bisikalo, D.V.; Gabdeev, M.M. Flow structure in the eclipsing polar V808 Aur. Results of 3D numerical simulations. *Astron. Rep.* **2019**, *63*, 751–777. [CrossRef]
84. Einfeldt, B. On Godunov-type methods for gas dynamics. *SIAM J. Numer. Anal.* **1988**, *25*, 294–318. [CrossRef]

95

Article

Particle Acceleration in Mildly Relativistic Outflows of Fast Energetic Transient Sources

Andrei Bykov *, Vadim Romansky and Sergei Osipov

Ioffe Institute, 194021 St. Petersburg, Russia; romanskyvadim@gmail.com (V.R.); osm2004@mail.ru (S.O.)
* Correspondence: byk@astro.ioffe.ru

Abstract: Recent discovery of fast blue optical transients (FBOTs)—a new class of energetic transient sources—can shed light on the long-standing problem of supernova—long gamma-ray burst connections. A distinctive feature of such objects is the presence of modestly relativistic outflows which place them in between the non-relativistic and relativistic supernovae-related events. Here we present the results of kinetic particle-in-cell and Monte Carlo simulations of particle acceleration and magnetic field amplification by shocks with the velocities in the interval between 0.1 and 0.7 c. These simulations are needed for the interpretation of the observed broad band radiation of FBOTs. Their fast, mildly to moderately relativistic outflows may efficiently accelerate relativistic particles. With particle-in-cell simulations we demonstrate that synchrotron radiation of accelerated relativistic electrons in the shock downstream may fit the observed radio fluxes. At longer timescales, well beyond those reachable within a particle-in-cell approach, our nonlinear Monte Carlo model predicts that protons and nuclei can be accelerated to petaelectronvolt (PeV) energies. Therefore, such fast and energetic transient sources can contribute to galactic populations of high energy cosmic rays.

Keywords: fast blue optical transients; non-thermal particle acceleration; particle-in-cell plasma modeling; high energy cosmic rays

Citation: Bykov, A.; Romansky, V.; Osipov, S. Particle Acceleration in Mildly Relativistic Outflows of Fast Energetic Transient Sources. *Universe* **2022**, *8*, 32. https://doi.org/10.3390/universe8010032

Academic Editors: Galina L. Klimchitskaya, Vladimir M. Mostepanenko and Nazar R. Ikhsanov

Received: 19 November 2021
Accepted: 3 January 2022
Published: 5 January 2022

Publisher's Note: MDPI stays neutral with regard to jurisdictional claims in published maps and institutional affiliations.

1. Introduction

The time domain astronomy operating now at all wavebands from radio to gamma rays has provided unique information on highly energetic processes in transient astrophysical objects such as supernovae (SNe), gamma-ray bursts (GRBs), fast radio bursts and others. The pioneer program of a dedicated search for supernovae in the optical band started by Fritz Zwicky in 1936, which has later allowed, in particular, obtaining fundamental results on the accelerated expansion of the Universe via spectroscopy of SN type Ia, is now ongoing with great perspectives [1]. The capabilities of fast and sensitive wide field imaging suggested for the forthcoming Large Synoptic Survey Telescope (LSST) would allow detecting many thousands of luminous SNe and tidal disruption events (TDEs) per year as well as studying other types of transient sources [2]. Together with the LSST, the currently operating wide field [3] and survey [4] X-ray observatories along with the future high energy missions [5–7], gravitational wave and neutrino observatories will allow revealing the physical nature of various types of energetic space transients.

The fast blue optical transients (FBOTs) [8–11] are among the most interesting recent discoveries. Their appearance is somewhat different from most of the core-collapse SNe [8,12]. Together with the low-luminosity GRBs they possibly belong to the intermediate class of phenomena filling the gap between non-relativistic SNe and "standard" long duration gamma-ray bursts, and which could have volumetric rates well above that of the GRBs [13,14]. The duration of both the energy-momentum release from the central engine and the interaction of the anisotropic ejecta with the outer layers of the progenitor star and its circumstellar matter determine the transient appearance (see, e.g., [15]).

Three recently studied powerful FBOT sources AT2018cow [9], CSS161010 [10] and ZTF18abvkwla [11] were characterized by a low ejected mass and fast outflows. Indeed,

Margutti et al. [9] found that to explain the fast rise of the optical and radio emission together with the persistent photosphere appearance in AT2018cow a wide range of velocities in the range from about 0.02 c to 0.2 c is needed, where c is the speed of light. The aspherical ejecta with the range of velocities had the estimated mass $\sim$0.1–1 $M_\odot$ [9]. The long low frequency *uGMRT* radio observations [16] allowed estimating the shock radius to be $R = (6.1 - 14.4) \times 10^{16}$ cm and the speed of the fast ejecta to be above 0.2 c at 257 days after the shock breakout. The mass loss rate of the progenitor star was found to be a few times 10^{-6} $M_\odot$ yr^{-1} during the period of 20–50 years assuming the wind velocity $\sim$1000 km s^{-1}, while it was possibly 100 times higher in a period of a few years just before the event [16].

Another interesting FBOT source is CSS161010 located in a dwarf galaxy at a distance about 150 Mpc [10]. On the basis of the synchrotron interpretation of its radio emission, which is peaked at about 100 days after the FBOT event, the authors suggested a presence of a mildly relativistic outflow of four-velocity 0.55 c driving a blast wave. They estimated the ejected mass to be in the range of 0.01–0.1 $M_\odot$. The outflow is faster than that estimated for AT2018cow and is similar to that in ZTF18abvkwla. The origin of the bright X-ray luminosity is attributed to an emission component, which is likely different from the primary one, which produced the synchrotron radio emission. Recent radio, millimeter wave and X-ray observations [17,18] of a short-duration luminous FBOT transient ZTF20acigmel (AT2020xnd) located at z = 0.2433 indicated a presence of a fast ejecta with a $\sim$0.2 c speed shock with estimated energy above 10^{49} ergs. AT2020xnd has shown high radio luminosity of $L_\nu \sim 10^{30}$ ergs^{-1} Hz^{-1} at 20 GHz almost 75 days after the event [18]. The observational data suggested a shock driven by a fast outflow of velocity 0.1–0.2 c interacting with the dense circumstellar matter shaped by an intense wind of $\dot{M} \approx 10^{-3}$ $M_\odot$ yr^{-1} with velocity of $v_w = 1000$ km s^{-1} from the progenitor star [18]; the presence of a steep density profile of $\rho(r) \propto r^{-3}$ in the wind was suggested by [17]. Similar to AT2018cow, the detected X-ray emission is in excess compared to the extrapolated synchrotron spectrum and constitutes a different emission component, possibly powered by accretion onto a newly formed black hole or neutron star.

The distinctive features of the four FBOT transients discovered so far are their high peak bolometric luminosity $L \gtrsim 10^{43}$ erg s^{-1} and a rapid timescale of a few days duration. Multi-wavelength observations of these objects uncovered the presence of powerful sub- or mildly relativistic outflows which are likely originated from either rare SN-type or TDE-type events with intermediate or stellar mass black holes (see, e.g., [19]). The kilonova type sources with the neutron stars mergers typically eject the masses in the range of 10^{-4}–10^{-2} $M_\odot$ (and even $\sim$0.1 $M_\odot$ for the black hole—neutron star mergers) with velocities $\sim$0.1–0.3 c [20]. It is interesting that the recent model of a supernova from a primordial population III star of a 55,500 $M_\odot$ mass with general relativistic instability [21] predicts the ejecta velocities of about 0.3 c. Earlier mildly and moderately relativistic ejecta outflows were found in a few broad line type Ic SNe (e.g., [22,23]). The geometry and structure of the outflows producing FBOTs depend on the source of their power and it is a subject of modeling [9]. Relativistic mass ejection in spherically symmetric shock outflows of core-collapse supernovae was studied in detail in [24–26]. Their high-velocity solutions demonstrated rather a steep dependence of the deposited kinetic energy E_k as a function of the ejecta four-speed $E_k \propto (\beta\Gamma)^{-5.2}$. Much flatter energy—ejecta velocity distribution $E_k \propto (\beta\Gamma)^{-2.4}$ can be obtained for engine-driven asymmetric supernovae with a powerful activity of compact stellar remnants [27–30]. Recently, numerical models of supernova explosions where the supernova ejecta interacts with the relativistic wind from the central engine were constructed in [31]. The formation of relativistic flows in the interaction of the powerful non-thermal radiation produced by the central machine with the supernova ejecta was considered in the papers [32,33].

Very luminous optical FBOT events with light curves extending to a couple of weeks can be expected in the case of shock breakout into a dense circumstellar shell produced by the dense progenitor wind a few years before the SN event [34–36]. The presence of an hour timescale central engine activity with a luminosity of about 10^{47} erg s^{-1} producing mildly relativistic jet was proposed in [15] to model SN 2006aj associated with a low-luminosity GRB. The bright high frequency radio emission of FBOTs and, possibly, the hard X-ray component detected in some events of this type, is produced by relativistic electrons accelerated by shocks driven by fast moderately relativistic outflows from the central engine. Moreover, the synchrotron self-absorption effects [37] are apparent in some FBOT spectra.

The SN shock breakout is usually accompanied by a bright ultraviolet (UV) flash, and it is likely that afterwards the shock enters a collisionless regime [38] with the X-ray dominated spectrum. Some models of relativistic shock breakout which consider a multifluid structure of a relativistic shock mediated by radiation in a cold electron-proton plasma are currently under discussion [39,40]. A MeV gamma-ray flash of the total energy $\sim 10^{48}$ erg lasting from a few seconds to a few hours was predicted in [41,42] for some SN events. At the later SN stages (after a few days from the event) the radiative shock is transforming into a collisionless plasma shock regulated by kinetic plasma instabilities. A specific feature of the collisionless shocks is their ability to create a powerful non-thermal particle population and to accelerate relativistic particles [43]. The collisionless shock structure and the efficiency of particle acceleration depend on the shock speed, on magnetic field inclination to the shock normal and on the plasma magnetization parameter [44,45]. Modeling of particle acceleration by non-relativistic shocks of velocities below 0.1 c in supernova remnants was discussed in [46] while studies of relativistic shocks was presented in [47].

Kinetic simulations of the efficiencies of the shock ram pressure conversion to magnetic fluctuations and relativistic particles are needed to provide an adequate interpretation of the observed non-thermal radiation. The mildly relativistic magneto-hydrodynamic (MHD) flows were shown (see, e.g., [48,49]) to be the most efficient environment providing the maximum energies of the accelerated nuclei for a given magnetic/kinetic luminosity of the power engine. Recent models suggested a possibility of cosmic ray acceleration to ultra high energies in the low-luminosity GRBs associated with SNe (see, e.g., [50,51]) and in relativistic SNe [52]. Fast outflows from SNe with dense circumstellar shells could accelerate cosmic rays up to the high energy regime on a few weeks timescale (e.g., [53,54]).

The structure and particle acceleration in the fast collisionless shocks can be successfully modeled with the kinetic particle-in-cell (PIC) technique [43,55] at many thousands of the particle gyro-scales. On the other hand, as the observed multi-wavelength spectra of supernova remnants and GRBs compellingly demonstrated that the spectra of accelerated particles extend to many decades in the particle momentum, a combination of both microscopic (see Section 2) and macroscopic kinetic models (see Section 3) is necessary to construct realistic models of such sources. Cosmic ray acceleration by supernova remnants is a subject of extensive modeling [46,52,56–59]. One of the most uncertain points is at what stage PeV regime cosmic rays can be accelerated. We discuss in Section 3 high energy cosmic ray acceleration in FBOT-type sources as potential pevatrons.

Here we will present 2D PIC simulation of sub and mildly relativistic shocks within the shock speed interval 0.1–0.7 c in proton-electron plasmas which could be applied to FBOT sources modeling. The fiducial case in our work is 0.3 c.

Earlier, Park et al. [60] presented a set of 1D PIC models of shocks in a range of velocities up to 0.1 c. They performed in the 1D case long simulation runs up to $5 \times 10^5 \, \omega_{pe}^{-1}$. For higher shock speed of 0.75 c Crumley et al. [61,62] published 2D PIC simulation results. In the quasi-parallel shock of a velocity 0.75 c simulation of about 4100 ω_{pi}^{-1} duration allowed them to model Bell-mediated shock. In particular, they noted that acceleration of electrons is likely initially associated with the shock drift acceleration and then switches to diffusive shock acceleration regime as it was seen in [60].

To model radio emission observed from CSS161010, where the shock velocity of about 0.3 c was suggested by observations, we make 2D PIC simulations with m_p/m_e ratio of 100, which are limited to the timescale $\sim 7 \times 10^4 \, \omega_{pe}^{-1}$. We discuss possible extrapolations of electron spectra to the energy range needed to model the radio emission.

2. Particle-in-Cell Simulations of Fast Mildly Relativistic Shocks

In this section, we present the results of particle-in-cell simulations of mildly relativistic shocks at gyro-scales with application to the observed non-thermal emission from FBOTs.

To simulate the structure and non-thermal particle acceleration by a mildly relativistic collisionless shock wave we have employed the publicly available code Smilei developed by Derouillat et al. [63]. This code is based on explicit Finite Differences Time Domain approach for solving Maxwell equations and on a relativistic solver for particle movement with charge-conserving algorithm proposed by [64].

For model shock initialization we employ a common approach: a homogeneous plasma flow collides with an ideal reflecting wall. The simulation box is two-dimensional, with the number of cells $Nx = 204{,}800$ in the direction along the flow velocity and $Ny = 400$ in the transverse direction, while particle velocities and the electromagnetic field are represented by full 3D vectors. Homogeneous plasma flows in through the right boundary and the reflecting wall is placed at the left boundary. Boundary conditions in the transverse direction are periodic. The initial magnetic field $\vec{B}$ lies in the plane of the simulation inclined by the angle θ to the velocity of the flow. The electric field is initialized to compensate the Lorentz force in the laboratory frame $\vec{E} = -\vec{v} \times \vec{B}/c$. The velocity of plasma flow v is 0.3 c and its Lorentz-factor γ is 1.05. The magnetization $\sigma = B^2/4\pi n\gamma m_p c^2$, where n is upstream concentration and m_p is mass of proton, in all the modeled configurations is about 10^{-4}. The electron mass m_e is artificially increased up to $m_p/m_e = 100$ in order to save computational resources. All the quantities obtained from the simulation can be scaled with respect to the plasma concentration, which can be chosen arbitrary during the data analysis. The spatial grid step is $dx = 0.2c/w_e$ and the time step is $dt = 0.09\ w_e^{-1}$, where w_e is electron plasma frequency $w_e = \sqrt{4\pi n e^2/m_e\gamma}$, e is the absolute value of the electron charge. Within such a setting the simulation box size along the x-axis corresponds to 500 gyroradii of protons in plasma flow $r_g = m_p vc\gamma/eB$ and along the y-axis—to 1 gyroradius. The maximum simulation time is $7 \times 10^4\ w_e^{-1}$ or about 100 inverse proton gyrofrequencies. For plasma concentration $n \approx 1$ cm^{-3}, this corresponds to timescale about 1 s, which is much smaller than the typical activity period of FBOTs.

The efficiency of particle acceleration depends on the inclination angle θ, especially in the case of relativistic flows [44,61,62,65]. To participate in the diffusive acceleration process, a particle needs to escape from the shock front and if it moves along the magnetic field, the maximum velocity along the x-axis is $c\cos(\theta')$, and it should be larger than the shock velocity v'_{sh} (all quantities here are measured in the upstream frame). As one can see in Figure 1, the high energy tail of the electron distribution is much higher for a quasiparallel shock. Such an angle dependence may lead to the presence of different electron distributions within one object, and this may possibly explain the difference between the spectral indices of synchrotron radio and inverse Compton X-ray radiation observed in FBOT AT2018cow [9].

For further modeling of synchrotron radiation from FBOT CSS161010 we have used a setup with parameters $v = 0.3$ c, $\sigma = 0.0002$ and $\theta = 30°$ (a quasiparallel shock). Time evolution of the concentration profile in such a shock averaged in the transverse direction is shown in Figure 2 and the corresponding magnetic field at the moment $7 \times 10^4\ w_e^{-1}$ is shown in Figure 3.

With the computed concentration profile we can determine the coordinate of the shock front: we consider that x_{sh} is the first point from the right where the concentration is two times larger than that in the far upstream. The average shock velocity is the ratio of the shock coordinate and the simulation time $v_{sh} = x_{sh}/t$. At later times the shock wave propagation is close to a stationary regime. The shock velocity measured in the downstream frame is close to the constant value $v_{sh} \approx 0.085$ c. In the upstream (observer) frame it corresponds to $v'_{sh} = (v_{sh} + v)/\left(1 + v_{sh}v/c^2\right) \approx 0.38$ c.

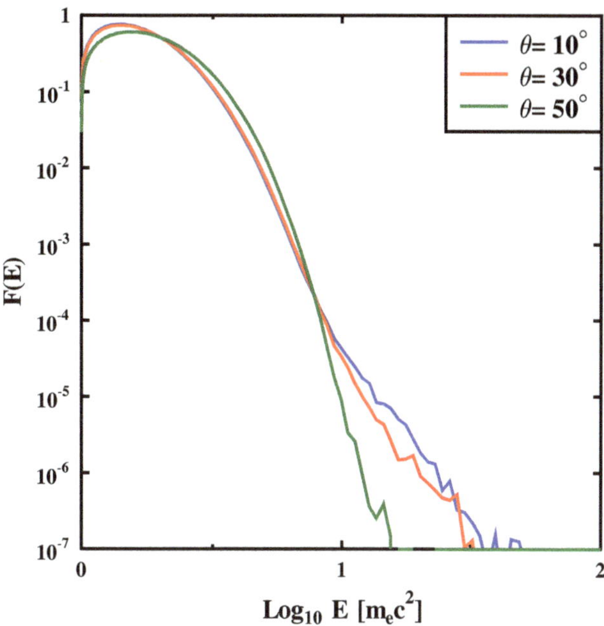

Figure 1. Electron distribution function in the shock downstream with initial parameters $v = 0.3$ c, $\sigma = 0.0002$ and different inclination angles.

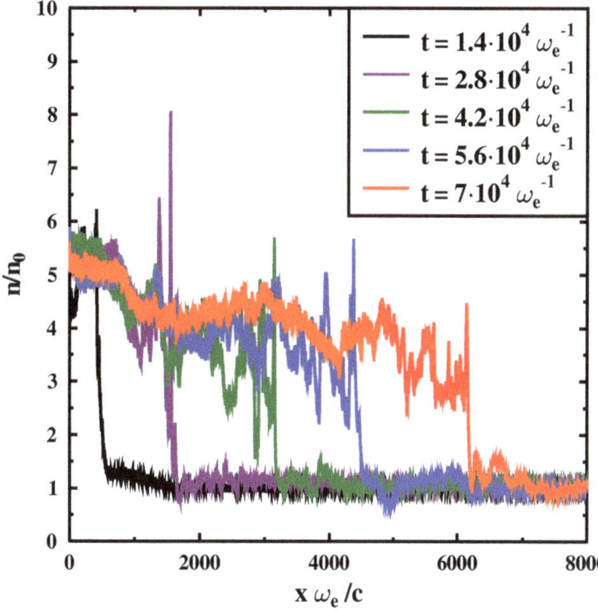

Figure 2. Time evolution of concentration normalized to the far upstream concentration.

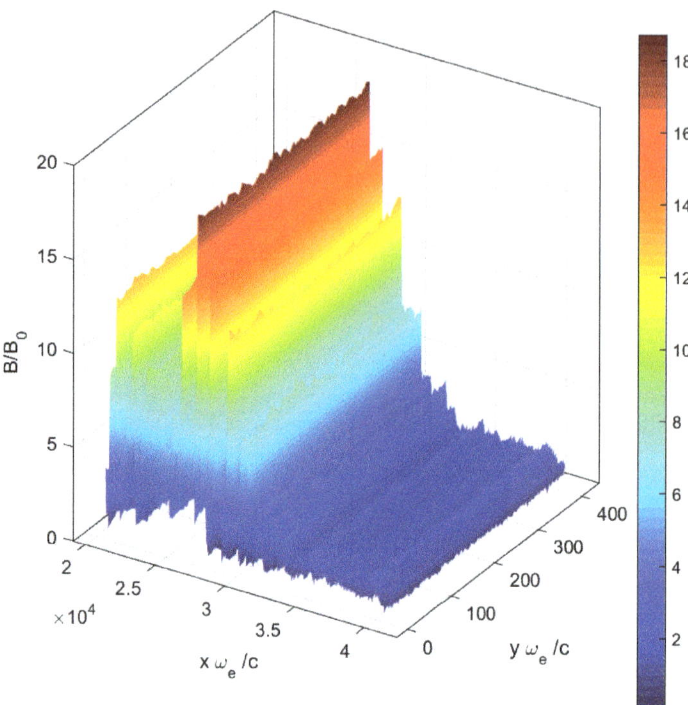

Figure 3. Magnetic field normalized to the far upstream magnetic field.

The simulations have allowed us to compute model distributions of electrons. In the downstream region such distributions have a complex shape, where one can see two main components: a thermal peak and a power law tail at high energies. At low energies ($E < 5\ m_e c^2$) we approximate the distribution with a Maxwell-Juttner function: an iterative process is employed to minimize the functional $f(T) = \int_{m_e c^2}^{5 m_e c^2} (F(E) - F_{mj}(E,T))^2 dE$, where $F(E)$ is simulated electron distribution function and $F_{mj}(E,T)$ is Maxwell-Juttner distribution function, and find the effective temperature. For high energies ($20\ m_e c^2 < E < 50\ m_e c^2$) we follow a least squares approach for linear regression in double logarithmic coordinates and obtain the power law spectral index. The electron spectrum in a close downstream of the shock (5000 grid cells behind the shock) at time $t = 70{,}000\ \omega_e^{-1}$ for the setup with $\sigma = 0.0002$ and $\theta = 30°$ and its approximation with temperature $T_e = 5 \times 10^{10}$ K and spectral index $s = 3.59$ are shown in Figure 4.

Time dependence of electron distribution function at later stages of simulation is shown in Figure 5. Distribution is not stationary and one can see the oscillation of the electron spectral distribution due to the influence of magneto-hydrodynamic instabilities, as described in [61,66]. So we have chosen the electron distribution at time $t = 70{,}000\ \omega_e^{-1}$ for the further modeling.

Recent observations of FBOTs [8–11] discussed above have shown that their synchrotron spectra are strongly influenced by synchrotron self-absorption. At the given time moment, observable radio fluxes rise as $F_\nu \propto \nu^{5/2}$ for low frequencies, then have a peak and fade with a power law tail, which depends on the particular electron distribution function. Additionally, the time dependence of such spectrum is very specific and its details could be used to imply a number of source parameters: the magnetic field and the shock radius can be determined from the measured maximum of the light-curve F_{max} at given frequency

ν, once the fractions of energy in accelerated electrons and magnetic field ϵ_e and ϵ_B are established, with the relations derived by Chevalier [37].

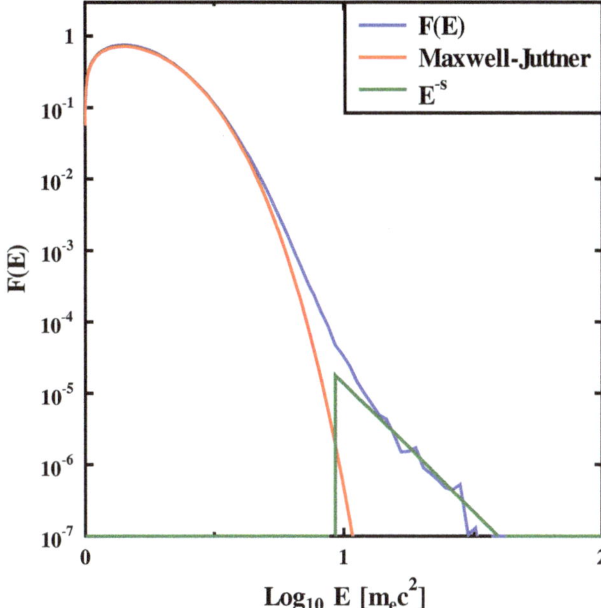

Figure 4. The electron distribution function in the shock downstream with initial parameters $v = 0.3\,\mathrm{c}$, $\sigma = 2 \times 10^{-4}$ and $\theta = 30°$.

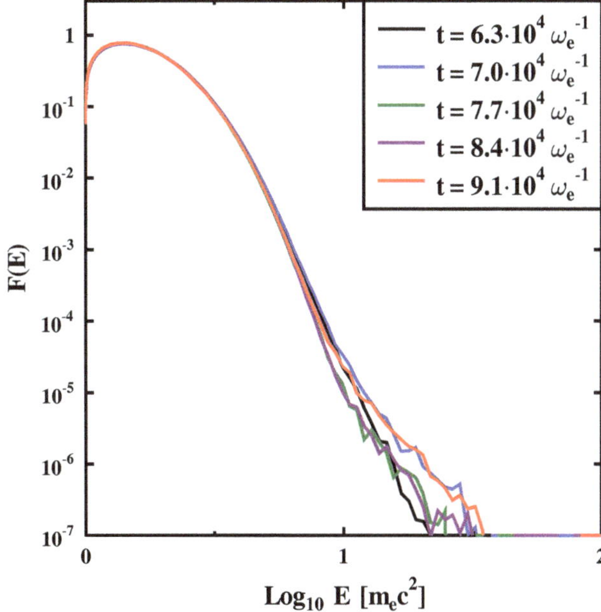

Figure 5. The electron distribution function in the shock downstream at different time moments.

$$R = \left(\frac{6\epsilon_B c_6^{s+5} F_{max}^{s+6} D^{2s+12}}{\epsilon_e f (s-2) \pi^{s+5} c_5^{s+6} E_1^{s-2}} \right)^{\frac{1}{2s+13}} \frac{2c_1}{v}, \tag{1}$$

$$B = \left(\frac{\epsilon_B^2 36 \pi^3 c_5}{\epsilon_e^2 f^2 (s-2)^2 c_6^3 E_1^{2s-4} F_{max} D^2} \right)^{\frac{2}{2s+13}} \frac{v}{2c_1}, \tag{2}$$

where s is the electron spectral index, derived from the power law tail of emissivity spectrum and its time dependence, c_1, c_5, c_6 are the Pacholczyk constants [67], depending on s, E_1 is the minimum energy of the electron power law distribution, which usually equals to the electron rest energy, f is the filling factor—the emitting fraction of the source volume, and D is the distance to the source. A problem of this method is that it needs a lot of assumptions about the shock structure and electron distribution. In this model the emitting region is considered to be a homogeneous flat disk with the radius r and depth $f \times r$. There is also a model modification for a spherical source with an account for source inhomogenuity [17]. The electron distribution is considered to be a power law. Additionally, fractions of energy ϵ_e and ϵ_B are unknown and are often chosen according to the equipartition rule $\epsilon_e = \epsilon_B = 1/3$. However the simulation results show, that these values are unlikely to be that large. Following Chevalier [68] we define the energy fractions in terms of the upstream kinetic energy density $(m_p + m_e) n_u v_{sh}'^2$, where n_u is the upstream concentration. In this notation $\epsilon_e = E_e / \rho_u v_{sh}'^2$ and $\epsilon_B = B^2 / \left(8\pi (m_p + m_e) n_u v_{sh}'^2 \right)$, E_e is the accelerated electron energy evaluated by subtracting from the total electron energy, that corresponds to the Maxwell–Juttner fit of the distribution function. The values obtained by particle-in-cell simulations in our fiducial setup are $\epsilon_e = 0.014$ and $\epsilon_B = 0.03$. They depend on the initial conditions, but for a wide variety of parameters we see that the fraction of energy in magnetic field is lower than 10%. Particle-in-cell simulations have rather small scales and cannot describe the influence of long wave upstream instabilities caused by high energy particles, but Monte Carlo simulations show similar values, as described below.

Using the electron distribution function obtained from the PIC simulation we can evaluate the spectral density of the energy flux of synchrotron radiation from the source, taking into account the effect of synchrotron self absorption. Standard formulae for that effect are described in detail in [69]. Emitted power per unit frequency per unit volume is

$$I(v) = \int_{E_{min}}^{E_{max}} dE \frac{\sqrt{3} e^3 n F(E) B \sin(\phi)}{m_e c^2} \frac{v}{v_c} \int_{\frac{v}{v_c}}^{\infty} K_{5/3}(x) dx, \tag{3}$$

where ϕ is the angle between the magnetic field and the line of sight, v_c is the critical frequency $v_c = 3e^2 B \sin(\phi) E^2 / 4\pi m_e^3 c^5$, and $K_{5/3}$ is the modified Bessel function. The absorption coefficient is

$$k(v) = \int_{E_{min}}^{E_{max}} dE \frac{\sqrt{3} e^3}{8\pi m_e v^2} \frac{n B \sin(\phi)}{E^2} \frac{d}{dE} E^2 F(E) \frac{v}{v_c} \int_{\frac{v}{v_c}}^{\infty} K_{5/3}(x) dx. \tag{4}$$

The obtained spectral density is further integrated over the volume of the source, which is described as a spherical shell, whose volume is determined by a filling factor $f = 0.5$. Hence the total emissivity and observable flux at distance D can be derived. The magnetic field B is considered constant and perpendicular to the line of sight. The concentration in the stellar wind depends on radius as $n \propto r^{-2}$, but in the downstream of the shock we assume it constant. The electron distribution function is also constant in the volume of the source. This modeling is further applied to explain the spectrum of the FBOT transient CSS161010. As described in [10] at time $t = 357$ days after explosion its shock velocity was $v_{sh} = 0.36c$, which corresponds to the modeled upstream plasma flow velocity $vs. = 0.3c$, measured in the downstream frame. At this moment, parameters derived with Equations (1) and (2), assuming the equipartition regime $\epsilon_e = \epsilon_B = 1/3$, are as follows:

the magnetic field $B = 0.052$ G, the outer radius $R = 3.3 \times 10^{17}$ cm, the concentration $n = 1.9$ cm^{-3}, the electron spectral index $s = 3.5$, which is similar to that obtained from a PIC simulation, and the minimum energy $E_1 = 4\, m_e c^2$.

The thermal electrons may contribute substantially to the observed synchrotron spectrum and it was shown in [70] that models with the power law electron distribution might be oversimplified. Additionally, equipartition regime is not very reasonable assumption. Thus, the use of the realistic kinetic simulations is important to model broad band radiation spectra of FBOTs. We have compared the observed fluxes to three models of source emission, evaluated with power law electron distribution and two distributions obtained from PIC simulation: with extrapolation of power law tail to higher energies ($E_{max} = 500\, m_e c^2$) and without it. For the first case, we have used the parameters described above. In order to evaluate parameters B and R of the simulated particle distribution, we minimized the functional $g(B, R) = \sum (F(\nu_i, B, R) - F_{obs}(\nu_i))^2$, where ν_i are the observed frequencies, $F_{obs}(\nu_i)$ are the observed fluxes and $F(\nu_i, B, R)$—the modeled fluxes, using the gradient descent algorithm. We employed six measurements from [10]: four made with VLA at day 357 at frequencies 1.5, 3.0, 6.05 and 10 GHz and two made by GMRT at day 350 at frequencies 0.33 and 0.61 GHz. Concentration in these equations is determined by the magnetic field ϵ_B, obtained from the PIC simulation. One can see that distribution, obtained directly from particle-in-cell simulation, cannot correctly fit the power law tail in observational data, so we had to extrapolate the simulated electron distribution to higher energies because the PIC approach requires a lot of computational resources and thus is not suitable to simulate the considered system up to timescales long enough to form a long tail of accelerated particles. The modeled parameters for extrapolated distribution are $B = 0.069$ G, $R = 3.0 \times 10^{17}$ cm and concentration $n = 210$ cm^{-3}. The values of the magnetic field and radius are rather close to the values from [10], while the concentration is much higher. These results are illustrated in Figure 6.

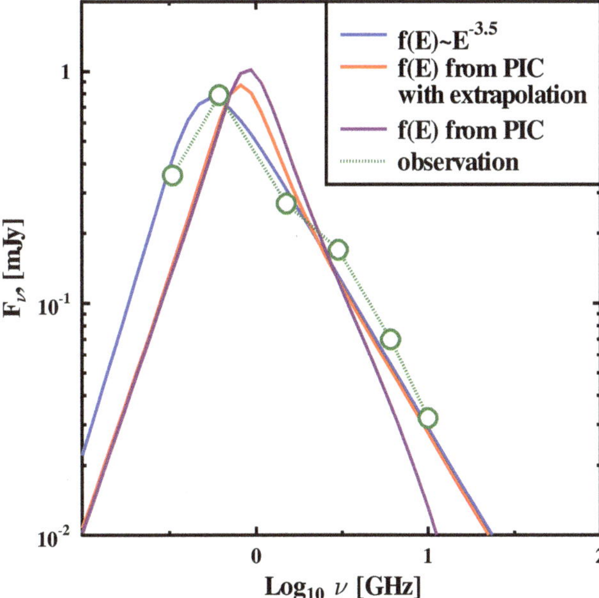

Figure 6. Observed (green circles) and modeled spectral energy distribution of CSS161010 at day 357 for various model electron distributions.

One can see that the parameters of the shock (especially the concentration), obtained from the radiation model strongly depend on the electron distribution function. This should be kept in mind during interpretation of the observational data.

3. A Monte Carlo Model of Cosmic Ray Acceleration in Fast Transient Sources

We have developed a Monte Carlo model of particle acceleration by collisionless shock waves. Acceleration occurs according to the first-order Fermi mechanism when particles are scattered by magnetic fluctuations and cross the shock front many times—the so-called diffusive shock acceleration (DSA). The distribution function of the accelerated particles has a significant anisotropy in the upstream of the shock, and this anisotropy leads to development of plasma instabilities in the upstream. The development of plasma instabilities also leads to a magnetic field amplification (MFA) in the upstream. The pressure of the accelerated particles can be on the order of the total momentum flux flowing onto the shock. In this case, the pressure gradient of the accelerated particles leads to modification of the plasma flow in the upstream.

The Monte Carlo code employed to describe the DSA and MFA is one-dimensional, stationary, nonlinear and plane-parallel. We consider acceleration of protons. The nonrelativistic model described in [71] has been modified to be applicable for the case of relativistic shocks. Within this model all the particles are divided into background and accelerated ones: the accelerated particles are those that have crossed the shock front from the downstream to the upstream at least once. The accelerated particles are treated individually. Background particles are described via macroscopic parameters under the assumption of their local Maxwell distribution function up to the injection point near the shock front in the upstream. After this point, the background particles are described as particles, which allows us not to add any additional parameters to describe the injection of particles into the acceleration process. During the propagation of particles, they are scattered elastically and isotropically in the rest frame of the scattering centers according to the pitch-angle scattering approach [72–76]. The reference frame of the scattering centers for background particles moves with a speed $u(x)$ relative to the rest frame of the shock. The rest frame of the scattering centers for accelerated particles moves with a speed $u(x) + v_{scat}(x)$ relative to the rest frame of the shock. $u(x)$ is the speed of the background plasma flow. The presence of $v_{scat}(x)$ is due to the fact that with the development of resonant instability, the modes propagating only in a certain direction relative to the background plasma are amplified. $v_{scat}(x)$ will be determined below.

Between the scatterings the particles uniformly move straightforward. The distance between particle scatterings is proportional to its mean free path, which we define as:

$$\lambda(x,p) = \frac{1}{\frac{1}{\lambda_{B,st}(x,p)} + \frac{1}{\lambda_{ss}(x,p)}},\qquad(5)$$

where k is the absolute value of the wavenumber of modes that make up magnetic fluctuations, p is the particle momentum, x is the particle coordinate that is counted from the front of the shock wave. Negative values of x correspond to the upstream, positive values correspond to the downstream.

$$\lambda_{B,st}(x,p) = \frac{pc}{eB_{ls,st}(x,k_{res})},\qquad(6)$$

$$B_{ls,st}(x,k) = \sqrt{4\pi \int_0^k W(x,k)dk},\qquad(7)$$

where $W(x,k)$ is the spectral density of the turbulence energy.

$$\lambda_{ss}(x,p) = \left(\frac{pc}{\pi e}\right)^2 \frac{1}{\int_{k_{res}}^{\infty} \frac{W(x,k)}{k}dk},\qquad(8)$$

$$\frac{k_{res}pc}{eB_{ls}(x,k_{res})} = 1,\tag{9}$$

B_{ls} is the large-scale magnetic field:

$$B_{ls}(x,k) = \sqrt{4\pi \int_0^k W(x,k)dk + B_0^2},\tag{10}$$

where B_0 is the constant longitudinal magnetic field. The particle propagates until it gets outside of the model box to the far downstream or crosses the free escape boundary (FEB) located at $x = x_{FEB}$ in the far upstream.

The iterative scheme of the employed numerical model allows us to keep the conservation laws of momentum and energy fluxes near the shock front. In the stationary relativistic case these are are formulated as follows. The law of conservation of particle flux (automatically kept here) has the form:

$$\gamma(x)\beta(x)n(x) = F_{n0},\tag{11}$$

where $\gamma(x) = 1/\sqrt{1 - \beta^2(x)}$ is the Lorentz factor of the background plasma flow, $\beta(x) = u(x)/c$, $n(x)$ is the background plasma number density, F_{n0} is the particle flux in the far upstream.

The conservation law of momentum flux takes the form:

$$\gamma^2(x)\beta^2(x)\left[m_p c^2 n(x) + \Phi_{th}(x) + \Phi_w(x)\right] + P_{th}(x) + P_w(x) + F_{px}^{cr}(x) = F_{px0} + Q_{px}^{ESC},\tag{12}$$

where $P_{th}(x)$ is the background plasma pressure, $\Phi_{th}(x) = \Gamma_{th}(x)P_{th}(x)/(\Gamma_{th}(x) - 1)$, Γ_{th} is the background plasma adiabatic index, $P_w(x)$ is the turbulence pressure, $P_w(x) = 0.5\int_{(k)} W(x,k)dk$, $\Phi_w(x) = 3P_w(x)$, F_{px0} is the momentum flux in the far unperturbed upstream, Q_{px}^{ESC} is the momentum flux carried away by the accelerated particles through the FEB ($Q_{px}^{ESC} = F_{px}^{cr}(x_{FEB})$), $F_{px}^{cr}(x)$ is the momentum flux of the accelerated particles.

The conservation law of energy flux takes the form:

$$\gamma^2(x)\beta(x)\left[m_p c^2 n(x) + \Phi_{th}(x) + \Phi_w(x)\right] + F_{en}^{cr}(x) = F_{en0} + Q_{en}^{ESC},\tag{13}$$

where F_{en0} is the energy flux in the far unperturbed upstream, Q_{en}^{ESC} is the energy flux carried away by accelerated particles through the FEB ($Q_{en}^{ESC} = F_{en}^{cr}(x_{FEB})$), $F_{en}^{cr}(x)$ is the energy flux of the accelerated particles.

To keep the conservation laws (12) and (13), it is necessary to determine the profile of the background plasma flow $u(x)$ in the upstream and the full compression by the shock R_{tot} by means of an iterative process. At the initial iteration, approximate profiles $u(x)$, $v_{scat}(x)$ in the upstream, $W(x,k)$ and the full shock compression R_{tot} are set. Then the particles are propagated and their distribution function is calculated accordingly. In the far downstream, the momentum distribution function of all the particles in the rest frame of the flow is isotropic. Thus, it is possible to determine the adiabatic index of the entire plasma in the downstream Γ_{p2} based on the obtained particle distribution function. Here and below, the subscript 0(2) denotes the values in the far upstream (downstream). In these designations $R_{tot} = \beta_0/\beta_2$.

To find the full compression, we write down the flux conservation laws (11), (12) and (13) for the far upstream and for the downstream.

$$\gamma_2\beta_2 n_2 = \gamma_0\beta_0 n_0,\tag{14}$$

$$\gamma_2^2\beta_2^2\left[m_p c^2 n_2 + \Phi_{p2} + \Phi_{w2}\right] + P_{p2} + P_{w2} = \gamma_0^2\beta_0^2\left[m_p c^2 n_0 + \Phi_{th0} + \Phi_{w0}\right] +$$
$$+ P_{th0} + P_{w0} + Q_{px}^{ESC}, \quad (15)$$

$$\gamma_2^2\beta_2\left[m_p c^2 n_2 + \Phi_{p2} + \Phi_{w2}\right] = \gamma_0^2\beta_0\left[m_p c^2 n_0 + \Phi_{th0} + \Phi_{w0}\right] + Q_{en}^{ESC}, \quad (16)$$

where P_{p2} is the pressure of the entire plasma in the downstream, $\Phi_{p2} = \Gamma_{p2}P_{p2}/(\Gamma_{p2}-1)$. We determine the current values Q_{en}^{ESC}, Q_{px}^{ESC} and Γ_{p2} at this iteration of the quantities from the distribution function obtained after particle propagation. The value P_{w2} is also calculated based on the equation given below. Thus, three unknowns R_{tot}, P_{p2} and n_2 remain in the Equations (14)–(16). Solving these equations, we find a new value of R_{tot}. The value R_{tot} for the next iteration is found by averaging the new value and the old one.

A new profile of the flow velocity in the upstream is determined according to the formula based on (12):

$$\gamma_{new}(x)\beta_{new}(x) = \gamma_{old}(x)\beta_{old}(x) + \frac{F_{px0} + Q_{px}^{ESC} - F_{px}^{cr}(x) - F_{px}^{th}(x) - F_{px}^{w}(x)}{\gamma_0\beta_0 m_p c^2 n_0}, \quad (17)$$

where the values obtained after propagation of particles are in the right part of the expression. The background plasma momentum flux is $F_{px}^{th}(x) = \gamma^2(x)\beta^2(x)\Phi_{th}(x) + P_{th}(x)$. The turbulence momentum flux is $F_{px}^{w}(x) = \gamma^2(x)\beta^2(x)\Phi_w(x) + P_w(x)$. $\gamma_{old}(x)\beta_{old}(x)$ is determined by the flow profile at the previous iteration. Selection of the flow profile based on the expression (17) in the area where the background flow is described in the form of particles, works well in the case of non-relativistic motion, when calculating in this area $F_{px}^{th}(x)$ based on the distribution function of background particles. In the relativistic case, as shown in [77], the momentum flow in the iterative process converges to a greater value than in the far upstream, when using (17) near the shock wave front. Hence, following [77] we smooth out the flow profile $u(x)$ near the shock front. The new speed profile is then averaged with the old one. Below we describe the equations that are used to calculate the values included in $F_{px}^{th}(x)$ and $F_{px}^{w}(x)$.

The turbulence energy spectrum defining $F_{px}^{w}(x)$ is found based on the solution of the following equation:

$$\gamma(x)u(x)\frac{\partial W(x,k)}{\partial x} + \frac{3}{2}\frac{\partial(\gamma(x)u(x))}{\partial x}W(x,k) + \frac{\partial\Pi(x,k)}{\partial k} =$$
$$= G(x,k)W(x,k) - \mathcal{L}(x,k), \quad (18)$$

where $\mathcal{L}(x,k)$ is the turbulent energy dissipation. The spectral flux of the turbulent energy (turbulent cascade) is:

$$\Pi(x,k) = -\frac{C^*}{\sqrt{\rho(x)}}k^{\frac{11}{2}}W(x,k)^{\frac{1}{2}}\frac{\partial}{\partial k}\left(\frac{W(x,k)}{k^2}\right), \quad (19)$$

where $\rho(x)$ is the background plasma density,

$$C^* = \frac{3}{11}C_{Kolm}^{-\frac{3}{2}}, \quad (20)$$

where C_{Kolm} is the Kolmogorov's constant, which is here taken equal to $C_{Kolm} = 1.6$. The expression (19) is derived from [78]. $G(x,k)$ is the growth rate of the turbulent energy in the background plasma rest frame due to plasma instabilities. Similar to [71], here we considered the growth rate of current instabilities—the Bell's non-resonant [79] and resonant instability. The accelerated particle current, which determines the growth rates, is calculated in the rest frame of the scattering centers, after propagation of particles.

Differentiating the Equations (12) and (13) by x, using (11), and excluding from the equations the term proportional to $n(x)$ we get the following relation:

$$\beta(x)\left(\frac{d\Phi_{th}(x)}{dx} + \frac{d\Phi_w(x)}{dx}\right) + \frac{dF_{en}^{cr}(x)}{dx} + \frac{1}{\gamma(x)}\frac{d(\gamma(x)\beta(x))}{dx}(\Phi_{th}(x) + \Phi_w(x)) =$$
$$= \beta(x)\left(\frac{dP_{th}(x)}{dx} + \frac{dP_w(x)}{dx} + \frac{dF_{px}^{cr}(x)}{dx}\right). \quad (21)$$

If we assume that each of the components of the system is affected only by the change of $(\gamma(x)\beta(x))$, that is, the change in energy and momentum flux occurs adiabatically due to the change of the flow velocity, we can divide the Equation (21) into separate adiabatic equations for the components:

$$\beta(x)\frac{d\Phi_{th}(x)}{dx} + \frac{1}{\gamma(x)}\frac{d(\gamma(x)\beta(x))}{dx}\Phi_{th}(x) = \beta(x)\frac{dP_{th}(x)}{dx}, \quad (22)$$

$$\beta(x)\frac{d\Phi_w(x)}{dx} + \frac{1}{\gamma(x)}\frac{d(\gamma(x)\beta(x))}{dx}\Phi_w(x) = \beta(x)\frac{dP_w(x)}{dx}, \quad (23)$$

$$\frac{dF_{en}^{cr}(x)}{dx} = \beta(x)\frac{dF_{px}^{cr}(x)}{dx}. \quad (24)$$

The Equation (23) in this case is equivalent to the Equation (18) integrated by k in the absence of MFA and dissipation. In the presence of MFA and dissipation after integration by k, the Equation (18) will take the form (similar to (23)):

$$\beta(x)\frac{d\Phi_w(x)}{dx} + \frac{1}{\gamma(x)}\frac{d(\gamma(x)\beta(x))}{dx}\Phi_w(x) = \beta(x)\frac{dP_w(x)}{dx} +$$
$$+ \frac{1}{c\gamma(x)}\int_{(k)}G(x,k)W(x,k)dk - \frac{1}{c\gamma(x)}L(x), \quad (25)$$

where the dissipation term is $L(x) = \int_{(k)}\mathcal{L}(x,k)dk$. Here we use the following expression for the turbulent energy dissipation:

$$\mathcal{L}(x,k) = v_\Gamma(x)\frac{k^2}{k_{th}}W(x,k), \quad (26)$$

$$v_\Gamma(x) = \frac{B_{ls}(x,k_{th})}{\sqrt{4\pi\rho(x)}}, \quad (27)$$

$$\frac{k_{th}c\sqrt{k_bT_{th}(x)}}{eB_{ls}(x,k_{th})} = 1, \quad (28)$$

where k_b is the Boltzmann's constant, $T_{th}(x)$ is the background plasma temperature ($P_{th}(x) = n(x)k_bT_{th}(x)$). In this model, it is assumed that in the downstream $\Pi(x,k) = 0$ and $\mathcal{L}(x,k) = 0$.

From the comparison of Equations (23) and (25) one can see that there are two additional terms in the right side of (25). To fulfill the total energy conservation Equation (21), these terms must be compensated by introducing additional terms into the equations for the remaining components of the system. We assume that the dissipation of the turbulent energy flow leads to heating of the background plasma. An increase in the energy flow turbulence due to MFA can be compensated by scattering accelerated particles in the frame of scattering centers moving with the speed $u(x) + v_{scat}(x)$. Then the equation for the energy flux of the accelerated particles will take the form:

$$c\frac{dF^{cr}_{en}(x)}{dx} = [u(x) + v_{scat}(x)]\frac{dF^{cr}_{px}(x)}{dx}. \tag{29}$$

Within the considered geometry we introduce $v_{scat}(x)$ as follows:

$$v_{scat}(x) = \min\left(v_{ampl}(x), v_{A,eff}(x)\right), \tag{30}$$

$$v_{ampl}(x) = -\frac{\int_{(k)} G(x,k)W(x,k)dk}{\gamma(x)\frac{dF^{cr}_{px}(x)}{dx}}, \tag{31}$$

$$v_{A,eff}(x) = -\frac{B_{eff}}{\sqrt{4\pi\rho}}, \tag{32}$$

$$B_{eff}(x) = \sqrt{4\pi\int_{(k)} W(x,k)dk + B_0^2}, \tag{33}$$

up to the injection point of the background particles in the upstream. After the injection point and before the shock front $v_{scat}(x) = v_{ampl}(x)$. In the downstream $v_{scat}(x) = 0$. In this model, we assume that part of the energy flux taken from the accelerated particle flux in the region where $\left|v_{ampl}(x)\right| < \left|v_{A,eff}(x)\right|$, goes to heat the background plasma. Thus, the equation determining the change in the energy flux of the background plasma has the form:

$$\beta(x)\frac{d\Phi_{th}(x)}{dx} + \frac{1}{\gamma(x)}\frac{d(\gamma(x)\beta(x))}{dx}\Phi_{th}(x) = \beta(x)\frac{dP_{th}(x)}{dx} + \frac{v_{diss}(x)}{c}\frac{dF^{cr}_{px}(x)}{dx} + $$
$$+ \frac{1}{c\gamma(x)}L(x), \tag{34}$$

where $v_{diss}(x) = \left|v_{A,eff}(x)\right| - \left|v_{ampl}(x)\right|$ if $\left|v_{ampl}(x)\right| < \left|v_{A,eff}(x)\right|$, in the opposite limit $v_{diss}(x) = 0$. After the injection point of background particles and in the downstream $v_{diss}(x) = 0$.

Substituting the expression for $\Phi_{th}(x)$ into the Equation (34), we obtain the following equation for the background plasma pressure:

$$\gamma(x)\beta(x)\frac{d}{dx}\frac{P_{th}(x)}{\Gamma_{th}(x) - 1} + \frac{d(\gamma(x)\beta(x))}{dx}\frac{\Gamma_{th}(x)P_{th}(x)}{\Gamma_{th}(x) - 1} = $$
$$= \gamma(x)\frac{v_{diss}(x)}{c}\frac{dF^{cr}_{px}(x)}{dx} + \frac{1}{c}L(x). \tag{35}$$

The solution of the Equation (35) defines $F^{th}_{px}(x)$ in the expression (17).

To solve the Equation (18), one needs to set the profile $W(x,k)$ on the FEB $x = x_{FEB}$. We define $W(x_{FEB}, k) \sim k^{-\frac{5}{3}}$—the Kolmogorov's spectrum at the FEB with the energy-carrying scale L_{en}. $W(x_{FEB}, k)$ is normalized as follows:

$$\int_{(k)} W(x_{FEB}, k)dk = \frac{B_0^2}{4\pi}. \tag{36}$$

Based on the developed Monte Carlo model, we have performed calculations of proton acceleration by mildly relativistic shocks, which are thought to come out during explosions of some SN kinds. In this case, the shocks often propagate into the wind of the pre-supernova star. The number density of particles (protons) in the stellar wind can be estimated as:

$$n_w(r) = \frac{\dot{M}}{4\pi v_w m_p r^2},$$ (37)

where $\dot{M}$ is the mass-loss rate of the pre-supernova, v_w is the speed of the stellar wind, r is the distance from the center of the star. For $\dot{M} = 10^{-4} M_\odot \mathrm{yr}^{-1}$, $v_w = 1000$ km s^{-1} and $r = 10^{15}$ cm: $n_w \approx 3 \times 10^6$ cm^{-3}. This estimate is used to estimated the number density n_0 in the far upstream. $L_{en} = 3 \times 10^{16}$ cm in all the calculations. We assume that the free escape boundary is at 0.2 of the current radius of the shock. The calculation parameters and their results are presented in Table 1.

$$\epsilon'_B = \frac{B^2_{eff,2}}{8\pi\gamma_0^2 u_0^2 m_p n_0}.$$ (38)

The spatial coordinate x in the figures is measured in $r_{g0} = m_p c u_0 / e B_0$. For the model A1: $x_{FEB} = -4.782 \times 10^6 r_{g0}$.

As can be seen from Figure 7, with an increase of the shock speed, the maximum momentum of accelerated particles increases. The value ϵ'_B also increases with an increase in the velocity of the shock wave in the range of simulated shock velocities 0.1–$0.5\,c$ and it decreases slightly for the shock velocity of $0.7c$ (see Table 1) where the transition to relativistic shock regime occurs. Particle-in-cell simulations of electron and proton spectra in the trans-relativistic regime were discussed in [61,62].

Table 1. The grid of Monte Carlo calculation parameters.

Model	u_0	n_0, cm^{-3}	x_{FEB}, cm	B_0, G	$B_{eff,2}$, G	ϵ'_B
A1	$0.1 \times c$	5×10^5	-5×10^{14}	3×10^{-3}	2.21	2.6×10^{-2}
A1n	$0.1 \times c$	1×10^5	-5×10^{14}	3×10^{-3}	1.00	2.7×10^{-2}
A2	$0.1 \times c$	2.5×10^3	-5×10^{15}	3×10^{-4}	1.57×10^{-1}	2.6×10^{-2}
A2b	$0.1 \times c$	2.5×10^3	-5×10^{15}	3×10^{-3}	1.55×10^{-1}	2.5×10^{-2}
A3	$0.1 \times c$	25	-5×10^{16}	3×10^{-5}	1.53×10^{-2}	2.5×10^{-2}
B1	$0.3 \times c$	5×10^5	-5×10^{14}	3×10^{-3}	8.57	3.9×10^{-2}
C1	$0.5 \times c$	5×10^5	-5×10^{14}	3×10^{-3}	16.7	4.4×10^{-2}
D1	$0.7 \times c$	5×10^5	-5×10^{14}	3×10^{-3}	26.3	3.8×10^{-2}

Figure 8 shows particle distribution functions used for evaluations of the models A1, A2 and A3, with varying values of x_{FEB} and the magnetic field at the FEB. Note, that with an increase of the FEB distance, the strength of the magnetic field on the FEB decreases proportionally. It can be seen that in these configurations the maximum momenta of the accelerated particles are almost the same. The current of the highest energy particles near the FEB amplify turbulent fluctuations due to the small-scale Bell's instability on scales much smaller than their own gyroradius. In Figure 9, the spectral energy density of the turbulence is shown at various upstream points for the model A1. Furthermore, it takes time to significantly amplify the magnetic field, and thus, the amplitude of the amplified turbulent field in a significant part of the upstream differs slightly from its value at the FEB (see Figure 10). The gyroradius of the highest energy particles near the FEB turns out to be significantly larger than the scale of the amplified fluctuations, which leads to a small contribution of the expression (8) to the free path (5). Thus, the free path (5) of the highest energy particles in a significant part of the upstream is determined by the contribution (6) with the initial turbulent field. That is, the free path (5) of the highest energy particles in a significant part of the upstream is $\lambda(x, p) \sim p/B_0$ according to the normalization (36). The non-relativistic theory of DSA gives a simple estimate of the maximum momentum p_{max} in the case of an upstream path independent of the coordinate: $\lambda(p_{max})c/3u_0 \approx x_{FEB}$. Accordingly, our model has a good estimate: $p_{max} \sim B_0 x_{FEB}$.

In Figure 11, the independence of the maximum particle momentum from the number density n_0 in the far upstream is illustrated. In Figure 12, it is shown, that with an increase of the magnetic field B_0 with the other parameters kept constant, the maximum particle momentum increases.

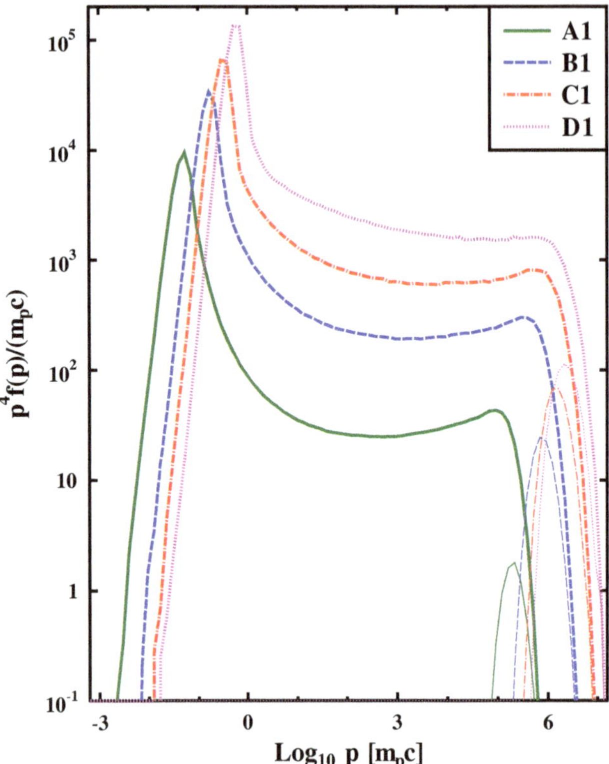

Figure 7. Particle distribution function in the shock rest frame, thick curves correspond to the point $x = 0$ (the front shock), thin curves correspond to a point $x = x_{FEB}$. The correspondence of the certain model Table 1 is reflected in the legend.

We have made the following estimate of the acceleration time τ_a of the particles to their maximum momentum. In the estimation, we assume that the mean free path of the highest energy particles with momentum p_{max} in most of the upstream can be estimated by the gyroradius in the magnetic field B_0. Thus, taking into account that the magnetic field in the downstream is much stronger and, accordingly, the diffusion coefficient of particles is much smaller than in the upstream $\tau_a \approx 3D(p_{max})/u_0^2$, where $D(p_{max}) = \lambda(p_{max})c/3 \approx p_{max}c^2/3eB_0$ is the diffusion coefficient of the highest energy particles in the upstream. Accordingly, the estimate of the particle acceleration time to the momentum p_{max} has the form:

$$\tau_a \approx 3.5 \times 10^5 \left(\frac{u_0}{0.1c}\right)^{-2}\left(\frac{B_0}{3 \times 10^{-3}\mathrm{G}}\right)^{-1}\left(\frac{p_{max}}{10^5 m_p c}\right)\mathrm{s}. \tag{39}$$

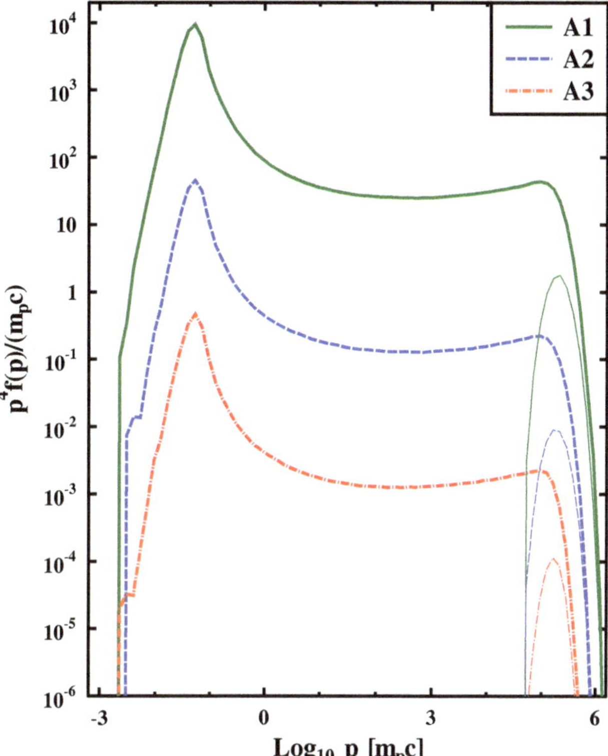

Figure 8. The particle distribution function in the shock rest frame, the thick curves correspond to the point $x = 0$ (the front shock), the thin curves correspond to the point $x = x_{FEB}$. The correspondence to Table 1 is reflected in the legend.

In the considered models we assume that the current radius of the shock is $R_f = 5|x_{FEB}|$. The expansion time of the supernova remnant can be estimated as the ratio of the current shock radius to the shock velocity. Accordingly, the estimate for the expansion time has the form:

$$\tau_{exp} \approx 8.3 \cdot 10^5 \left(\frac{|x_{FEB}|}{5 \cdot 10^{14} \text{cm}} \right) \left(\frac{u_0}{0.1c} \right)^{-1} \text{s.} \tag{40}$$

It can be seen from the expressions (39), (40) that the expansion time is longer than the acceleration time for the model (see Table 1), the maximum momentum can be estimated from Figure 7, which confirms the self-consistency of the stationary model we have employed.

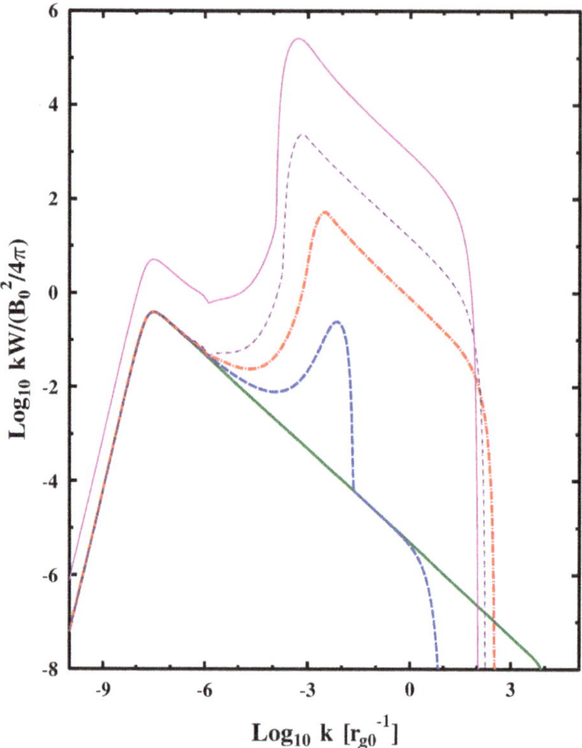

Figure 9. Spectral turbulent energy density for the model A1. The green solid curve corresponds to the point $x = x_{FEB}$. The blue dashed curve corresponds to the point $x \approx 0.75 \cdot x_{FEB}$. The red dash-point curve corresponds to the point $x \approx 0.5 \cdot x_{FEB}$. The thin purple dashed curve corresponds to the point $x \approx 0.1 \cdot x_{FEB}$. The thin magenta solid curve corresponds to the point $x = 0$ (the shock front).

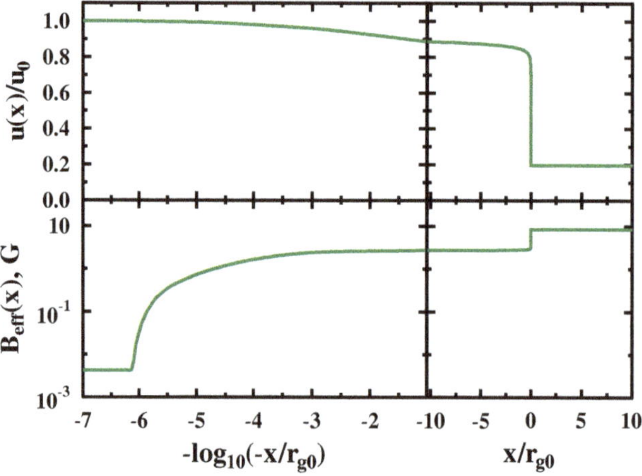

Figure 10. Background plasma velocity profile (upper panel) and magnetic field profile (bottom panel) for the model B1.

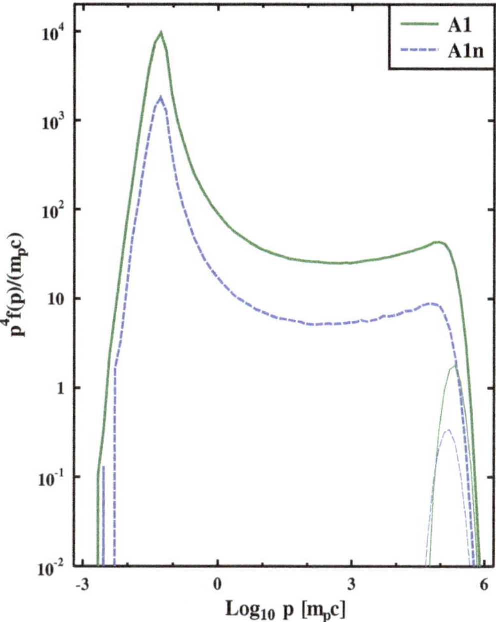

Figure 11. Particle distribution function in the shock rest frame, the thick curves correspond to the point $x = 0$ (the front shock), the thin curves correspond to the point $x = x_{FEB}$. The correspondence to Table 1 is reflected in the legend.

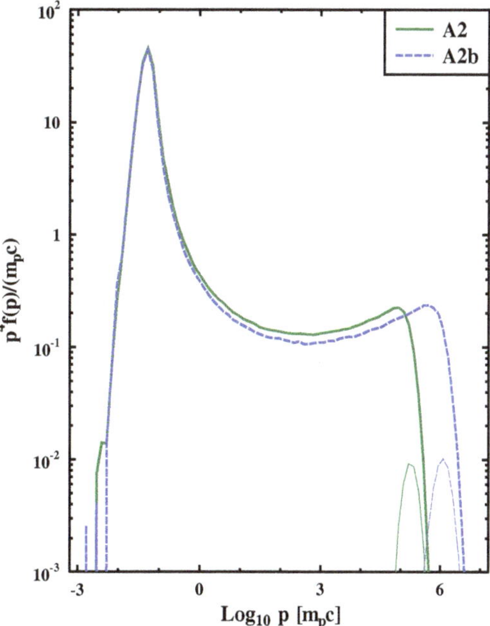

Figure 12. Particle distribution function in the shock rest frame, the thick curves correspond to the point $x = 0$ (the front shock), the thin curves correspond to the point $x = x_{FEB}$. The correspondence to Table 1 is reflected in the legend.

4. Discussion and Conclusions

In order to understand the processes of non-thermal particle acceleration at work in fast and energetic transients we have carried out and presented above a microscopic gyro-scale modeling of the collisionless shock structure and non-thermal particle spectra, which show a strong dependence of both electron and proton acceleration efficiency on the shock obliquity.

In this context one should keep in mind that the particle-in-cell simulation assumed a fixed homogeneous magnetic field and cold flow of the incoming particles in the shock upstream at the boundary of the simulation box. On the other hand, fast non-thermal particles can penetrate into the upstream plasma flow. It is clear from the results of the Monte Carlo modeling presented in Section 3 (see, in particular, Figure 10) that the efficient cosmic ray acceleration at the shock may provide a strong magnetic field amplification and modification of the upstream plasma flow. The cosmic ray pressure gradient in the shock upstream decelerates the incoming plasma flow and the cosmic ray driven instabilities may highly amplify the fluctuating magnetic fields well outside the particle-in-cell simulation box which is limited by a few hundred of the proton gyroradii around the plasma sub-shock. Such simulations show the potential importance of the feedback effects which cannot be modeled with the microscopic particle-in-cell simulations so far. Therefore we have used here the particle-in-cell model to simulate the electron injection and acceleration in sub and mildly relativistic shocks of fast energetic transients together with Monte Carlo modeling. While the Monte Carlo technique describes the structure of upstream flow modified by the accelerated particles at scales well above the proton gyro-scales, it cannot be used to simulate the electron injection where particle-in-cell approach is required. Thus, we used the combination of the two techniques to model radio emission.

In our study, of electron spectrum convergence in quasi-parallel shock of $0.3c$ speed we observed some non-monotonic temporal behavior of the electron spectra as it is shown in Figure 5. This could be due to the development of Bell's instability as it was found earlier in [61] and which mediate the quasi-parallel shock structure at proton gyro-scales. The maximum electron energy achievable in our simulation was about 50 $m_e c^2$ and we extrapolate the spectral slope to larger energy. We extended the spectra from 50 to 500 $m_e c^2$ using the same spectral slope, it is enough to model radio spectra. This is an assumption in our case. However, the extrapolation seems to be justified by the presence of a power-law such as a spectral component with a similar slope right after the thermal peak in Crumley et al. simulation (see the electron spectra presented in Figure 7 [61] in the electron momentum range between 0.1 and 30 $m_p c$). A similar spectral component is also apparent right after the peak of the spectrum in our steady Monte Carlo simulations (see, e.g., our Figure 8).

We applied 2D kinetic PIC simulations for sub-relativistic flows with velocity of 0.3 c to model the spectra of electrons and protons. The dynamical range of the full kinetic PIC simulations is limited but the electron spectrum extrapolated with account for the results of Monte Carlo simulations is enough to model the observed radio emission in the fast transients where such fast outflows were found. In the Monte Carlo model the particle-in-cell simulated domain where the electron injection occurs correspond to the sub-shock structure which is apparent at $x = 0$ in Figure 10. The plasma compression ratio at the sub-shock of the cosmic ray modified shock is about 3 and the spectra of particles accelerated at the sub-shock in the low energy regime (c.f. the proton spectral slope right after the peak in Figure 8) are consistent with the ones obtained within the particle-in-cell simulation dynamical range. The gyroradius of electrons with Lorentz-factor of 1000 in our simulations is about the gyroradius of protons with Lorentz-factor of 10. The spectral index of proton distribution simulated with Monte Carlo model at Lorentz-factor of 10 is consistent with the PIC electron spectrum extrapolated to $\sim 500\ m_e c^2$.

This allows to model the non-thermal emission and to understand the origin of the synchrotron radio emission of FBOTs. Additionally, one may assume an explanation of their hard X-ray spectrum using the dependence of the electron power law distribution index on the shock obliquity. As can be seen in Figure 1 it leads to the presence of relativistic electron populations of comparable intensities within a range of spectral indices along the curved shock surface expected in a wide angle outflow. In this case the synchrotron radio emission could have rather steep spectral indices, while the X-ray component produced by the inverse Compton scattering of the intense optical radiation by the relativistic electrons with the harder spectral index may appear to be flatter.

The simplified macroscopic Monte Carlo kinetic model was used in Section 3 to estimate the maximum energies of the accelerated cosmic ray nuclei. The model accounted for both the nonlinear feedback effects of the flow modification by the cosmic ray pressure gradient and magnetic turbulence amplification by cosmic ray driven instabilities in the upstream of a plain mildly relativistic shock. The model demonstrated a possibility of PeV regime proton acceleration in the shocks driven by mildly relativistic outflows of fast energetic transients on a few weeks timescale.

Transient objects of different types such as gamma-ray bursts and tidal disruption events [80–82] are widely discussed as possible sources of the observed high energy neutrinos. The search of PeV gamma-ray sources have recently revealed a number of potential galactic pevatron sources (see, e.g., [83–86]). The analysis presented in [59] allowed the authors to conclude that the extended galactic supernova remnants are not likely to be pevatrons, while some other types of galactic sources associated with the SN events in the compact clusters of young massive stars [87] and gamma-ray binaries with compact relativistic stars [88,89] can produce PeV photons and neutrinos. While PeV photons undergo strong attenuation and can be detected mostly from the galactic sources, high energy neutrinos from extragalactic energetic transients can be detected with currently operating and future neutrino observatories simultaneously with LSST optical transients detection.

Author Contributions: All of the authors have equally contributed into the paper. A.B. constructed the model concept. PiC simulations were performed by V.R. Monte Carlo simulations were performed by S.O. All authors have read and agreed to the published version of the manuscript.

Funding: This research received funding from Russian Science Foundation grant 21-72-20020.

Institutional Review Board Statement: The study did not involve humans or animals.

Informed Consent Statement: The study did not involve humans.

Acknowledgments: We are honored to dedicate this paper to the memory of professor Yury Nikolaevich Gnedin (1935–2018). His great scientific enthusiasm, competence and the unfailing goodwill for many years were extremely important for us. We are grateful to the referees for constructive comments and suggestions. The authors were supported by the RSF grant 21-72-20020. Some of the modeling was performed at the Joint Supercomputer Center JSCC RAS and some—at the "Tornado" subsystem of the Peter the Great Saint-Petersburg Polytechnic University Supercomputing Center.

Conflicts of Interest: The authors declare no conflict of interest.

References

1. Kulkarni, S.R. Towards An Integrated Optical Transient Utility. *arXiv* **2020**, arXiv:2004.03511.
2. Abell, P.A.; Allison, J.; Anderson, S.F.; Andrew, J.R.; Angel, J.R.P.; Armus, L.; Arnett, D.; Asztalos, S.; Axelrod, T.S.; Bailey, S.; et al. LSST Science Book, Version 2.0. *arXiv* **2009**, arXiv:0912.0201.
3. Gehrels, N.; Chincarini, G.; Giommi, P.e.; Mason, K.; Nousek, J.; Wells, A.; White, N.; Barthelmy, S.; Burrows, D.; Cominsky, L.; et al. The Swift Gamma-Ray Burst Mission. *Astrophys. J.* **2004**, *611*, 1005–1020. [CrossRef]
4. Sunyaev, R.; Arefiev, V.; Babyshkin, V.; Bogomolov, A.; Borisov, K.; Buntov, M.; Brunner, H.; Burenin, R.; Churazov, E.; Coutinho, D.; et al. The SRG X-ray orbital observatory, its telescopes and first scientific results. *arXiv* **2021**, arXiv:2104.13267.
5. Greiner, J.; Mannheim, K.; Aharonian, F.; Ajello, M.; Balasz, L.G.; Barbiellini, G.; Bellazzini, R.; Bishop, S.; Bisnovatij-Kogan, G.S.; Boggs, S.; et al. GRIPS—Gamma-Ray Imaging, Polarimetry and Spectroscopy. *Exp. Astron.* **2012**, *34*, 551–582. [CrossRef]
6. De Angelis, A.; Tatischeff, V.; Argan, A.; Brandt, S.; Bulgarelli, A.; Bykov, A.; Costantini, E.; Silva. R.C.d.; Grenier, I.A.; Hanlon, L.; et al. Gamma-ray astrophysics in the MeV range. *Exp. Astron.* **2021**, *51*, 1225–1254. [CrossRef]
7. Mereghetti, S.; Balman, S.; Caballero-Garcia, M.; Del Santo, M.; Doroshenko, V.; Erkut, M.; Hanlon, L.; Hoeflich, P.; Markowitz, A.; Osborne, J.; et al. Time Domain Astronomy with the THESEUS Satellite. *arXiv* **2021**, arXiv:2104.09533.

8.   Drout, M.R.; Chornock, R.; Soderberg, A.M.; Sanders, N.E.; McKinnon, R.; Rest, A.; Foley, R.J.; Milisavljevic, D.; Margutti, R.; Berger, E.; et al. Rapidly Evolving and Luminous Transients from Pan-STARRS1. *Astrophys. J.* **2014**, *794*, 23. [CrossRef]

9.   Margutti, R.; Metzger, B.; Chornock, R.; Vurm, I.; Roth, N.; Grefenstette, B.; Savchenko, V.; Cartier, R.; Steiner, J.; Terreran, G.; et al. An Embedded X-ray Source Shines through the Aspherical AT 2018cow: Revealing the Inner Workings of the Most Luminous Fast-evolving Optical Transients. *Astrophys. J.* **2019**, *872*, 18. [CrossRef]

10.   Coppejans, D.L.; Margutti, R.; Terreran, G.; Nayana, A.; Coughlin, E.; Laskar, T.; Alexander, K.; Bietenholz, M.; Caprioli, D.; Chandra, P.; et al. A Mildly Relativistic Outflow from the Energetic, Fast-rising Blue Optical Transient CSS161010 in a Dwarf Galaxy. *Astrophys. J. Lett.* **2020**, *895*, L23. [CrossRef]

11.   Ho, A.Y.; Perley, D.A.; Kulkarni, S.; Dong, D.Z.; De, K.; Chandra, P.; Andreoni, I.; Bellm, E.C.; Burdge, K.B.; Coughlin, M.; et al. The Koala: A Fast Blue Optical Transient with Luminous Radio Emission from a Starburst Dwarf Galaxy at z = 0.27. *Astrophys. J.* **2020**, *895*, 49. [CrossRef]

12.   Nicholl, M.; Blanchard, P.K.; Berger, E.; Chornock, R.; Margutti, R.; Gomez, S.; Lunnan, R.; Miller, A.A.; Fong, W.-f.; Terreran, G.; et al. An extremely energetic supernova from a very massive star in a dense medium. *Nat. Astron.* **2020**, *4*, 893–899. [CrossRef]

13.   Weiler, K.W; Panagia N.; Montes, M.J.; Sramek, R.A. Radio Emission from Supernovae and Gamma-Ray Bursters. *Ann. Rev. Astron. Astrophys.* **2002**, *40*, 387–438. [CrossRef]

14.   Soderberg, A.M.; Kulkarni, S.R.; Nakar, E.; Berger, E.; Cameron, P.; Fox, D.; Frail, D.; Gal-Yam, A.; Sari, R.; Cenko, S.; et al. Relativistic ejecta from X-ray flash XRF 060218 and the rate of cosmic explosions. *Nature* **2006**, *442*, 1014–1017. [CrossRef]

15.   Irwin, C.M.; Chevalier, R.A. Jet or shock breakout? The low-luminosity GRB 060218. *Mon. Not. R. Astron. Soc.* **2016**, *460*, 1680–1704. [CrossRef]

16.   Nayana, A.J.; Chandra P. uGMRT Observations of a Fast and Blue Optical Transient—AT 2018cow. *Astrophys. J. Lett.* **2021**, *912*, L9.

17.   Ho, A.Y.; Margalit, B.; Bremer, M.; Perley, D.A.; Yao, Y.; Dobie, D.; Kaplan, D.L.; O'Brien, A.; Petitpas, G.; Zic, A. Luminous Millimeter, Radio, and X-ray Emission from ZTF20acigmel (AT2020xnd). *arXiv* **2021**, arXiv:2110.05490.

18.   Bright, J.S.; Margutti, R.; Matthews, D.; Brethauer, D.; Coppejans, D.; Wieringa, M.H.; Metzger, B.D.; DeMarchi, L.; Laskar, T.; Romero, C.; et al. Radio and X-ray observations of the luminous Fast Blue Optical Transient AT2020xnd. *arXiv* **2021**, arXiv:2110.05514.

19.   Kremer, K.; Lu, W.; Piro, A.L.; Chatterjee, S.; Rasio, F.A.; Claire, S.Y. Fast Optical Transients from Stellar-mass Black Hole Tidal Disruption Events in Young Star Clusters. *Astrophys. J.* **2021**, *911*, 104. [CrossRef]

20.   Metzger, B.D. Kilonovae. *Living Rev. Relat.* **2019**, *23*, 1.

21.   Moriya, T.J.; Chen K.J.; Nakajima, K.; Tominaga, N.; Blinnikov S.I. Observational properties of a general relativistic instability supernova from a primordial supermassive star. *Mon. Not. R. Astron. Soc.* **2021**, *503*, 1206–1213. [CrossRef]

22.   Soderberg, A.; Chakraborti, S.; Pignata, G.; Chevalier, R.; Chandra, P.; Ray, A.; Wieringa, M.; Copete, A.; Chaplin, V.; Connaughton, V.; et al. A relativistic type Ibc supernova without a detected γ-ray burst. *Nature* **2010**, *463*, 513–515. [CrossRef]

23.   Corsi, A.; Gal-Yam, A.; Kulkarni, S.; Frail, D.; Mazzali, P.; Cenko, S.; Kasliwal, M.; Cao, Y.; Horesh, A.; Palliyaguru, N.; et al. Radio Observations of a Sample of Broad-line Type IC Supernovae Discovered by PTF/IPTF: A Search for Relativistic Explosions. *Astrophys. J.* **2016**, *830*, 42. [CrossRef]

24.   Imshennik, V.S.; Nadezhin, D.K. Supernova 1987A in the Large Magellanic Cloud: Observations and theory. *Sov. Sci. Rev. Sect. E* **1989**, *8*, 1–147.

25.   Matzner, C.D.; McKee, C.F. The Expulsion of Stellar Envelopes in Core-Collapse Supernovae. *Astrophys. J.* **1999**, *510*, 379–403. [CrossRef]

26.   Tan, J.C.; Matzner, C.D.; McKee, C.F. Trans-Relativistic Blast Waves in Supernovae as Gamma-Ray Burst Progenitors. *Astrophys. J.* **2001**, *551*, 946–972. [CrossRef]

27.   MacFadyen, A.I.; Woosley, S.E.; Heger, A. Supernovae, Jets, and Collapsars. *Astrophys. J.* **2001**, *550*, 410–425. [CrossRef]

28.   Lazzati, D.; Morsony, B.J.; Blackwell, C.H.; Begelman, M.C. Unifying the Zoo of Jet-driven Stellar Explosions. *Astrophys. J.* **2012**, *750*, 68. [CrossRef]

29.   Margutti, R.; Milisavljevic, D.; Soderberg, A.; Guidorzi, C.; Morsony, B.; Sanders, N.; Chakraborti, S.; Ray, A.; Kamble, A.; Drout, M.; et al. Relativistic Supernovae have Shorter-lived Central Engines or More Extended Progenitors: The Case of SN 2012ap. *Astrophys. J.* **2014**, *797*, 107. [CrossRef]

30.   Corsi, A.; Lazzati, D. Gamma-ray burst jets in supernovae. *New Astron. Rev.* **2021**, *92*, 101614. [CrossRef]

31.   Suzuki, A.; Maeda, K. Supernova ejecta with a relativistic wind from a central compact object: A unified picture for extraordinary supernovae. *Mon. Not. R. Astron. Soc.* **2017**, *466*, 2633–2657. [CrossRef]

32.   Suzuki, A.; Maeda, K. Three-dimensional Hydrodynamic Simulations of Supernova Ejecta with a Central Energy Source. *Astrophys. J.* **2019**, *880*, 150. [CrossRef]

33.   Suzuki, A.; Maeda, K. Two-dimensional Radiation-hydrodynamic Simulations of Supernova Ejecta with a Central Power Source. *Astrophys. J.* **2021**, *908*, 217. [CrossRef]

34.   Grassberg, E.K.; Imshennik, V.S.; Nadyozhin, D.K. On the Theory of the Light Curves of Supernovae. *Astroph. Space Sci.* **1971**, *10*, 28–51. [CrossRef]

35.   Chevalier, R.A.; Irwin C.M. Shock Breakout in Dense Mass Loss: Luminous Supernovae. *Astrophys. J. Lett.* **2011**, *729*, L6. [CrossRef]

36. Tolstov, A.G.;. Blinnikov, S.I; Nadyozhin, D.K. Coupling of matter and radiation at supernova shock breakout. *Mon. Not. R. Astron. Soc.* **2013**, *429*, 3181–3199. [CrossRef]

37. Chevalier, R.A. Synchrotron Self-Absorption in Radio Supernovae. *Astrophys. J.* **1998**, *499*, 810–819. [CrossRef]

38. Waxman, E.; Katz, B. Shock Breakout Theory. In *Handbook of Supernovae*; Alsabti, A.W., Murdin, P., Eds.; Springer: New York, NY, USA, 2017; p. 967.

39. Budnik R.; Katz, B.; Sagiv A.; Waxman, E. Relativistic Radiation Mediated Shocks. *Astrophys. J.* **2010**, *725*, 63–90. [CrossRef]

40. Levinson, A. Plasma kinetic effects in relativistic radiation-mediated shocks. *Phys. Rev. E* **2020**, *102*, 063210. [CrossRef] [PubMed]

41. Granot, A.; Nakar, E.; Levinson, A. Relativistic shock breakout from a stellar wind. *Mon. Not. R. Astron. Soc.* **2018**, *476*, 5453–5463. [CrossRef]

42. Ito, H.; Levinson, A.; Nakar, E. Monte Carlo simulations of fast Newtonian and mildly relativistic shock breakout from a stellar wind. *Mon. Not. R. Astron. Soc.* **2020**, *499*, 4961–4971. [CrossRef]

43. Marcowith, A.; Bret, A.; Bykov, A.; Dieckman, M.E.; Drury, L.O.C.; Lembège, B.; Lemoine, M.; Morlino, G.; Murphy, G.; Pelletier, G.; et al. The microphysics of collisionless shock waves. *Rep. Prog. Phys.* **2016**, *79*, 046901. [CrossRef] [PubMed]

44. Sironi, L.; Spitkovsky, A. Particle Acceleration in Relativistic Magnetized Collisionless Pair Shocks: Dependence of Shock Acceleration on Magnetic Obliquity. *Astrophys. J.* **2009**, *698*, 1523–1549. [CrossRef]

45. Pelletier, G.; Bykov, A.; Ellison, D.; Lemoine, M. Towards Understanding the Physics of Collisionless Relativistic Shocks. Relativistic Collisionless Shocks. *Space Sci. Rev.* **2017**, *207*, 319–360. [CrossRef]

46. Bell, A.R.; Schure, K.M.; Reville, B.; Giacinti G. Cosmic-ray acceleration and escape from supernova remnants. *Mon. Not. R. Astron. Soc.* **2013**, *431*, 415–429. [CrossRef]

47. Bell, A.R.; Araudo, A.T.; Matthews, J.H.; Blundell, K.M. Cosmic-ray acceleration by relativistic shocks: Limits and estimates. *Mon. Not. R. Astron. Soc.* **2018**, *473*, 2364–2371. [CrossRef]

48. Lemoine, M.; Waxman E. Anisotropy vs. chemical composition at ultra-high energies. *J. Cosmol. Astropart. Phys.* **2009**, *2009*, 009. [CrossRef]

49. Bykov, A.; Gehrels, N.; Krawczynski, H.; Lemoine, M.; Pelletier, G.; Pohl, M. Particle Acceleration in Relativistic Outflows. *Space Sci. Rev.* **2012**, *173*, 309–339. [CrossRef]

50. Zhang, B.T.; Murase, K. Ultrahigh-energy cosmic-ray nuclei and neutrinos from engine-driven supernovae. *Phys. Rev. D* **2019**, *100*, 103004. [CrossRef]

51. Samuelsson, F.; Bégué, D.; Ryde, F.; Pe'er, A.; Murase, K. Constraining Low-luminosity Gamma-Ray Bursts as Ultra-high-energy Cosmic Ray Sources Using GRB 060218 as a Proxy. *Astrophys. J.* **2020**, *902*, 148. [CrossRef]

52. Bykov, A.M.; Ellison, D.C.; Marcowith, A.; Osipov, S.M. Cosmic Ray Production in Supernovae. *Space Sci. Rev.* **2018**, *214*, 41. [CrossRef]

53. Zirakashvili, V.N.; Ptuskin, V.S. Type IIn supernovae as sources of high energy astrophysical neutrinos. *Astropart. Phys.* **2016**, *78*, 28–34. [CrossRef]

54. Murase, K.; Franckowiak, A.; Maeda, K.; Margutti, R.; Beacom, J.F. High-energy Emission from Interacting Supernovae: New Constraints on Cosmic-Ray Acceleration in Dense Circumstellar Environments. *Astrophys. J.* **2019**, *874*, 80. [CrossRef]

55. Sironi, L.; Keshet, U.; Lemoine, M. Relativistic Shocks: Particle Acceleration and Magnetization. *Space Sci. Rev.* **2015**, *191*, 519–544. [CrossRef]

56. Ptuskin, V.;Zirakashvili, V.; Seo, E.-S. Spectra of Cosmic-Ray Protons and Helium Produced in Supernova Remnants. *Astrophys. J.* **2013**, *763*, 47. [CrossRef]

57. Blasi, P. The origin of galactic cosmic rays. *Astron. Astrophys. Rev.* **2013**, *21*, 70. [CrossRef]

58. Amato, E. The origin of galactic cosmic rays. *Int. J. Mod. Phys. D* **2014**, *23*, 1430013. [CrossRef]

59. Cristofari, P. The Hunt for Pevatrons: The Case of Supernova Remnants. *Universe* **2021**, *7*, 324. [CrossRef]

60. Park, J.; Caprioli, D.; Spitkovsky, A. Simultaneous Acceleration of Protons and Electrons at Nonrelativistic Quasiparallel Collisionless Shocks. *Phys. Rev. Lett.* **2015**, *114*, 085003. [CrossRef]

61. Crumley, P.; Caprioli, D.; Markoff, S.; Spitkovsky, A. Kinetic simulations of mildly relativistic shocks—I. Particle acceleration in high Mach number shocks. *Mon. Not. R. Astron. Soc.* **2019**, *485*, 5105–5119. [CrossRef]

62. Romansky, V.I.; Bykov, A.M.; Osipov, S.M. Electron and ion acceleration by relativistic shocks: particle-in-cell simulations. *J. Phys. Conf. Ser.* **2018**, *1038*, 012022. [CrossRef]

63. Derouillat, J.; Beck, A.; Pérez, F.; Vinci, T.; Chiaramello, M.; Grassi, A.; Flé, M.; Bouchard, G.; Plotnikov, I.; Aunai, N.; et al. SMILEI : A collaborative, open-source, multi-purpose particle-in-cell code for plasma simulation. *Comput. Phys. Commun.* **2018**, *222*, 351–373. [CrossRef]

64. Esirkepov, T.Z. Exact charge conservation scheme for Particle-in-Cell simulation with an arbitrary form-factor. *Comput. Phys. Commun.* **2001**, *135*, 144–153. [CrossRef]

65. Guo, X.; Sironi, L.; Narayan, R. Non-thermal Electron Acceleration in Low Mach Number Collisionless Shocks. I. Particle Energy Spectra and Acceleration Mechanism *Astrophys. J.* **2014**, *794*, 153. [CrossRef]

66. Romansky, V.I.; Bykov, A.M.; Osipov, S.M. On electron acceleration by mildly-relativistic shocks: PIC simulations. *J. Phys. Conf. Ser.* **2021**, *2103*, 012009. [CrossRef]

67. Pacholczyk, A.G. *Radio Astrophysics. Nonthermal Processes in Galactic and Extragalactic Sources*; Freeman: San Francisco, CA, USA, 1970.

68. Chevalier, R.A.; Fransson, C. Thermal and Non-thermal Emission from Circumstellar Interaction. In *Handbook of Supernovae*; Alsabti, A.W., Murdin, P., Eds.; Springer: New York, NY, USA, 2017; p. 875.

69. Ghisellini, G. *Radiative Processes in High Energy Astrophysics*; Springer International Publishing: Cham, Switzerland, 2013; Volume 873.

70. Margalit, B.; Quataert, E. Thermal Electrons in Mildly Relativistic Synchrotron Blast Waves. *Astrophys. J. Lett.* **2021**, *923*, L14. [CrossRef]

71. Bykov, A.M.; Ellison, D.C.; Osipov, S.M.; Vladimirov, A.E. Magnetic Field Amplification in Nonlinear Diffusive Shock Acceleration Including Resonant and Non-resonant Cosmic-Ray Driven Instabilities. *Astrophys. J.* **2014**, *789*, 137. [CrossRef]

72. Jones, F.C.; Ellison, D.C. The plasma physics of shock acceleration. *Space Sci. Rev.* **1991**, *58*, 259–346. [CrossRef]

73. Ellison, D.C.; Baring, M.G.; Jones, F.C. Nonlinear Particle Acceleration in Oblique Shocks. *Astrophys. J.* **1996**, *473*, 1029. [CrossRef]

74. Vladimirov, A.E.; Bykov, A.M.; Ellison, D.C. Turbulence Dissipation and Particle Injection in Nonlinear Diffusive Shock Acceleration with Magnetic Field Amplification. *Astrophys. J.* **2008**, *688*, 1084–1101. [CrossRef]

75. Vladimirov, A.E.; Bykov, A.M.; Ellison, D.C. Spectra of Magnetic Fluctuations and Relativistic Particles Produced by a Nonresonant Wave Instability in Supernova Remnant Shocks. *Astrophys. J. Lett.* **2009**, *703*, L29–L32. [CrossRef]

76. Vladimirov, A. Modeling Magnetic Field Amplification in Nonlinear Diffusive Shock Acceleration. Ph.D. Thesis, North Carolina State University, Raleigh, NC, USA 2009.

77. Ellison, D.C.; Warren, D.C.; Bykov, A.M. Monte Carlo Simulations of Nonlinear Particle Acceleration in Parallel Trans-relativistic Shocks. *Astrophys. J.* **2013**, *776*, 46. [CrossRef]

78. Matthaeus, W.H.; Oughton, S.; Zhou, Y. Anisotropic magnetohydrodynamic spectral transfer in the diffusion approximation. *Phys. Rev. E* **2009**, *79*, 035401. [CrossRef]

79. Bell, A.R. Turbulent amplification of magnetic field and diffusive shock acceleration of cosmic rays. *Mon. Not. R. Astron. Soc.* **2004**, *353*, 550–558. [CrossRef]

80. Farrar, G.R.; Piran, T. Tidal disruption jets as the source of Ultra-High Energy Cosmic Rays. *arXiv* **2014**, arXiv:1411.0704.

81. Murase, K.; Waxman, E. Constraining high-energy cosmic neutrino sources: Implications and prospects. *Phys. Rev. D* **2016**, *94*, 103006. [CrossRef]

82. Murase, K.; Kimura, S.S.; Zhang, B.T.; Oikonomou, F.; Petropoulou, M. High-energy Neutrino and Gamma-Ray Emission from Tidal Disruption Events. *Astrophys. J.* **2020**, *902*, 108. [CrossRef]

83. Aharonian, F.; Yang, R.; de Oña Wilhelmi, E. Massive stars as major factories of Galactic cosmic rays. *Nat. Astron.* **2019**, *3*, 561–567. [CrossRef]

84. Amenomori, M.; Bao, Y.; Bi, X.; Chen, D.; Chen, T.; Chen, W.; Chen, X.; Chen, Y.; Cui, S.; Ding, L.; et al. First Detection of sub-PeV Diffuse Gamma Rays from the Galactic Disk: Evidence for Ubiquitous Galactic Cosmic Rays beyond PeV Energies. *Phys. Rev. Lett.* **2021**, *126*, 141101. [CrossRef] [PubMed]

85. Cao, Z.; Aharonian, F.; An, Q.; Bai, L.; Bai, Y.; Bao, Y.; Bastieri, D.; Bi, X.; Bi, Y.; Cai, H.; et al. Ultrahigh-energy photons up to 1.4 petaelectronvolts from 12 γ-ray Galactic sources. *Nature* **2021**, *594*, 33–36. [CrossRef]

86. Dzhappuev, D.; Afashokov, Y.Z.; Dzaparova, I.; Dzhatdoev, T.; Gorbacheva, E.; Karpikov, I.; Khadzhiev, M.; Klimenko, N.; Kudzhaev, A.; Kurenya, A.; et al. Observation of Photons above 300 TeV Associated with a High-energy Neutrino from the Cygnus Region. *Astrophys. J. Lett.* **2021**, *916*, L22. [CrossRef]

87. Bykov, A.M.; Ellison, D.C.; Gladilin, P.E.; Osipov S.M. Ultrahard spectra of PeV neutrinos from supernovae in compact star clusters. *Mon. Not. R. Astron. Soc.* **2015**, *453*, 113–121. [CrossRef]

88. Kimura, S.S.; Sudoh, T.; Kashiyama, K.; Kawanaka, N. Magnetically Arrested Disks in Quiescent Black Hole Binaries: Formation Scenario, Observable Signatures, and Potential PeVatrons. *Astrophys. J.* **2021**, *915*, 31. [CrossRef]

89. Bykov, A.M.; Petrov, A.E.; Kalyashova, M.E.; Troitsky, S.V. PeV photon and neutrino flares from galactic gamma-ray binaries. *Astrophys. J. Lett.* **2021**, *921*, L10. [CrossRef]

Review

Studies of Magnetic Chemically Peculiar Stars Using the 6-m Telescope at SAO RAS

Iosif Romanyuk

Special Astrophysical Observatory, 369167 Nizhnij Arkhyz, Russia; roman@sao.ru; Tel.: +7-87822-93359

Abstract: We present a survey of the most important results obtained in observations with the 6-m telescope in the studies of magnetic fields of chemically peculiar stars. It is shown that we have found more than 200 new magnetic chemically peculiar stars, which is more than 30% of their total known number. Observations of ultra-slow rotators (stars with rotation periods of years and decades) have shown that there are objects with strong fields among them, several kG in magnitude. In the association of young stars in Orion, it has been found that the occurrence and strength of magnetic fields of chemically peculiar stars decrease sharply with age in the interval from 2 to 10 Myr. These data indicate the fossil nature of magnetic fields of chemically peculiar stars. About 10 magnetic stars were found based on ultra-accurate photometry data obtained from the Kepler and TESS satellites. A new effective method of searching for magnetic stars was developed. In addition, the exact rotation periods make it possible to build reliable curves of the longitudinal field component variability with the phase of the star's rotation period, and hence to create its magnetic model. The survey is dedicated to the memory of Prof. Yuri Nikolaevich Gnedin.

Keywords: magnetic field; chemically peculiar stars; observation

Citation: Romanyuk, I. Studies of Magnetic Chemically Peculiar Stars Using the 6-m Telescope at SAO RAS. *Universe* **2021**, *7*, 465. https://doi.org/10.3390/universe7120465

Academic Editors: Nazar R. Ikhsanov, Galina L. Klimchitskaya, Vladimir M. Mostepanenko and Lorenzo Iorio

Received: 25 October 2021
Accepted: 28 November 2021
Published: 29 November 2021

Publisher's Note: MDPI stays neutral with regard to jurisdictional claims in published maps and institutional affiliations.

1. Yuri Gnedin, a Researcher of Stellar Magnetism

Prof. Yuri Nikolaevich Gnedin is a prominent Russian theoretical physicist. In the field of astrophysics, his studies of the polarized radiation transfer in the inhomogeneous environment, the study of objects with a super strong magnetic field, the polarization and magnetic field of quasars and active galaxy nuclei, and the search for axions and other unusual objects are widely known.

From 1993 to 2015, Yuri Gnedin was the Chairman of the Russian Telescope Time Allocation Committee (RTTAC)[1]. The Committee members develop the scientific program of our several largest telescopes. The RTTAC includes the Russian leading astronomers working in various fields.

The author of this paper worked together with Yuri Gnedin for many years as the scientific secretary of the above Committee. As a rule, RTTAC worked twice a year in SAO RAS and the arrival of the Committee members was an event for the astronomers of the observatory. Each time, open sessions were held, during which leading scientists talked about the latest science news, and telescope time applicants reported on their observation programs. Yuri Gnedin regularly made reports on the most important events in science that have taken place in recent years. His open mindedness allowed him to estimate the importance of the already completed works and the prospects for future research.

Research on stellar magnetism was one of the main areas of his activity. We will mention here some of the main studies: Gnedin et al. [1], Gnedin [2], Gnedin et al. [3–5], and others. Yuri Gnedin's scientific interest mainly was in the studies of objects with strong and very strong magnetic fields (including magnetic white dwarfs and polars).

However, he also solved more general problems associated with stellar magnetism. The classical monograph by Dolginov et al. [6] "Propagation and polarization of radiation in cosmic media" discusses the magnetism of stars of different types, and the more recent paper by Gnedin and Silant'ev [7] summarizes the main theories of the light polarization mechanism by cosmic particles.

His colleagues will be able to write about Yuri Gnedin's papers in the field of theoretical astrophysics in more detail. But he obviously made a great contribution to the development of the origin and evolution of stellar magnetic fields theory. Prof. Gnedin has always supported stellar magnetism studies with the 6-m telescope.

Thanks to the efforts of the staff of the Special Astrophysical Observatory of the Russian Academy of Sciences (SAO RAS), the 6-m telescope was equipped for measuring stellar magnetic fields which allowed our scientists to carry out observations of various objects and be among the leaders in this field of science.

2. Studies of Magnetic Stars with the 6-m Telescope

2.1. Some General Issues in the Stellar Magnetism Studies

The study of stellar magnetism is one of the most important areas of research in modern observational astrophysics. The world's largest telescopes are equipped with the appropriate instruments for this purpose. The magnetic field is responsible for a variety of outbursts, explosions, and other manifestations of non-stationarity in our Galaxy and beyond. Techniques for measuring magnetic fields in various objects are still imperfect and insufficiently developed. Often, even the world's largest telescopes are not powerful enough to obtain results of the required accuracy. Therefore, our knowledge of stellar magnetic fields is fairly superficial.

As a theoretician, Yuri Gnedin wondered: how can one find observation evidence of the efficiency of some or other mechanisms of the stellar magnetic field formation and evolution? This question is one of the key questions in modern observational astrophysics.

It is well known that the magnetic field of the Sun has a two-component structure: Hale [8] discovered a strong (2 kG) magnetic field in spots; and only after almost half a century, Babcock [9] found its weak (1–2 G) total surface magnetic field. The process of formation, evolution, and dissipation of the local magnetic field of the Sun spots is perfectly observed and explained in detail. Fields in the spots arise due to convective motions of the magnetized plasma in the Solar atmosphere. The duration of one cycle from occurrence to disappearance of a field takes days or weeks [10,11]. It must be assumed that Solar-like and other stars with convective atmospheres have powerful local magnetic fields formed in their atmospheres.

The nature of formation of the stellar global magnetic field is completely different. Obviously, these are not atmospheric phenomena; they are the result of overall processes occurring in the entire star. But if, by analogy with the Sun, they are weak (of about several G), then the accuracy of modern measurement methods is still too low for a detailed study of such a magnetic field. Therefore, to solve the problem of the large-scale magnetic field formation, it is necessary to find such objects, where global, but strong, fields are observed. Fortunately, such objects exist: chemically peculiar (CP) stars. Babcock [12] discovered strong (measured in kG) coherent magnetic fields in these objects, which were measured with a sufficiently good accuracy making it possible to analyze them.

The methods and techniques of measuring stellar magnetic fields stepped forward compared to those of the time, when Babcock worked. But Babcock's basic ideas are relevant to this day. To study the stellar magnetism, he designed a special differential circular polarization analyzer that split the light beam, entering the spectrograph slit, into two circularly polarized components [9]. Using the longitudinal Zeeman effect, Babcock synchronously obtained two spectra with the opposite orientation of circular polarization. The lines in these spectra are shifted relative to each other in proportion to the magnetic field strength and the Lande factor of the line. By measuring several tens of lines on a photographic plate, Babcock found their average shift and calculated the magnetic field longitudinal component B_e averaged over the entire visible surface. The fact that the spectra were registered in the same photographic plate made it possible to remove numerous small instrumental effects and sharply (by an order of magnitude) increased the accuracy of measurements.

Observing the polarized spectra in different phases of the rotation period, it was possible to build the B_e variations with time and build a field model (if it was strong and close to dipole). When working with photographic plates, it was impossible to achieve more. But with the introduction of CCDs, the accuracy of spectrum measurements has increased dramatically; the signal-to-noise ratio has increased typically by 1–2 orders. It became possible to obtain high-accuracy polarized profiles of spectral lines with details corresponding to the spots of chemical abundance. This made it possible to analyze the line profiles and develop methods of the Doppler and then magnetic imaging [13,14] of chemically peculiar stars.

The Mapping Methods

The mapping methods were first used in the study of chemically peculiar stars, but now they have become widespread in the study of other objects. More information on the history and methods of magnetic field measurements can be found in the series of papers by Romanyuk [15–17].

2.2. Magnetic CP-Stars

It turned out that about 10–15% of all A and B stars of the Main Sequence (MS) have anomalous chemical abundance. Anomalies consist in a significant increase of abundances in some of the selected chemical elements heavier than iron and in a weakening of calcium and light elements. But the most prominent increase is observed in the lines of chromium, strontium, and especially rare-earth elements. Chemically peculiar stars have strong stable magnetic fields that have not changed for decades. How the chemical abundance anomalies and the presence of a magnetic field are related is not completely clear. Do all CP stars have a global magnetic field?

Chemically peculiar stars are easily distinguished by their spectra. Antonia Maury first classified them in 1897 when compiling the Henry Drapper (HD) catalog by eye viewing of low-quality (by modern standards) photographic plates. Subsequently, astronomers, using different methods, found new chemically peculiar stars and the most complete catalog of such objects (which currently has over 8000 stars) was compiled by Renson and Manfroid [18].

Surface chemical anomalies are not uniformly located but in the form of spots or rings. This leads to the observed photometric and spectral variability of the star. Brightness and spectral features of such stars vary repeatedly with a period of axial rotation. Having a limited set of observational characteristics, it becomes possible to describe the star within the frame of the oblique rotator model [19].

The existence of CP stars with strong magnetic fields makes it possible, from an observational point of view, to set the task of studying the formation and evolution of the large-scale magnetic field of these and related objects. In fact, there are many options for the formation of magnetic fields in stars, but two of them stand out particularly [20]:

- the fossil theory, according to which no field is currently generated but we observe only the consequences of earlier processes (it does not matter how and when the field appeared) before the arrival of the star on the Main Sequence [21–24];
- different types of the dynamo theory, according to which the generation also occurs during the life of a star on the Main Sequence [25–27];
- merger scenario, there is growing evidence that interaction in binaries or multiple stars plays a role in the generation in magnetic fields [28].

If the field generation on the Main Sequence does not occur, then magnetic fields of young stars should be greater than those of old ones. And if the dynamo mechanism works, then the field strength should depend on the rotation velocity: it should be higher for fast rotators. All observation tests are built based on these principles [20,29].

Attempts to find correlations of the magnetic field strength with age or with the rotation velocity have been made several times before [30–32] and many others. But no reliable results were obtained. Borra [33] found the field weakening with age from the study of magnetic stars in the Orion OB1 association. Other results were obtained in the paper by Thompson et al. [32], in which the magnetic fields of 13 stars in the Sco–Cen association were measured and no differences were found in two samples of stars of different ages.

The reason for this is the relatively low determination accuracy of the field strength and the lack of magnetic stars with known rotation periods, as well as very large errors in determining the age of stars. To begin with, the process of measuring a magnetic field itself is far from an ordinary procedure. In order to perform such work, a large telescope and special spectropolarimetric instruments are needed. Therefore, magnetic measurements are quite rare. The relative accuracy of measurements rarely exceeds 20% which is an obstacle to the detection of the required weak effects. In addition, the measurement method allows one to obtain the data on the longitudinal field component only. To compare the results for different stars, it is necessary: to obtain a series of observations for each object in different phases of the rotation period and to build the phase variability curve of the longitudinal field B_e. And finally, to build a magnetic field model. Only in this case, it is possible to compare the magnetic fields of different objects. The fields are usually compared at the dipole pole B_p. It is an extremely time-consuming task that requires long-term observations to cover the entire rotation period of a star and complex modeling of a magnetic field, since often the magnetic field variations cannot be described by a simple dipole.

The main spectrographs that perform spectropolarimetric measurements in the world are: FORS1/2[2] at the 8-m VLT telescope, HARPSpol[3] at the 3.6 m ESO telescope, the same-type NARVAL[4] at the 2-m Telescope Bernard Lyot and ESPaDOnS[5] at the 3.6-m CFHT, and Main Stellar Spectrograph (MSS[6] at the 6-m BTA.

For our research at the 6-m BTA telescope, the Main Stellar Spectrograph (MSS) [34] with a circular polarization analyzer has been actively used for more than 40 years [35]. The observation and data reduction methods are described, for example, in the papers by Romanyuk et al. [36], Kudryavtsev et al. [37]. The exposure time is chosen so that the signal-to-noise ratio (S/N) in the spectra exceeds 150–200. The main bulk of the spectra covers the 4450–4950 Å wavelength region with the mean resolution $R = 15,000$. The instrument allows us to observe the stellar magnetic fields of the magnitude $m_V = 11^m$. In addition to the magnetic field, the circularly polarized spectra obtained with it can be used to measure the radial velocity V_R and the projected rotation velocity $v_e \sin i$. But the low resolution of MSS does not permit to estimate $v_e \sin i$ values for slowly rotating stars: $v_e \sin i < 20$ km s^{-1}. More than 20 thousand spectra were received with this spectrograph.

The FORS1/2 spectropolarimeter with the spectral resolution $R = 2000$ is very effectively used to search for faint magnetic stars in clusters of different ages [38]. To date, this instrument allows studying the magnetic fields of the weakest objects, up to the magnitude $m_V = 13^m$, which is a record for permeability. With other telescopes, it is impossible to observe objects weaker than those observed with FORS1/2.

Besides these instruments, magnetic measurements are being carried out at several other observatories in the world (e.g., [39]), but the amount of data obtained with them is much smaller.

In order to obtain high-quality observed data required for building magnetic and chemical maps of the surface of CP stars, it is necessary to obtain a series of high-resolution spectra in different phases of the rotation period. The HARPSpol ($R > 100,000$), ESPaDOnS, and NARVAL high-resolution spectrographs ($R = 65,000$) are intended for observing the four Stokes parameters of magnetic stars for measuring very weak magnetic fields of bright stars with an accuracy of about 1 G [40–42]. The results of observations with these spectropolarimeters are successfully used for magnetic mapping not only of chemically peculiar stars but also of other types of objects [43]. Mapping technique requires long and periodic observations to exclude data redundancy for each object. Such job is not possible to realize with the largest telescopes, because of high pressure to their time. So the work can actually be done with moderate-sized telescopes.

Theoretical calculations have shown that the ohmic decay of a field, if there is no field generation on the Main Sequence, occurs very slowly, the decay time is comparable to the lifetime of a star on it [29]. Since a slow decrease in the magnetic field strength with age is assumed, a large number of high-accuracy observations of MS stars of different ages are required to reveal the effect. So far, the field measurement errors are so large that trends cannot be detected.

Another problem is determining the age of stars. For single stars of the galactic field, errors in the age determination can exceed 100% [44,45]. The age of field stars is determined from the evolutionary tracks based on the effective temperature and surface gravity found using the spectra. The initial data, especially for stars with anomalous chemical abundance, are not very reliably determined. This, in turn, results in large errors in determining the age of CP stars. Since both components (the field and the age) are found with large errors, it has not been possible to find any reliable relationship between these values until now.

The problem of determining the age of CP stars was repeatedly analyzed by John Landstreet and he came to the conclusion that the best solution is to carry out observations of stars in clusters of different ages; in this case, the accuracy and reliability increase significantly. The star is assigned the age of the cluster, in which it is located. Of course, the reliability depends on cluster membership, i.e. on the astrometric data. Results of GAIA essentially improved the quality of astrometric data and in many cases, known members were excluded from clusters.

The search for correlations between the field and the rotation velocity, which is expected by the dynamo mechanism, also presents great difficulties. The number of magnetic CP stars with the determined periods and, hence, the rotation velocities, is still insufficient.

Some tendencies (the faster the rotation is, the stronger the field is) are visible, but since fast rotators are mainly young hot stars, it was not possible to finally give preference to any hypothesis [20]. Summarizing the issue (formed 30 years ago) with studies of the evolution of a large-scale magnetic field, we proposed a large program for studying these objects with the 6-m telescope.

Systematic integrated observations were supposed to be carried out in three main directions.

- Searching for new magnetic CP stars. This is necessary because the available sample of stars was not enough for statistical studies.
- Searching for magnetic stars with very long rotation periods and building the variability curves of the longitudinal field component with the rotation period phase. This is necessary to test the assertion that slow rotators cannot have a strong field, according to the dynamo mechanism. Nevertheless, that could have worked at some other stage of the stellar evolution, so it cannot be excluded.
- Performing a large number of measurements of magnetic fields of young and old stars, members of open clusters of different ages. If there is no generation on the Main Sequence, then the fields of young stars should be systematically large.

This program has been running for over 30 years. Of course, our work was not done in isolation from the rest of the world. Astronomers world-wide have received new impressive results that have made it possible to clarify a lot in the evolution theory of the stellar magnetic field. Below, we present an overview of long-term observations at the 6-m telescope with the results obtained at other telescopes and the conclusions, to which they led.

3. Research Results Obtained from Observations at the 6-m BTA Telescope

Measurements of the magnetic field at the BTA have been carried out since 1977, from the very beginning of the telescope scientific programs. The instruments for magnetic field measurements were developed, manufactured, and implemented in observations by the SAO RAS staff: G. A. Chountonov, I. D. Naidenov, and V. G. Shtol under the leadership of Yu. V. Glagolevskij [46].

By the mid-70s of the last century, Babcock had already completed his observations. Significant contribution to understanding the structure of the magnetic field of CP stars was made by Preston. Angel and Landstreet built and implemented Balmer magnetometer, and its team carried out effective studies of magnetic stars. Magnetic fields of white dwarfs [47] were detected; it was found that stars with anomalous helium lines also have large-scale magnetic fields. An overview of normal A and B stars showed the absence of magnetic fields in them. In this context, observations of magnetic fields began at the 6-m telescope in 1977.

3.1. Search for New Magnetic Stars

By the mid-80s of the twentieth century, 126 magnetic chemically peculiar stars were found [48]. This amount was very small for building various statistical dependencies. Therefore, we started a large program at the BTA with the aim of searching for new magnetic CP stars. However, finding new magnetic stars is not easy. The problem is that the spectra of magnetic CP stars do not actually differ from the spectra of non-magnetic such objects. The Zeeman effect with kilo-Gauss fields has too little influence on spectral lines leading to their insignificant broadening invisible in the moderate spectral resolution observations. Experience has shown that the large fraction of chemically peculiar stars has magnetic fields below the limit of detection [49].

Observing all CP stars indiscriminately with a large telescope to search for their magnetic fields is extremely ineffective. We have proposed a new approach to solving this issue. The point is that using not only the chemical abundance anomalies, but also the energy distribution anomalies of CP stars is necessary. Kodaira and Unno [50] found that wide and short depressions are observed in the continuum of CP stars. The strongest of them is located at the wavelength about $\lambda = 5200\,\text{Å}$ which has a depth of up to 5%.

The most successful, in our opinion, photometric measurements were carried out in the mid-band Geneva system. One of the filters of this system was centered on the region $\lambda = 5200\,\text{Å}$, which made it possible to carry out numerous observations of different stars. The Swiss astronomers Cramer and Maeder [51] found a correlation between the so-called Z parameter of the Geneva photometric system, which characterizes the depression depth, and the magnetic field strength on the surface of CP stars. The reason for the appearance of such a correlation is not entirely clear; most likely, the effect of magnetic amplification of the lines takes place. Observations of about 30 thousand objects were carried out in the seven-color Geneva system [52], the Z parameters are determined for several thousand CP stars.

A similar system, the Δa photometry, was developed at the Vienna observatory. Mid-band special filters were made, one of them was centered at the wavelength $\lambda = 5200\,\text{Å}$, and the other two were in the continuous spectrum at a distance of about 200 Å from the center of the depression. A large series of observations was carried out by Maitzen et al. [53] in the specified region. The applicability of photometric methods strongly depends on the spectral type of the studied stars.

Using the photometric data, we have selected more than 100 previously unstudied stars with a strong depression in the continuum, believing that the proportion of magnetic stars among them is more typical. The approach proved to be efficient. In the period from 1998 to 2005, we carried out observations of 96 CP stars with a circular polarization analyzer at the 6-m telescope; and magnetic fields were found for 72 of them for the first time [37]. Thus, 75% of the candidates appeared to be actually magnetic stars. The efficiency of our searches has tripled compared to that with the commonly used method. This allowed our group to become leading in the number of magnetic stars found. More than 10 objects turned out to be stars with very strong magnetic fields, the longitudinal component of which was greater than 3 kG [37,54–57].

By about 2010, we have completed observations of all stars with strong depressions. Unfortunately, our list has been completed. At both Geneva and Vienna observatories, mid-band photometric observations were discontinued. We carried out observations of all objects with strong depressions, the data for which were obtained from observations of the mid-band photometry. Nevertheless, we continued to search for new magnetic stars. We proposed other criteria: strong anomalies and variability of helium lines in hot stars, very strong lines of rare-earth elements for cool ones, the presence of periodic photometric variability, etc. The efficiency of these criteria is not so high, the fraction of successful searches for new magnetic stars has decreased; nevertheless, it amounts to 40–50% of all the observed magnetic star candidates.

We discover about 10 new magnetic CP stars every year. The total number of such objects found by us is more than 200, i.e., more than 1/3 [58] of the total number of currently known magnetic stars. The results are published annually in a series of papers; currently the latest paper is by Romanyuk et al. [36].

Of course, our observatory is not the only in the world, in which searches for magnetic stars are carried out. The world largest telescopes were involved in this work. First of all, the very extensive MiMeS survey should be noted, the purpose of which is to search for the magnetic fields of massive stars. More than 500 objects were observed, 35 new magnetic stars were found [59]. For the first time, large-scale magnetic fields have been found in O stars, similar in structure to the fields of A and B stars. It was found that all five known Of?p stars have magnetic fields. The first, unfortunately unsuccessful attempt was made to detect a magnetic field outside of our Galaxy, in the Magellanic clouds [60].

Another program, "B fields in OB stars (BOB)", is in many ways similar to the previous one. With FORS2 and HARPSpol, they carried out carried out observations of about 70 stars and found that the proportion of magnetic stars is 7%, which is in full agreement with the results of the previous studies [61].

From earlier papers, let us note the paper by Bagnulo et al. [38], in which a large number of magnetic stars were found with the FORS1 VLT in open clusters of different ages. Note that the observations with FORS1 were carried out for stars up to magnitude 13, which are extremely weak and unobservable for other telescopes. This made it possible to study magnetic stars features in dozens of open cluster. It should be noted that the new GAIA data revised membership of large number of stars in clusters. Thus, we can see that the new magnetic CP stars were discovered mainly with the 8-m VLT telescope in the southern hemisphere and with the 6-m BTA telescope in the northern one.

Searches for magnetic stars were also carried out with other telescopes but in much smaller volumes. The number of known magnetic CP stars has now exceeded 600 [58], which allows one to carry out a variety of statistical studies. Resumption the mid-band photometric observations in the depression region at $\lambda = 5200\,\text{Å}$ would help us efficiently speed up the search for new magnetic stars. But new possibilities of spectroscopy have appeared.

The paper by Hümmerich et al. [62] reports the identification of chemically peculiar stars based on the LAMOST spectroscopic survey[7]. Candidates were identified by the spectra based on the presence of a depression at $\lambda = 5200\,\text{Å}$. The GAIA parallaxes were used to analyze the spatial distribution and their properties in the color-magnitude diagram. The final sample by the authors contains 1002 CP stars, most of them are new discoveries: only 59 objects are in the catalogs of peculiar stars. The ages of all the objects range from 100 million to 1 billion years, the masses range from 2 to 3 $M_\odot$. As a result of this study, the sample of the known CP stars in our Galaxy has been significantly increased. Although, the amount of the found magnetic stars is still insufficient to build their spatial distribution in the Galaxy, few of them have yet been found in open clusters of different ages. But it is already clear that the upper limit of the surface field of these objects is about 34 kG.

More than 60 years ago, Babcock [63] found the star HD 215441 with a magnetic field of 34 kG on the surface. Despite all attempts, it was not possible to find a star with a stronger field. Apparently, fields of greater magnitude cannot be generated during the CP star formation. Magnetic stars are observed in all regions of the Galaxy: in open clusters and in a field of up to 1 kpc. There is no reason to believe that they can not exist farther away, just modern technical capabilities do not allow their observations to be carried out.

To date, no particular regions with a great or small number of magnetic stars in the Galaxy have been found.

The next task to be solved is to find out whether there is a connection between the distribution of magnetic stars and their field strengths with the structure of the Galactic magnetic field. For this purpose, it is necessary to continue the search for new magnetic stars especially distant and, therefore, weak.

3.2. CP Stars with Very Long Rotation Periods

Chemically peculiar stars rotate slower than normal ones, as was shown by Preston [64] and was certainly confirmed later. In normal stars without any spectral variability, one can find only the projected rotation velocities on the line of sight using the line broadening due to rotation.

The chemical abundance anomalies of CP stars are located non-uniformly over their surface. When the star rotates, regions with different chemical abundances fall on the line of sight; therefore, the spectral variability is observed. The variability proved to be periodic and undoubtedly associated with the rotation of the star. For many stars, the period of this variability was determined, and hence the actual rotation velocities on its surface, since the radii of the stars are well known.

After the magnetic field was discovered in CP stars, the large-scale measurements of the longitudinal field component B_e began. It turned out that in a number of cases, the periodic variability indicates very long variability periods: months and years. Preston [65], Preston and Wolff [66] noticed this fact for the first time. The spectral and photometric variability with periods of years and decades cannot be detected. Very small variations that last for many years can be masked by observation systematic errors. The only reliable option is to measure the longitudinal field component. It has been found, in this way, that the rotation period of the star γ Equ is very large: more than 70 years [67].

From the point of view of the dynamo theory of generation of the field at the MS should be a dependence: the faster the star rotates, the stronger the field can be generated. Theoretical calculations are complex and their results depend on many unobservable boundary conditions. Therefore, the theory provides an opportunity for verification in observations: it is necessary to carry out observations of fast and slow rotators and compare their magnetic fields. Such studies have been performed repeatedly before, but have not led to reliable definite conclusions. The main reason is that fast rotators are predominantly young hot stars; there are no slow rotators among them. And slow rotators are older cool stars, among which there are no fast rotators. The second reason is that there are still few stars with known rotation periods. The periods found are often erroneous.

In recent years, owing to spacecraft launches, the situation has been improving; however, there is still little data. Yes, in general, a certain trend is visible: the field decrease with the decrease in the rotation velocity, but it is possible that this is a dependence on temperature and age, and not on the parameters of rotation. To solve this problem, we proposed observing the magnetic field of stars with very long rotation periods: years and decades. If a field is found in such stars, then this will contradict the dynamo theory. The rotation energy in this case is too weak to generate a strong magnetic field.

There were assumptions for such a task. Even Babcock [67] found that the longitudinal magnetic field of the cool peculiar star γ Equ varies very slowly with a probable period of about 70 years. Later, dozens of magnetic observations of this object were carried out. A very long rotation period is confirmed [68,69]; however, an interesting phenomenon was found: over time, observations show an increase in the period. Since even one observation cycle of its magnetic field has not been completed yet, it is not clear whether the rotation of the star slows down or whether the B_e variability curve is anharmonic; and until the full cycle is completed, the exact rotation period of γ Equ is unknown.

The recent rotation periods and the literature on the period estimates can be found in the papers by Savanov et al. [68], Bychkov et al. [69]. The star's rotation period is estimated as 95 years; nevertheless, it has a strong magnetic field, the longitudinal component of which exceeds 1 kG. From the split components, due to the Zeeman effect, it was found that the surface field of the star reaches 4 kG [70]. The presence of such a star was a strong argument against the efficiency of the dynamo mechanism in magnetic CP stars. For a long time, this star was the only one with such characteristics. It was necessary to check whether there were more similar objects. More than 20 years ago, we began a program of searches for and further observations of stars with very long rotation periods. We took the works by astronomer Mathys as a basis, in which he was looking for CP stars with split Zeeman components. The results of his many years of studies are summarized in the comprehensive paper by Mathys [28]. Overall, he found 43 stars with split Zeeman components. This phenomenon can be observed only in stars with a strong magnetic field and very narrow lines in the spectra: the projected rotation velocity on the line of sight $v_e \sin i$ (does not exceed 10 km s^{-1}). Very narrow lines mean either a small value of $\sin i$ (the rotation pole is facing the observer) or very slow rotation.

We have carried out observations of about 15 stars with narrow lines from the lists of the paper by Wade et al. [71]. Among them, three stars with very long rotation periods were discovered. We have carried out observations covering the whole rotation cycle with a period phase. For two more stars, the observations continue. Another star, γ Equ, has been observed for over 70 years all over the world, but the full rotation period has not yet been obtained.

Long-term observations were required to build the variability curve of the longitudinal field component B_e with the phase of the rotation period. The presence of such a phase curve, along with the data on the surface magnetic field obtained from the split Zeeman components, allows one to obtain the data necessary for building a stellar magnetic field model.

We carried out observations with a rotation period phase for three stars, each of which has the longitudinal field component greater than 1 kG. For HD 18078, a rotation period of 1358 days was found. The geometric structure of its magnetic field is in agreement with the first-order collinear model; the dipole axis does not pass through the center of the star. The phase curve of the longitudinal component field B_e is strongly anharmonic which indicates an unusual magnetic structure (for more detail, see the paper by Mathys et al. [72]).

The next star, HD 50169, has been observed in various observatories for 30 years starting with the paper by Babcock [73]. The longitudinal field component was measured with the 6-m telescope; the modulus of the magnetic field from the split Zeeman components was determined at ESO. We found that the rotation period of the star is 29 years. This is the star with the longest period, the observations of which lasted more than one rotation period. In this star, the magnetic field is not symmetrical according to the axis passing through the center of the star [74].

The magnetic field of the chemically peculiar star HD 965 was discovered by us [75]. Further magnetic monitoring showed that the star is a very slow rotator [76]. The results of the HD 965 study are presented in the paper by Mathys et al. [77]. The rotation period of the star is shown to be 16.5 years. The magnetic field is symmetrical to the axis passing through the center of the star. It is well represented as a superposition of collinear dipole, quadrupole, and octupole.

In addition to these three stars, we are measuring two more with very long rotation periods, but it is necessary to carry out observations in all phases in order to build the curve. Recently, Mathys et al. [78] published the results of a study of the rapidly oscillating roAp star HD 166473, whose rotation period is 10.5 years. In all these cases, a magnetic field is observed, the longitudinal component of which exceeds 1 kG. All the objects are cool magnetic stars with effective temperatures of about 7000–8000 K [79]. At least some of them (γ Equ [80], HD 166473) exhibit fast brightness and spectrum oscillations.

We can see that the presence of ultra-slow rotators is a rare phenomenon but not accidental. The fact that we observe exactly ultra-slow rotation and not a cyclic process (like the Solar 11-yr cycle) has been proven repeatedly, for example, in the paper by Mathys [28].

The obtained data indicate that the reference system of magnetic measurements is stable: the data obtained with different telescopes, spectrographs, and using detectors of different types do not lead to principal differences in the data obtained from observations. The standard system of magnetic measurements in SAO is stable for decades and corresponds to the international one.

Available data unambiguously indicate that the generation mechanism cannot create a strong field with such slow rotation: several degrees per year for γ Equ. Thus, the hypothesis of the fossil magnetic field formation receives new weighty evidence.

3.3. Magnetic Fields of Stars in the Association in Orion

Another direction of research, that is actively developing at the 6-m telescope, is observations of chemically peculiar stars in open clusters of different ages. We have selected 17 clusters ranging in age from several million to hundreds of millions of years, of which at least three CP stars are the members. The program implies the magnetic field measurements of about 200 stars, the members of these clusters, in order to study the dependences of the field strength on age. We have completed the observations and analyzed the data obtained for the Orion OB1 association.

In the Orion constellation, at a distance of about 400 pc, there is one of the most famous groups of early-type stars: the Orion OB1 association. The structure of the association is very complex, it consists of various star clusters, gaseous and dust nebulae. Millions of different objects were discovered in it: maser sources, protostars of the τ Tau and Ae/Be Herbig type, OBA stars of the Main Sequence, various emission objects, etc.

Blaauw [81] made the first attempt to analyze the association structure and identified four subgroups: 1a, 1b, 1c, and 1d which differed in age and stellar composition. Subgroup 1a belongs to the northern region of the association, subgroup 1b is the Orion's belt, 1c is the region south of the Orion's belt, and 1d is a very compact area in the center of the association within subgroup 1c.

3.3.1. CP Stars in the Association

The most complete catalog of the association's stellar population is presented in the paper by Brown et al. [82]; it has 814 objects. The number of A and B stars in it is approximately the same (about half) and there is a small number of O stars.

Using various literary sources, in particular the catalog by Renson and Manfroid [18], we selected 85 chemically peculiar stars in the association. We can see that the proportion of such objects in it is 10.4% and corresponds to the average for the whole sample of chemically peculiar stars. Learn more about the star selection from the paper by Romanyuk et al. [83].

The bulk of the selected objects are B stars with strong or weak (in comparison with the Solar abundance) helium lines, with strong silicon lines, as well as early A stars with strong lines of metals and rare-earth elements. The paper by Brown et al. [82] was published before the publication of the results of the HIPPARCOS mission, so we decided check the belonging of objects to the association using parallaxes. The average distance to it is about 400 pc. Using the parallaxes of the HIPPARCOS and GAIA missions, we found that the 23 coolest peculiar stars with strong metal lines do not belong to the association, they are foreground objects and are located at distances from 100 to 300 pc. Thus, for observations with the 6-m telescope, we have selected 62 potentially magnetic chemically peculiar stars.

The association in Orion includes a large number of stars with anomalous helium lines. The researchers could not disregard this fact. Borra and Landstreet [30] first discovered several magnetic stars among objects with strong helium lines in the association. Borra [33] found magnetic braking and magnetic field decay; the evidence was obtained from observations of stars in the association. Based on observations of 13 chemically peculiar stars in the association in Orion, he found that magnetic stars in Orion have magnetic fields stronger than older stars by a factor of the order of three. He found a short decay time (about 100 million years), which he explained by the movements of masses inside the star. But in the paper by Thompson et al. [32], based on studies of magnetic stars in another young association, Sco–Cen, no significant differences were found in the magnetic field strength of young and old stars. Thus, it is unclear whether a rapid decrease of the field strength with age is characteristic only of the association in Orion, or there are some systematic errors in these studies.

Note that both results were obtained with the Balmer magnetometer [47]; classical spectropolarimetry using metal lines was not performed. In any case, these results turned out to be too small in number to solve the problem of the origin of the strong large-scale field and anomalies in the chemical abundance of CP stars. Therefore, we decided to carry out a large-scale program to search for and study magnetic fields in the Orion OB1 association.

3.3.2. Magnetic CP Stars in the Association

The first question that arises concerns the completeness of the sample of chemically peculiar stars in the association. The average distance to it is about 400 pc, which corresponds to the distance module $|V - M_V| \sim 8^m$ magnitudes. In the association, the vast majority are B stars with the absolute magnitude $M_V = -3 \ldots 0$ and only three are early A stars. Even taking into account the interstellar and circumstellar extinction, they are all brighter than $m_V > 10$ magnitude. Such bright objects are well studied, presented in different catalogs, and the omission of some peculiar star is unlikely. Therefore, we consider the completeness of the sample to be sufficient.

We decided to perform a magnetic survey of all 62 potentially magnetic CP stars in the association. The observations were carried out between 2013 and 2021. The observation results and analysis by subgroups are published in the papers by Romanyuk et al. [84–86]. They present the results of measurements of each CP star in the association subgroups, and describe in detail the methods of observation, data reduction, and analysis. An important feature that should be taken into account is that in most cases only the longitudinal component of the field B_e can be obtained in observations averaged over the whole visible surface of the star. In order to estimate the actual field strength in some region on the stellar surface, simulation is required, which cannot be carried out without knowing the rotation period. Therefore, it was decided to use the root-mean-square magnetic field and use χ^2/n (see Romanyuk et al. [85], section: Comments on Individual Stars, formulas 1, 2, 3).

Based on the results of the analysis of the association CP stars, it has been found that the root-mean-square magnetic fields in subgroups 1a and 1c are on average the same, but the reliability of identifying the magnetic stars in subgroup 1c is much higher. And in the Orion's belt in the young 1b subgroup, the magnetic field is almost three times stronger (see Table 1).

Table 1. Root-mean-square fields of magnetic stars in different subgroups of the Orion OB1 association.

Subgroup	$\log t$	$\langle B_e \rangle$, G	σ, G	χ^2/n
1a	7.0	1286	229	29.8
1b	6.2	3014	212	266.6
1c	6.6	1074	145	92.5

It was also found that the proportion of CP stars in the association relative to all A and B stars drops sharply with age. The youngest 1d subgroup is a separate issue; it contains only three CP stars [87]. It is possible that all these objects are Ae/Be Herbig stars. From one-dimensional spectra, it is difficult to distinguish the indicated objects with the H_α emission from a hot star simply located in the nebula, and emission in H_α is not a signature of a star but the radiation of the nebula itself. To solve this problem, additional high-resolution observations are needed in the H_α region.

We found that the magnetic fields of stars in the Orion's belt (subgroup 1b) are about three times stronger than those in the older 1a and 1c subgroups [86]. The result is consistent with that obtained by Borra [33]. Thus, not only the proportion of magnetic stars in the youngest part of the association, the Orion's Belt, but also the field strength in it is significantly higher than those in other parts of the association. We also note that, contrary to our expectations, the strongest magnetic field for the Orion OB1 association stars is observed not in its center, in the star formation regions, but on its periphery. Two stars with the strongest fields, HD 34736 and HD 37776, are on the periphery. In the Orion Nebula, in the star formation center, there is only one magnetic star HD 37017 with a strong field, the longitudinal component of which is 2 kG.

While working on the magnetic survey of the association, we noticed that among the peculiar stars included in the catalog by Parenago [88], there were very few magnetic stars. The catalog contains the stars from the Great Orion Nebula, all are in subgroups 1c and 1d of the association. One can expect that in the regions with high polarization and circumstellar extinction, in which star formation is apparently taking place, stars with the strongest magnetic fields could be found. However, the situation is exactly the opposite: 12 out of 24 stars of subgroup 1c are included in the Orion Nebula, 3 of them are magnetic and 9 are non-magnetic; 3 CP stars of subgroup 1d are included in the Orion Nebula, and none of them has a strong field. Since there are only 13 magnetic stars in subgroup 1c, it turns out that outside the Nebula we have 9 magnetic and 2 non-magnetic stars. And magnetic stars form in the regions with weak linear polarization and extinction (Romanyuk et al. [86]).

One way or another, we obtain the data about the reliable difference in the magnetic properties of stars inside and outside the Orion Nebula. The proportion of magnetic stars in the nebula is four times smaller and the average magnetic field is two times weaker than those of objects outside it.

By the mid-80s of the last century, mono-channel digital instruments were actively introduced in observations: first reticons and then CCDs. This made it possible not only to study a stellar magnetic field as a whole but also to analyze the polarized profiles of metal lines and, thus, the stellar magnetic field structure.

The paper by Thompson and Landstreet [89] was one of the first, in which they report the discovery of a very strong quadrupole field of the star HD 37776 with anomalously strong helium lines. Subsequently, this star was repeatedly studied and a strong complex field was confirmed. Here, let us note the analysis performed by Khokhlova et al. [90] based on the data obtained from long-term (about 15 years) observations at the 6-m telescope using the Doppler–Zeeman imaging method. A new analysis of the magnetic field structure of the star based on our observations was published in the paper by Kochukhov et al. [91], which also contains the distribution maps of some chemical elements over the surface of the star.

Within the MiMeS project, a new magnetosphere model of the rapidly rotating star σ Ori E was built [92]. The high spectral resolution observations were carried out with the NARVAL and ESPaDOnS spectropolarimeters. The field model built shows significant differences from a simple dipole configuration. Bohlender and Landstreet [93] performed observations and built a field model of the star δ Ori C with strong helium lines.

A total of 31 magnetic stars have been found in the association in Orion. However, the variability curves of the longitudinal component with the rotation period phase were obtained only for the following 25 objects (see Table 2).

For these stars, the rotation periods and fundamental parameters were found and magnetic models were built. Therefore, the Orion OB1 association already has more or less reliable data for making generalization.

Table 2. List of objects, for which the phase curve of the magnetic field variability is built in the Orion OB1 association.

Star	Citation
HD 34736	Semenko et al. [94]
HD 34859	Romanyuk et al. [84]
HD 35177	Romanyuk et al. [84]
HD 35298	Yakunin [95]
HD 35456	Romanyuk et al. [96]
HD 35502	Sikora et al. [97]
HD 36313	Borra [33], Romanyuk et al. [85,96]
HD 36485	Romanyuk et al. [85], Bohlender et al. [98], Yakunin et al. [99]
HD 36526	Borra [33], Romanyuk et al. [96]
HD 36668	Borra [33], Romanyuk et al. [85,100]
HD 36916	Romanyuk et al. [86,100]
HD 36955	Romanyuk et al. [85]
HD 36997	Romanyuk et al. [86]
HD 37017	Borra and Landstreet [30], Bohlender et al. [101]
HD 37058	Romanyuk et al. [86]
HD 37140	Romanyuk et al. [86]
HD 37333	Romanyuk et al. [85]
HD 37479	Bohlender et al. [101]
HD 37633	Romanyuk et al. [85]
HD 37687	Romanyuk et al. [86]
HD 37776	Thompson and Landstreet [89], Kochukhov et al. [91], Romanyuk et al. [102]
HD 40146	Romanyuk et al. [85]
HD 40759	Romanyuk et al. [86]
HD 290665	Romanyuk et al. [85]
HD 294046	Romanyuk et al. [84]

3.4. Mapping the Surface of CP Stars

As we wrote above, after the introduction of CCDs in observations, which allowed obtaining high-accuracy spectroscopic data, methods for analyzing spotted stars using the Doppler–Zeeman imaging methods were developed. Since our data were obtained at the 6-m telescope with a moderate-resolution spectrograph, it can be used for mapping only the stars with very strong magnetic fields, for example, HD 37776. Since new results have been obtained in the analysis with mapping methods, we present here a brief overview of the most important papers in this direction.

The pioneer papers belong to Khokhlova [13], Khokhlova et al. [90], where the main ideas of using methods to solve the inverse problem for reconstructing the maps of the chemical element distribution over the stellar surface using the spectral line profiles are presented. Further, Piskunov and Khokhlova [14] developed the Zeeman–Doppler imaging methods. Piskunov and Wehlau [103] showed that, when a very high S/N ratio is obtained, the stellar surface mapping can also be performed using the moderate-resolution spectra.

Significant contribution to the development of spectroscopy and spectropolarimetry was the creation of the Vienna Atomic Line Data Base (VALD[8]) [104]. This made it possible to unify the analyses performed by different groups of researchers. The further development of the "Spectroscopy made easy" software package by Piskunov et al. [105] made it possible to significantly speed up the estimation of the physical parameters of stars using the method of atmospheric models.

The papers by Piskunov and Kochukhov [106], Kochukhov and Piskunov [107] outline the basic principles of mapping used by the authors. It is shown that calculations require using a cluster of parallel computers. As examples, we present several papers, in which the Doppler–Zeeman imaging of different magnetic CP stars was performed.

In the paper by Kochukhov et al. [108], mapping of the field and distribution of elements over the surface of the star α^2 CVn was fulfilled. It was found that the magnetic field of the star is dominated by the dipole component with small contribution from the quadrupole component. The distribution of elements over the surface is related to the configuration of the magnetic field. Despite critics by Stift and Leone [109], Zeeman-Doppler Imaging (ZDI) remains to be the main method of surface mapping in the case of magnetic CP stars. In the paper by Kochukhov et al. [110], mapping field of 53 Cam based on the results of observations of 4 Stokes parameters is performed. The magnetic field of the star is found to be complex and cannot be described by low-order multi polar expansion. The distributions of spots of some chemical elements over the stellar surface were reconstructed. The paper by Kochukhov et al. [91] presented mapping of magnetic field and some chemical elements of the star HD 37776 with an extremely complex and strong magnetic field. New results were obtained when modeling 4 Stokes parameters. The basic principles and application to specific stars are presented in the paper by Silvester et al. [111].

At the end of the section, let us mention a paper by Yakunin et al. [112] on magnetic and chemical mapping of the star HD 184927 with anomalous helium lines. The fundamental parameters of the star were found, the maps of the distribution of chemical elements and magnetic field were built.

The magnetic and chemical mapping requires a large number of high-accuracy observations with the rotation period phase of the star under study. Therefore, for each object under study, as a rule, maps are built once with very rare exceptions. For the most famous star α^2 CVn, the maps were built several times. However, the accuracy of earlier maps is significantly worse than that of modern ones. Therefore, it is impossible to assess whether there is a measurement of the field structure or the migration of chemical abundance spots over the surface. Most likely there is no migration, because the brightness variability curve of, e.g., α^2 CVn has been perfectly described by the Farnsworth [113] ephemerides for 90 years.

3.5. High-Accuracy Photometry and Variability Searches

Photometric studies are much more common than spectroscopic and magnetic studies. This is mainly due to the greater availability of instruments: it is much easier to build a photometer than a spectrograph, and there is also no need to observe stars with large telescopes. The photometric studies are less informative than spectroscopic ones; however, high-accuracy photometry makes it possible to study the weak variability typical of chemically peculiar stars. The light curves can be used to determine whether the spots of certain chemical elements are hot or cold.

Very slow rotation period variations were found for a number of stars. For the repeatedly mentioned star HD 37776, Mikulášek et al. [114], found that the increase in the rotation period was equal to 18 s over 31 years of observations. In the paper, the data of magnetic observations at the 6-m telescope were used. Soon the second such star HR 7355 was found [115]. New data showed that the rotation period of HD 37776 peaked in 2003 and then began to shorten [116]. The authors explain the phenomenon by the presence of a magnetic outer shell.

Paunzen et al. [117] found the first brightest spotted CP stars in Large Magellanic Cloud (LMC).

A new epoch in the photometry of CP stars began after the successful Kepler, MOST, ASAS, and especially TESS missions. Here, we will not describe many new results obtained from these satellites. Let us turn our attention only to the studies carried out under cooperative agreement with magnetic measurements at the 6-m telescope. In the paper by Mikulášek et al. [118], two variable chemically peculiar stars are studied based on the MOST and Kepler photometry. From the spectra obtained with the 6-m telescope, both were found to be magnetic.

In the fields of the Kepler satellite, (in the paper by Hümmerich et al. [119]) many variable stars have been found, whose light curves resemble those for CP stars. As a result, a sample of 41 stars was obtained, in which the chemical peculiarity was confirmed; 39 stars are new CP stars. The authors consider the stability of the light curve to be the main criterion for selecting CP stars among many thousands of targets.

The paper by Romanyuk et al. [120] presents the results of measurements of the magnetic field of 8 CP stars and one candidate from the Kepler field obtained with the spectropolarimeter of the 6-m telescope. A strong magnetic field has been found in 5 stars, the status of 3 stars is not yet quite clear: additional observations are needed.

The data obtained with the TESS mission intended to search for exoplanets are of particular interest. The MOBSTER collaboration was created to study the variability of massive magnetic stars and intermediate-mass stars using the high-accuracy photometry data from this satellite. David-Uraz et al. [121] presented the first results of studies of magnetic OBA stars carried out by the MOBSTER collaboration. Observations of 19 already known magnetic OBA stars were carried out in sectors 1 and 2 of the TESS mission. The authors determined the exact periods from the newly obtained light curves and compared them with the previously published ones. The advantages of using high-accuracy TESS data were demonstrated. Mathys et al. [122] found long-period CP candidates based on the TESS data. The authors have found 60 such objects in the southern hemisphere, 31 of which are already known to have a long rotation period, and 23 are new discoveries. In the paper by Mikulášek et al. [116], new results obtained for the unique star HD 37776 are discussed. Very high-accuracy data from the TESS satellite showed that the light curve is difficult to reproduce using a standard model with chemical photometric spots and solid-body rotation. It seems that HD 37776 is a unique target among magnetic chemically peculiar stars with no analogs yet found.

4. Conclusions

We demonstrate that the 6-m telescope has been carrying out intense observations of stars of different types for the last 40 years. In this survey, we consider only magnetic chemically peculiar stars.

We have found over 200 new magnetic CP stars. Together with the data obtained in other observatories, there is currently the material on more than 600 such objects. This allowed one to significantly expand the understanding of the magnetic field strength of chemically peculiar stars, their spatial distribution in the Galaxy, in the field, and in open clusters of different ages.

It was found that, starting with a field of approximately 1 kG, the number of magnetic stars decreases with the field increase according to a log-normal law [123]. The upper limit of the field is about 34 kG. The lower limit depends on the determination accuracy and on the instruments and technique used.

Studies of several stars with very long rotation periods have shown that some of them have strong magnetic fields, the longitudinal component of which exceeds 1 kG. For three objects, the phase curves of the longitudinal field component variability were obtained covering more than one rotation period of the star. Therefore, ultra-slow rotation was reliably confirmed. This result is consistent with the theory of the fossil magnetic field of CP stars, which states that the field was generated before the exit of the stars on the Main Sequence.

At the 6-m telescope, extensive observed material was collected on the peculiar stars of the Orion OB1 association: more than 600 moderate-resolution spectra were obtained with a circular polarization analyzer. New important, mainly unexpected, results have been received. We have found that the total number of chemically peculiar stars with strong fields reaches 31, which is 55% of the total number of chemically peculiar stars in the association, which is 2 times greater than usual for the whole sample of these objects. A sharp decrease of the proportion of chemically peculiar stars in the association was found in the age interval from 2 million to 10 million years, as well as a significant decrease in the field strength with age was found. We also observe a sharp fall in the proportion of magnetic stars in the Orion Nebula. However, no strong magnetic field was found in the association for the youngest objects with an age of smaller than 1 million years. It is obvious that the formation of large-scale magnetic fields of chemically peculiar stars occurs in a complex manner. On the whole, our data support the theory of the fossil magnetic field formation in these objects, but the speed of field decay turned out to be unexpectedly large.

We performed a cycle of magnetic observations of stars, photometry of which was carried out at the Kepler and TESS satellites. About 10 new magnetic stars were found. Their phase curves of the longitudinal component variability with the rotation period were built.

Of course, we conduct our research in broad cooperation with scientists from different countries. The conclusions presented here are not only ours. We publish them regularly, the results are widely discussed at Russian and international conferences. Therefore, we believe that the observation results obtained with the 6-m telescope make significant contribution to solving the issue of the origin and evolution of stellar magnetic fields.

Funding: This research was funded by the Russian Science Foundation (grant number 21-12-00147).

Data Availability Statement: The data supporting reported results about Orion OB1 association can be found at VIZIER database: subgroup 1a (accessed on 25 October 2021): https://cdsarc.cds.unistra.fr/viz-bin/cat/J/other/AstBu/74.55; subgroup 1b (accessed on 25 October 2021): https://cdsarc.cds.unistra.fr/viz-bin/cat/J/other/AstBu/76.39; subgroup 1c, 1d (accessed on 25 October 2021): https://cdsarc.cds.unistra.fr/viz-bin/cat/J/other/AstBu/76.163.

Acknowledgments: Yuri Nikolaevich Gnedin always supported research on stellar magnetism. The author would like to thank his colleagues at the Stellar Magnetism Studies Laboratory of the SAO RAS for many years of fruitful collaboration. Special thanks are due to Anastasiya V. Moiseeva for her help in preparing the paper. The author is also grateful to the Russian Telescope Time Allocation Committee for allocating the observation time at the 6-m telescope. Observations with the SAO RAS telescopes are supported by the Ministry of Science and Higher Education of the Russian Federation (including agreement No 05.619.21.0016, project ID RFMEFI61919X0016). The renovation of the telescope equipment is currently provided within the "Science" national project.

Conflicts of Interest: The author declare no conflict of interest.

Abbreviations

The following abbreviations are used in this manuscript:

RTTAC	Russian Telescope Time Allocation Committee
SAO RAS	Special Astrophysical Observatory of the Russian Academy of Sciences
BTA	Big Telescope Alt-azimuthal
FORS	FOcal Reducer and low dispersion Spectrograph
VLT	Very Large Telescope
ESPaDOnS	Echelle SpectroPolarimetric Device for the Observation of Stars
MSS	Main Stellar Spectrograph
TESS	Transiting Exoplanet Survey Satellite
MS	Main Sequence
HD catalog	Henry Drapper Catalogue
CP star	Chemically peculiar star
HIPPARCOS	High Precision Parallax Collecting Satellite
HARPSpol	High Accuracy Radial velocity Planet Searcher
LAMOST	The Large Sky Area Multi-Object Fiber Spectroscopic Telescope
GAIA	Global Astrometric Interferometer for Astrophysics
MOBSTER	Magnetic OBA Stars with TESS:probing their Evolutionary and Rotational properties

Notes

[1] The Committee website (accessed on 25 October 2021): https://www.sao.ru/hq/Komitet/about-en.html.

[2] The FORS1/2 WEB-page (accessed on 25 October 2021): https://www.eso.org/sci/facilities/paranal/instruments/fors.html.

[3] The HARPSpol WEB-page (accessed on 25 October 2021): https://www.eso.org/public/teles-instr/lasilla/36/harps/.

[4] The NARVAL WEB-page (accessed on 25 October 2021): https://tbl.omp.eu/platform/.

[5] The ESPaDOnS WEB-page (accessed on 25 October 2021): https://www.cfht.hawaii.edu/Instruments/Spectroscopy/ESPADONS/.

[6] The MSS WEB-page (accessed on 25 October 2021): https://www.sao.ru/hq/lizm/mss/en/index.html.

[7] The LAMOST Web-page (accessed on 25 October 2021): http://www.lamost.org/public/?locale=en.

[8] The VALD Web-page (accessed on 25 October 2021): http://vald.astro.uu.se/.

References

1. Gnedin, Y.N.; Redkina, N.P.; Tarasov, K.V. Influence of the Magnetic Field on the Spectrum of Linear Polarization of the Emission of T-Tauri Stars. *Sov. Astron.* **1988**, *32*, 186.
2. Gnedin, I.N. Axion Decay in a Strong Magnetic Field and Radio Fluxes from Magnetic White Dwarfs. *Astrophys. Space Sci.* **1990**, *169*, 271–274. [CrossRef]
3. Gnedin, Y.N.; Natsvlishvili, T.M.; Shtol', V.G.; Valyavin, G.G.; Shakhovskoi, N.M. SS Cygni: A white dwarf with a sub-megagauss magnetic field. *Astron. Lett.* **1995**, *21*, 118–121.
4. Gnedin, Y.N.; Nagovitsyn, Y.A.; Natsvlishvili, T.M. Quasi-periodic brightness oscillations of the dwarf nova SS Cyg and their magnetic origin. *Astron. Rep.* **1999**, *43*, 462–470.
5. Gnedin, Y.N.; Borisov, N.V.; Larionov, V.M.; Natsvlishvili, T.M.; Piotrovich, M.Y.; Arkharov, A.A. Spectropolarimetry and IR photometry of magnetic white dwarfs: Vacuum polarization or rydberg states in their magnetic fields? *Astron. Rep.* **2006**, *50*, 553–561. [CrossRef]
6. Dolginov, A.Z.; Gnedin, Y.N.; Silant'ev, N.A. *Propagation and Polarisation of Radiation in Cosmic Media*; Gordon and Breach Publisher: Moscow, Russia, 1995; p. 382.
7. Gnedin, Y.N.; Silant'ev, N.A. Basic Mechanisms of Light Polarization in Cosmic Media. *Astrophys. Space Phys. Rev.* **1997**, *10*, 1–47.
8. Hale, G.E. On the Probable Existence of a Magnetic Field in Sun-Spots. *Astrophys. J.* **1908**, *28*, 315. [CrossRef]
9. Babcock, H.W. The Solar Magnetograph. *Astrophys. J.* **1953**, *118*, 387. [CrossRef]
10. Sokoloff, D.D.; Shibalova, A.S.; Obridko, V.N.; Pipin, V.V. Shape of solar cycles and mid-term solar activity oscillations. *Mon. Not. R. Astron. Soc.* **2020**, *497*, 4376–4383. [CrossRef]
11. Sokoloff, D. Dynamo theory and perspectives of forecasting solar cycles. *J. Atmos.-Sol.-Terr. Phys.* **2018**, *176*, 10–14. [CrossRef]
12. Babcock, H.W. Zeeman Effect in Stellar Spectra. *Astrophys. J.* **1947**, *105*, 105. [CrossRef]
13. Khokhlova, V.L. Mapping of "spots" on the surface of Ap stars by means of line profiles. *Sov. Astron.* **1976**, *19*, 576.
14. Piskunov, N.E.; Khokhlova, V.L. Numerical Modeling of Circular Polarization Profiles for Magnetic Ap-Stars. *Sov. Astron. Lett.* **1983**, *9*, 346–349.
15. Romanyuk, I.I. Magnetic CP stars of the main sequence. I. Diagnostic techniques for magnetic field. *Bull. Spec. Astrophys. Obs. RAS* **2005**, *58*, 63–89.
16. Romanyuk, I.I. Main-sequence magnetic CP stars: II. Physical parameters and chemical composition of the atmosphere. *Astrophys. Bull.* **2007**, *62*, 62–89. doi:10.1134/S1990341307010063. [CrossRef]
17. Romanyuk, I.I. Main-sequence magnetic CP stars III. Results of magnetic field measurements. *Astrophys. Bull.* **2010**, *65*, 347–380. [CrossRef]
18. Renson, P.; Manfroid, J. Catalogue of Ap, HgMn and Am stars. *Astron. Astrophys.* **2009**, *498*, 961–966. [CrossRef]
19. Monaghan, J.J. The oblique rotator model of magnetic stars. *Mon. Not. R. Astron. Soc.* **1973**, *163*, 423. [CrossRef]
20. Moss, D. Magnetic Fields in the Ap and Bp Stars: A Theoretical Overview. In *Magnetic Fields Across the Hertzsprung-Russell Diagram*; Astronomical Society of the Pacific Conference Series; Mathys, G., Solanki, S.K., Wickramasinghe, D.T., Eds.; Astronomical Society of the Pacific Press: San Francisco, CA, USA, 2001; Volume 248, p. 305.
21. Mestel, L.; Spitzer, L.J. Star formation in magnetic dust clouds. *Mon. Not. R. Astron. Soc.* **1956**, *116*, 503. [CrossRef]
22. Moss, D. The survival of fossil magnetic fields during pre-main sequence evolution. *Astron. Astrophys.* **2003**, *403*, 693–697. [CrossRef]
23. Alecian, E.; Wade, G.A.; Catala, C.; Folsom, C.; Grunhut, J.; Donati, J.F.; Petit, P.; Bagnulo, S.; Marsden, S.C.; Ramirez Velez, J.C.; et al. Magnetism in pre-MS intermediate-mass stars and the fossil field hypothesis. *Contrib. Astron. Obs. Skaln. Pleso* **2008**, *38*, 235–244.
24. Dudorov, A.E.; Khaibrakhmanov, S.A. Theory of fossil magnetic field. *Adv. Space Res.* **2015**, *55*, 843–850. [CrossRef]
25. Moss, D. A Gailitis-type dynamo in the magnetic CP stars? *Mon. Not. R. Astron. Soc.* **1990**, *243*, 537–542.
26. Hubrig, S.; North, P.; Mathys, G. Magnetic AP Stars in the Hertzsprung-Russell Diagram. *Astrophys. J.* **2000**, *539*, 352–363. [CrossRef]

27. Arlt, R.; Rüdiger, G. Amplification and stability of magnetic fields and dynamo effect in young A stars. *Mon. Not. R. Astron. Soc.* **2011**, *412*, 107–119. [CrossRef]
28. Mathys, G. Ap stars with resolved magnetically split lines: Magnetic field determinations from Stokes I and V spectra⋆. *Astron. Astrophys.* **2017**, *601*, A14. [CrossRef]
29. Mestel, L. Magnetic Fields across the H-R Diagram. In *Magnetic Fields Across the Hertzsprung-Russell Diagram*; Astronomical Society of the Pacific Conference Series; Mathys, G., Solanki, S.K., Wickramasinghe, D.T., Eds.; Astronomical Society of the Pacific Press: San Francisco, CA, USA, 2001; Volume 248, p. 3.
30. Borra, E.F.; Landstreet, J.D. The magnetic field of the helium-strong stars. *Astrophys. J.* **1979**, *228*, 809–816. [CrossRef]
31. Borra, E.F.; Landstreet, J.D. The magnetic fields of the AP stars. *Astrophys. J. Suppl.* **1980**, *42*, 421–445. [CrossRef]
32. Thompson, I.B.; Brown, D.N.; Landstreet, J.D. The Evolution of the Magnetic Fields of AP Stars: Magnetic Observations of Stars in the Scorpius-Centaurus Association. *Astrophys. J. Suppl.* **1987**, *64*, 219. [CrossRef]
33. Borra, E.F. Decaying stellar magnetic fields, magnetic braking: Evidence from magnetic observations in Orion OB1. *Astrophys. J. Lett.* **1981**, *249*, L39–L42. doi:10.1086/183654. [CrossRef]
34. Panchuk, V.E.; Chuntonov, G.A.; Naidenov, I.D. Main stellar spectrograph of the 6-meter telescope. Analysis, reconstruction, and operation. *Astrophys. Bull.* **2014**, *69*, 339–355. [CrossRef]
35. Chountonov, G.A. Dichroic circular polarization analyzer for the Main Stellar Spectrograph of the 6-m telescope. *Astrophys. Bull.* **2016**, *71*, 489–495. [CrossRef]
36. Romanyuk, I.I.; Moiseeva, A.V.; Semenko, E.A.; Kudryavtsev, D.O.; Yakunin, I.A. Results of Magnetic-Field Measurements with the 6-m Telescope. VI. Observations in 2012. *Astrophys. Bull.* **2020**, *75*, 294–310. [CrossRef]
37. Kudryavtsev, D.O.; Romanyuk, I.I.; Elkin, V.G.; Paunzen, E. New magnetic chemically peculiar stars. *Mon. Not. R. Astron. Soc.* **2006**, *372*, 1804–1828. [CrossRef]
38. Bagnulo, S.; Landstreet, J.D.; Mason, E.; Andretta, V.; Silaj, J.; Wade, G.A. Searching for links between magnetic fields and stellar evolution. I. A survey of magnetic fields in open cluster A- and B-type stars with FORS1. *Astron. Astrophys.* **2006**, *450*, 777–791. [CrossRef]
39. Leone, F.; Giarrusso, M.; Cecconi, M.; Cosentino, R.; Munari, M.; Ghedina, A.; Ambrosino, F.; Boschin, W. Twenty-year monitoring of the surface magnetic field of peculiar stars. *arXiv* **2021**, arXiv:2108.12527.
40. Donati, J.F.; Catala, C.; Wade, G.A.; Gallou, G.; Delaigue, G.; Rabou, P. A dedicated polarimeter for the MuSiCoS échelle spectrograph. *Astron. Astrophys. Suppl. Ser.* **1999**, *134*, 149–159. [CrossRef]
41. Wade, G.A.; Donati, J.F.; Landstreet, J.D.; Shorlin, S.L.S. High-precision magnetic field measurements of Ap and Bp stars. *Mon. Not. R. Astron. Soc.* **2000**, *313*, 851–867. [CrossRef]
42. Bagnulo, S.; Wade, G.A.; Donati, J.F.; Landstreet, J.D.; Leone, F.; Monin, D.N.; Stift, M.J. A study of polarized spectra of magnetic CP stars: Predicted š. observed Stokes IQUV profiles for beta CrB and 53 Cam. *Astron. Astrophys.* **2001**, *369*, 889–907. [CrossRef]
43. Donati, J.F.; Achilleos, N.; Matthews, J.M.; Wesemael, F. ZEBRA—ZEeman BRoadening Analysis of magnetic white dwarfs I. Maximum entropy reconstruction of Hα profiles. *Astron. Astrophys.* **1994**, *285*, 285–299.
44. Landstreet, J.D.; Bagnulo, S.; Andretta, V.; Fossati, L.; Mason, E.; Silaj, J.; Wade, G.A. Searching for links between magnetic fields and stellar evolution: II. The evolution of magnetic fields as revealed by observations of Ap stars in open clusters and associations. *Astron. Astrophys.* **2007**, *470*, 685–698. [CrossRef]
45. Landstreet, J.D.; Silaj, J.; Andretta, V.; Bagnulo, S.; Berdyugina, S.V.; Donati, J.F.; Fossati, L.; Petit, P.; Silvester, J.; Wade, G.A. Searching for links between magnetic fields and stellar evolution. III. Measurement of magnetic fields in open cluster Ap stars with ESPaDOnS. *Astron. Astrophys.* **2008**, *481*, 465–480. [CrossRef]
46. Glagolevskij, Y.V.; Chuntonov, G.A.; Najdjonov, I.D.; Romanjuk, I.I.; Rjadchenko, V.P.; Borisenko, A.N.; Drabek, S.V. First measurements of magnetic fields of stars with the photoelectric magnetometer 6-meter telescope. *Soobshcheniya Spetsial'Noj Astrofiz. Obs.* **1979**, *25*, 5.
47. Angel, J.R.P.; Landstreet, J.D. Magnetic Observations of White Dwarfs. *Astrophys. J. Lett.* **1970**, *160*, L147. [CrossRef]
48. Didelon, P. Catalog of magnetic field measurements. *Astron. Astrophys. Suppl. Ser.* **1983**, *53*, 119–137.
49. Bychkov, V.D.; Bychkova, L.V.; Madej, J. Catalogue of averaged stellar effective magnetic fields—II. Re-discussion of chemically peculiar A and B stars. *Mon. Not. R. Astron. Soc.* **2009**, *394*, 1338–1350. [CrossRef]
50. Kodaira, K.; Unno, W. New Evidence for the Oblique-Rotator Model for α$\hat{}$2 Canum Venaticorum. *Astrophys. J.* **1969**, *157*, 769. [CrossRef]
51. Cramer, N.; Maeder, A. Relation between surface magnetic field intensities and Geneva photometry. *Astron. Astrophys.* **1980**, *88*, 135–140.
52. Rufener, F. Catalogue of stars measured in the Geneva Observatory photometric system (fourth edition). *Astron. Astrophys. Suppl. Ser.* **1989**, *78*, 469–481.
53. Maitzen, H.M.; Schneider, H.; Weiss, W.W. Photoelectric search for CP2-stars in open clusters. XIII. NGC 3114 and IC 2602. *Astron. Astrophys. Suppl. Ser.* **1988**, *75*, 391–398.
54. El'Kin, V.G.; Kudryavtsev, D.O.; Romanyuk, I.I. New Magnetic Chemically Peculiar Stars. *Astron. Lett.* **2002**, *28*, 169–173. [CrossRef]
55. El'Kin, V.G.; Kudryavtsev, D.O.; Romanyuk, I.I. Eight New Magnetic Stars with Large Continuum Depressions. *Astron. Lett.* **2003**, *29*, 400–404. [CrossRef]

56. Ryabchikova, T.; Kochukhov, O.; Kudryavtsev, D.; Romanyuk, I.; Semenko, E.; Bagnulo, S.; Lo Curto, G.; North, P.; Sachkov, M. HD 178892—A cool Ap star with extremely strong magnetic field. *Astron. Astrophys.* **2006**, *445*, L47–L50. [CrossRef]

57. Semenko, E.A.; Kudryavtsev, D.O.; Ryabchikova, T.A.; Romanyuk, I.I. HD 45583—A chemically peculiar star with an unusual curve of longitudinal magnetic field variations. *Astrophys. Bull.* **2008**, *63*, 128–138. [CrossRef]

58. Shultz, M.E. Australia. Private communication, 2021.

59. Wade, G.A.; Neiner, C.; Alecian, E.; Grunhut, J.H.; Petit, V.; Batz, B.d.; Bohlender, D.A.; Cohen, D.H.; Henrichs, H.F.; Kochukhov, O.; et al. The MiMeS survey of magnetism in massive stars: Introduction and overview. *Mon. Not. R. Astron. Soc.* **2016**, *456*, 2–22. [CrossRef]

60. Bagnulo, S.; Nazé, Y.; Howarth, I.D.; Morrell, N.; Vink, J.S.; Wade, G.A.; Walborn, N.; Romaniello, M.; Barbá, R. First constraints on the magnetic field strength in extra-Galactic stars: FORS2 observations of Of?p stars in the Magellanic Clouds. *Astron. Astrophys.* **2017**, *601*, A136. [CrossRef]

61. Schöller, M.; Hubrig, S.; Fossati, L.; Carroll, T.A.; Briquet, M.; Oskinova, L.M.; Järvinen, S.; Ilyin, I.; Castro, N.; Morel, T.; et al. B fields in OB stars (BOB): Concluding the FORS 2 observing campaign. *Astron. Astrophys.* **2017**, *599*, A66. [CrossRef]

62. Hümmerich, S.; Paunzen, E.; Bernhard, K. A plethora of new, magnetic chemically peculiar stars from LAMOST DR4. *Astron. Astrophys.* **2020**, *640*, A40. [CrossRef]

63. Babcock, H.W. The 34-KILOGAUSS Magnetic Field of HD 215441. *Astrophys. J.* **1960**, *132*, 521. [CrossRef]

64. Preston, G.W. A List of AP Stars That May Have Long Periods. *Publ. Astron. Soc. Pac.* **1970**, *82*, 878. [CrossRef]

65. Preston, G.W. The Large Variable Magnetic Field of HD 126515 and its Implications for the Rigid-Rotator Model of Magnetic Stars. *Astrophys. J.* **1970**, *160*, 1059. [CrossRef]

66. Preston, G.W.; Wolff, S.C. The Very Slow Spectrum, Magnetic, and Photometric Variations of HD 9996. *Astrophys. J.* **1970**, *160*, 1071. [CrossRef]

67. Babcock, H.W. The Magnetic Field of γ Equulei. *Astrophys. J.* **1948**, *108*, 191. [CrossRef]

68. Savanov, I.S.; Romanyuk, I.I.; Dmitrienko, E.S. Long-Term Variability in the Magnetic Field of the Ap Star γ Equ. *Astrophys. Bull.* **2018**, *73*, 463–465. [CrossRef]

69. Bychkov, V.D.; Bychkova, L.V.; Madej, J. Periods of magnetic field variations in the Ap star γ Equulei (HD 201601). *Mon. Not. R. Astron. Soc.* **2016**, *455*, 2567–2572. [CrossRef]

70. Mathys, G. AP stars with resolved Zeeman split lines. *Astron. Astrophys.* **1990**, *232*, 151,

71. Wade, G.A.; Smolkin, S.; Romanyuk, I.I.; Kudryavtsev, D. Monitoring magnetic fields of sharp-lined Ap stars with the 6 m telescope. *Magn. Stars* **2004**, *121*, 127,

72. Mathys, G.; Romanyuk, I.I.; Kudryavtsev, D.O.; Landstreet, J.D.; Pyper, D.M.; Adelman, S.J. HD 18078: A very slowly rotating Ap star with an unusual magnetic field structure. *Astron. Astrophys.* **2016**, *586*, A85. [CrossRef]

73. Babcock, H.W. A Catalog of Magnetic Stars. *Astrophys. J. Suppl.* **1958**, *3*, 141. [CrossRef]

74. Mathys, G.; Romanyuk, I.I.; Hubrig, S.; Kudryavtsev, D.O.; Landstreet, J.D.; Schöller, M.; Semenko, E.A.; Yakunin, I.A. Variation of the magnetic field of the Ap star HD 50169 over its 29-year rotation period. *Astron. Astrophys.* **2019**, *624*, A32. [CrossRef]

75. Elkin, V.G.; Kurtz, D.W.; Mathys, G.; Wade, G.A.; Romanyuk, I.I.; Kudryavtsev, D.O.; Smolkin, S. High time resolution spectroscopy and magnetic variability of the cool Ap star HD965*. *Mon. Not. R. Astron. Soc.* **2005**, *358*, 1100–1104. [CrossRef]

76. Romanyuk, I.I.; Kudryavtsev, D.O.; Semenko, E.A.; Yakunin, I.A. Magnetic field monitoring of the very slowly rotating CP star HD 965. *Astrophys. Bull.* **2015**, *70*, 456–459. [CrossRef]

77. Mathys, G.; Romanyuk, I.I.; Hubrig, S.; Kudryavtsev, D.O.; Schöller, M.; Semenko, E.A.; Yakunin, I.A. HD 965: An extremely peculiar A star with an extremely long rotation period. *Astron. Astrophys.* **2019**, *629*, A39. [CrossRef]

78. Mathys, G.; Khalack, V.; Landstreet, J.D. The 10.5 year rotation period of the strongly magnetic rapidly oscillating Ap star HD 166473. *Astron. Astrophys.* **2020**, *636*, A6. [CrossRef]

79. Moiseeva, A.V.; Romanyuk, I.I.; Semenko, E.A.; Yakunin, I.A.; Kudryavtsev, D.O. Physical Parameters of Long-Period CP Stars. In *Ground-Based Astronomy in Russia. 21st Century, Proceedings of the All-Russian Conference, Nizhnij Arkhyz, Russia, 21–25 September 2020*; Romanyuk, I.I., Yakunin, I.A., Valeev, A.F., Kudryavtsev, D.O., Eds.; SNEG Press: Pyatigorsk, Russia, 2020; pp. 304–306. [CrossRef]

80. Hubrig, S.; Järvinen, S.P.; Ilyin, I.; Strassmeier, K.G.; Schöller, M. The rapidly oscillating Ap star γ Equ: Linear polarization as an enhanced pulsation diagnostic? *Mon. Not. R. Astron. Soc.* **2021**, *508*, L17–L21. [CrossRef]

81. Blaauw, A. The O Associations in the Solar Neighborhood. *Annu. Rev. Astron. Astrophys.* **1964**, *2*, 213, doi:10.1146/annurev.aa.02. 090164.001241. [CrossRef]

82. Brown, A.G.A.; de Geus, E.J.; de Zeeuw, P.T. The Orion OB1 association. I. Stellar content. *Astron. Astrophys.* **1994**, *289*, 101–120.

83. Romanyuk, I.I.; Semenko, E.A.; Yakunin, I.A.; Kudryavtsev, D.O. Chemically peculiar stars in the orion OB1 association. I. occurrence frequency, spatial distribution, and kinematics. *Astrophys. Bull.* **2013**, *68*, 300–337. [CrossRef]

84. Romanyuk, I.I.; Semenko, E.A.; Moiseeva, A.V.; Yakunin, I.A.; Kudryavtsev, D.O. Magnetic Fields of CP Stars in the Orion OB1 Association. III. Stars of Subgroup (a). *Astrophys. Bull.* **2019**, *74*, 55–61. [CrossRef]

85. Romanyuk, I.I.; Semenko, E.A.; Moiseeva, A.V.; Yakunin, I.A.; Kudryavtsev, D.O. Magnetic Fields of CP Stars in the Orion OB1 Association. IV. Stars of Subgroup 1b. *Astrophys. Bull.* **2021**, *76*, 39–54. [CrossRef]

86. Romanyuk, I.I.; Semenko, E.A.; Moiseeva, A.V.; Yakunin, I.A.; Kudryavtsev, D.O. Magnetic Fields of CP Stars in the Orion OB1 Association. V. Stars of Subgroups (c) and (d). *Astrophys. Bull.* **2021**, *76*, 163–184. [CrossRef]

87. Romanyuk, I.I.; Semenko, E.A.; Kudryavtsev, D.O.; Yakunin, I.A. Magnetic field of massive chemically peculiar stars in the Orion OB1 association. *Contrib. Astron. Obs. Skaln. Pleso* **2018**, *48*, 208–212.

88. Parenago, P.P. Issledovaniia zvezd v oblasti tumannosti Oriona. *Tr. Gos. Astron. Instituta* **1954**, *25*, 3.

89. Thompson, I.B.; Landstreet, J.D. The extraordinary magnetic variation of the helium-strong star HD 37776: A quadrupole field configuration. *Astrophys. J. Lett.* **1985**, *289*, L9–L13. [CrossRef]

90. Khokhlova, V.L.; Vasilchenko, D.V.; Stepanov, V.V.; Romanyuk, I.I. Doppler-Zeeman Mapping of the Rapidly Rotating Magnetic CP Star HD37776. *Astron. Lett.* **2000**, *26*, 177–191. [CrossRef]

91. Kochukhov, O.; Lundin, A.; Romanyuk, I.; Kudryavtsev, D. The Extraordinary Complex Magnetic Field of the Helium-strong Star HD 37776. *Astrophys. J.* **2011**, *726*, 24. [CrossRef]

92. Oksala, M.E.; Wade, G.A.; Townsend, R.H.D.; Owocki, S.P.; Kochukhov, O.; Neiner, C.; Alecian, E.; Grunhut, J. Revisiting the Rigidly Rotating Magnetosphere model for σ Ori E—I. Observations and data analysis. *Mon. Not. R. Astron. Soc.* **2012**, *419*, 959–970. [CrossRef]

93. Bohlender, D.A.; Landstreet, J.D. The Peculiar Helium-Strong Star Delta Orionis C. *Bull. Am. Astron. Soc.* **1988**, *20*, 699.

94. Semenko, E.A.; Romanyuk, I.I.; Kudryavtsev, D.O.; Yakunin, I.A. On discovery of strong magnetic field in the binary system HD34736. *Astrophys. Bull.* **2014**, *69*, 191–197. doi:10.1134/S1990341314020060. [CrossRef]

95. Yakunin, I.A. Magnetic field measurement of the star HD 35298. *Astrophys. Bull.* **2013**, *68*, 214–218. doi:10.1134/S1990341313020090. [CrossRef]

96. Romanyuk, I.I.; Semenko, E.A.; Yakunin, I.A.; Kudryavtsev, D.O.; Moiseeva, A.V. Magnetic field of CP stars in the Ori OB1 association. I. HD35456, HD35881, HD36313 A, HD36526. *Astrophys. Bull.* **2016**, *71*, 436–446. [CrossRef]

97. Sikora, J.; Wade, G.A.; Bohlender, D.A.; Shultz, M.; Adelman, S.J.; Alecian, E.; Hanes, D.; Monin, D.; Neiner, C.; MiMeS Collaboration; et al. HD 35502: A hierarchical triple system with a magnetic B5IVpe primary. *Mon. Not. R. Astron. Soc.* **2016**, *460*, 1811–1828. [CrossRef]

98. Bohlender, D.A.; Bolton, C.T.; Walker, G.A.H. Variable Hα emission in the helium-strong star δ orionis C. In *Proceedings of a Workshop Organized Jointly by the UK SERC's Collaborative Computational Project No. 7 and the Institut für Theoretische Physik und Sternwarte, University of Kiel Held at the University of Kiel, Kiel, Germany, 18–20 September 1991*; Heber, U., Jeffery, C.S., Eds.; Springer-Verlag Berlin Heidelberg GmbH Press: Berlin, Germany, 1992; Volume 401, pp. 221–223. [CrossRef]

99. Yakunin, I.; Romanyuk, I.; Kudryavtsev, D.; Semenko, E. Results of magnetic field observations of stars with helium anomalies with the 6-m telescope. *Astron. Nachrichten* **2011**, *332*, 974. [CrossRef]

100. Romanyuk, I.I.; Semenko, E.A.; Yakunin, I.A.; Kudryavtsev, D.O.; Moiseeva, A.V. Magnetic field of CP stars in the Ori OB1 association. II. HD36540, HD36668, HD36916, HD37058. *Astrophys. Bull.* **2017**, *72*, 165–177. [CrossRef]

101. Bohlender, D.A.; Brown, D.N.; Landstreet, J.D.; Thompson, I.B. Magnetic Field Measurements of Helium-strong Stars. *Astrophys. J.* **1987**, *323*, 325. [CrossRef]

102. Romanyuk, I.I.; Elkin, V.G.; Wade, G.A.; Landstreet, J.D. The very strong and complex magnetic field of the helium-strong star HD 37776. In *Stellar Magnetic Fields, Proceedings of the International Conference, Nizhnij Arkhyz, Russia, 13–18 May, 1996*; Glagolevskij, Y., Romanyuk, I., Eds.; Zvezdochet Press: Moscow, Russia, 1997; pp. 101–105.

103. Piskunov, N.E.; Wehlau, W.H. Mapping stellar surfaces from spectra of medium resolution. *Astron. Astrophys.* **1990**, *233*, 497–502.

104. Piskunov, N.E.; Kupka, F.; Ryabchikova, T.A.; Weiss, W.W.; Jeffery, C.S. VALD: The Vienna Atomic Line Data Base. *Astron. Astrophys. Suppl. Ser.* **1995**, *112*, 525.

105. Piskunov, N.; Ryabchikova, T.; Pakhomov, Y.; Sitnova, T.; Alekseeva, S.; Mashonkina, L.; Nordlander, T. Program Package for the Analysis of High Resolution High Signal-To-Noise Stellar Spectra. In *Stars: From Collapse to Collapse*; Astronomical Society of the Pacific Conference Series; Balega, Y.Y., Kudryavtsev, D.O., Romanyuk, I.I., Yakunin, I.A., Eds.; Astronomical Society of the Pacific: San Francisco, CA, USA, 2017; Volume 510, pp. 509–513.

106. Piskunov, N.; Kochukhov, O. Doppler Imaging of stellar magnetic fields. I. Techniques. *Astron. Astrophys.* **2002**, *381*, 736–756. [CrossRef]

107. Kochukhov, O.; Piskunov, N. Doppler Imaging of stellar magnetic fields. II. Numerical experiments. *Astron. Astrophys.* **2002**, *388*, 868–888. [CrossRef]

108. Kochukhov, O.; Piskunov, N.; Ilyin, I.; Ilyina, S.; Tuominen, I. Doppler Imaging of stellar magnetic fields. III. Abundance distribution and magnetic field geometry of alpha^2 CVn. *Astron. Astrophys.* **2002**, *389*, 420–438. [CrossRef]

109. Stift, M.J.; Leone, F. Zeeman Doppler Maps: Always Unique, Never Spurious? *Astrophys. J.* **2017**, *834*, 24. [CrossRef]

110. Kochukhov, O.; Bagnulo, S.; Wade, G.A.; Sangalli, L.; Piskunov, N.; Landstreet, J.D.; Petit, P.; Sigut, T.A.A. Magnetic Doppler imaging of 53 Camelopardalis in all four Stokes parameters. *Astron. Astrophys.* **2004**, *414*, 613–632. [CrossRef]

111. Silvester, J.; Wade, G.A.; Kochukhov, O.; Bagnulo, S.; Folsom, C.P.; Hanes, D. Stokes IQUV magnetic Doppler imaging of Ap stars—I. ESPaDOnS and NARVAL observations. *Mon. Not. R. Astron. Soc.* **2012**, *426*, 1003–1030. [CrossRef]

112. Yakunin, I.; Wade, G.; Bohlender, D.; Kochukhov, O.; Marcolino, W.; Shultz, M.; Monin, D.; Grunhut, J.; Sitnova, T.; Tsymbal, V.; et al. The surface magnetic field and chemical abundance distributions of the B2V helium-strong star HD 184927. *Mon. Not. R. Astron. Soc.* **2015**, *447*, 1418–1438. [CrossRef]

113. Farnsworth, G. The Period of 12 $\alpha\{2\}$ Canum Venaticorum. *Astrophys. J.* **1932**, *76*, 313. [CrossRef]

114. Mikulášek, Z.; Krtička, J.; Henry, G.W.; Zverko, J.; Žižǎovský, J.; Bohlender, D.; Romanyuk, I.I.; Janík, J.; Božić, H.; Korčáková, D.; et al. The extremely rapid rotational braking of the magnetic helium-strong star HD 37776. *Astron. Astrophys.* **2008**, *485*, 585–597. [CrossRef]
115. Mikulášek, Z.; Krtička, J.; Henry, G.W.; de Villiers, S.N.; Paunzen, E.; Zejda, M. HR 7355—Another rapidly braking He-strong CP star? *Astron. Astrophys.* **2010**, *511*, L7. [CrossRef]
116. Mikulášek, Z.; Krtička, J.; Shultz, M.E.; Henry, G.W.; Prvák, M.; David-Uraz, A.; Janík, J.; Zejda, M.; Romanyuk, I.I. What's New with Landstreet's Star HD 37776 (V901 Ori)? In *Stellar Magnetism: A Workshop in Honour of the Career and Contributions of John D. Landstreet*; Wade, G., Alecian, E., Bohlender, D.; Sigut, A., Eds.; Astronomical Society of the Pacific Press: San Francisco, CA, USA, 2020; Volume 11, pp. 46–53.
117. Paunzen, E.; Mikulášek, Z.; Poleski, R.; Krtička, J.; Netopil, M.; Zejda, M. The (non-)variability of magnetic chemically peculiar candidates in the Large Magellanic Cloud. *Astron. Astrophys.* **2013**, *556*, A12. [CrossRef]
118. Mikulášek, Z.; Paunzen, E.; Zejda, M.; Semenko, E.; Bernhard, K.; Hümmerich, S.; Zhang, J.; Hubrig, S.; Kuschnig, R.; Janík, J.; et al. Fine detrending of raw Kepler and MOST photometric data of KIC 6950556 and HD 37633. *Bulg. Astron. J.* **2016**, *25*, 19.
119. Hümmerich, S.; Mikulášek, Z.; Paunzen, E.; Bernhard, K.; Janík, J.; Yakunin, I.A.; Pribulla, T.; Vaňko, M.; Matěchová, L. The Kepler view of magnetic chemically peculiar stars. *Astron. Astrophys.* **2018**, *619*, A98. [CrossRef]
120. Romanyuk, I.I.; Mikulášek, Z.; Hümmerich, S.; Yakunin, I.; Moiseeva, A.; Janík, J.; Bernhard, K.; Krtička, J.; Paunzen, E.; Jagelka, M.; et al. Magnetic field measurements of Kepler Ap/CP2 stars. In *Stars and Their Variability Observed from Space*; Neiner, C., Weiss, W.W., Baade, D., Griffin, R.E., Lovekin, C.C., Moffat, A.F.J., Eds.; University of Vienna Press: Wien, Austria, 2020; pp. 197–198.
121. David-Uraz, A.; Neiner, C.; Sikora, J.; Bowman, D.M.; Petit, V.; Chowdhury, S.; Handler, G.; Pergeorelis, M.; Cantiello, M.; Cohen, D.H.; et al. Magnetic OB[A] Stars with TESS: Probing their Evolutionary and Rotational properties (MOBSTER)—I. First-light observations of known magnetic B and A stars. *Mon. Not. R. Astron. Soc.* **2019**, *487*, 304–317. [CrossRef]
122. Mathys, G.; Kurtz, D.W.; Holdsworth, D.L. Long-period Ap stars discovered with TESS data. *Astron. Astrophys.* **2020**, *639*, A31. [CrossRef]
123. Medvedev, A.S.; Kholtygin, A.F.; Hubrig, S.; Schöller, M.; Fabrika, S.; Valyavin, G.G.; Chountonov, G.A.; Milanova, Y.V.; Tsiopa, O.A.; Yakovleva, V.A. Statistics of magnetic field measurements in OBA stars and the evolution of their magnetic fields. *Astron. Nachrichten* **2017**, *338*, 910–918. [CrossRef]

 *universe*

Review

Searching for Magnetospheres around Herbig Ae/Be Stars

Mikhail Pogodin [1,*], Natalia Drake [2,3], Nina Beskrovnaya [1,*], Sergei Pavlovskiy [1], Swetlana Hubrig [4], Markus Schöller [5], Silva Järvinen [4], Olesya Kozlova [6] and Ilya Alekseev [6]

1   Central Astronomical Observatory at Pulkovo, 196140 Saint-Petersburg, Russia; sergpavlovsky@gmail.com
2   Laboratory of Observational Astrophysics, Saint Petersburg State University, Universitetski pr. 28, 198504 Saint-Petersburg, Russia; natalia.drake.2008@gmail.com
3   Laboratório Nacional de Astrofisica/MCTI, Rua dos Estados Unidos 154, Itajuba 37504-364, Brazil
4   Leibniz-Institut für Astrophysik Potsdam (AIP), An der Sternwarte 16, 14482 Potsdam, Germany; shubrig@aip.de (S.H.); sjarvinen@aip.de (S.J.)
5   European Southern Observatory (ESO), Karl-Schwarzschild-Str. 2, 85748 Garching, Germany; mschoell@eso.org
6   Crimean Astrophysical Observatory, Russian Academy of Sciences, 298409 Nauchny, Russia; oles_kozlova@mail.ru (O.K.); ilya-alekseev@mail.ru (I.A.)
*   Correspondence: mikhailpogodin@mail.ru (M.P); beskrovnaya@yahoo.com (N.B.)

Citation: Pogodin, M.; Drake, N.; Beskrovnaya, N.; Pavlovskiy, S.; Hubrig, S.; Schöller, M.; Järvinen, S.; Kozlova, O.; Alekseev, I. Searching for Magnetospheres around Herbig Ae/Be Stars. *Universe* 2021 , 7 , 489. https://doi.org/10.3390/universe7120489

Academic Editors: Galina L. Klimchitskaya, Vladimir M. Mostepanenko and Nazar R. Ikhsanov

Received: 1 November 2021
Accepted: 9 December 2021
Published: 12 December 2021

Abstract: We describe four different approaches for the detection of magnetospheric accretion among Herbig Ae/Be stars with accretion disks. Studies of several unique objects have been carried out. One of the objects is the Herbig Ae star HD 101412 with a comparatively strong magnetic field. The second is the early-type Herbig B6e star HD 259431. The existence of a magnetosphere in these objects was not recognized earlier. In both cases, a periodicity in the variation of some line parameters, originating near the region of the disk/star interaction, has been found. The third object is the young binary system HD 104237, hosting a Herbig Ae star and a T Tauri star. Based on the discovery of periodic variations of equivalent widths of atmospheric lines in the spectrum of the primary, we have concluded that the surface of the star is spotted. Comparing our result with an earlier one, we argue that these spots can be connected with the infall of material from the disk onto the stellar surface through a magnetosphere. The fourth example is the Herbig Ae/Be star HD 37806. Signatures of magnetospheric accretion in this object have been identified using a different method. They were inferred from the short-term variability of the He I λ5876 line profile forming in the region of the disk/star interaction.

Keywords: Herbig Ae/Be stars; disk accretion; magnetosphere; individual: HD 10141; HD 259431; HD 104237; HD 37806

1. Introduction

Herbig Ae/Be stars (HAeBes) are widely recognized as pre-main sequence (PMS) objects with pronounced emission line features and a far-infrared (FIR) excess indicative of cool dust in their accretion disks [1–3]. These stars are intermediate-mass (2–8 $M_\odot$) analogues of T Tauri K–M stars. According to current views, the HAeBes have no convective interiors that could support classical dynamo actions as found in the fully convective T Tauri stars (e.g., [4]). For this reason, strong magnetic fields (such as $\sim 10^3$ G for T Tauri stars) are not expected in HAeBes. On the other hand, over the last years a number of spectropolarimetric studies revealed that about 20% of the Herbig Ae/Be stars have globally organized magnetic fields of the order of 100 G ([5–11], and some other similar works).

The question of the origin of magnetic fields in intermediate mass stars with radiative envelopes is still under debate. It has been argued that these fields could be fossil relics of fields that are present in the interstellar medium from which the stars were formed (e.g., Moss [12]). However, the fossil field hypothesis also faces problems (see, e.g., the review by Hubrig et al. [13]).

Notably, the traditional spectropolarimetric method of stellar magnetic field diagnostics has some drawbacks:

- It is effective only for objects with small projected rotation velocity $v \sin i$, because the method is based on the measurement of the fine Zeeman splitting in the presence of a magnetic field, which is difficult to detect if the line broadening due to stellar rotation is significant.

- This method allows one to measure only the mean longitudinal magnetic field $\langle B_z \rangle$, which is the projection of the magnetic field vector on the line of sight averaged over the stellar limb. As a rule, this value is more than three times smaller than the modulus of the surface magnetic field B. It is necessary to repeat the spectropolarimetric measurements on different phases of the stellar rotation, and to use modeling to reconstruct the complete configuration of the global magnetic field of the object and to estimate its strength.

This work presents a review of several of our investigations, in which other methods to study magnetic fields in the Herbig Ae/Be stars were proposed. It is based on a search for any signs of magnetospheric accretion from a disk onto a star. These methods can be applied in cases where a direct spectropolarimetric measurement of the magnetic field is inefficient, in particular, if the star is rapidly rotating.

2. Magnetospheres of Herbig Ae/Be Stars

We propose an alternative method to study the magnetism of stars with accretion disks such as T Tauri and Herbig Ae/Be stars. This method is related to the search of magnetospheres around such objects, where a specific structure of the circumstellar (CS) environment is observed in the region of the disk/star interaction.

According to the magnetospheric accretion (MA) scenario, if a star possesses a sufficiently strong magnetic field, the transfer of accreted material from the disk to the star is stopped by the magnetic field at a certain distance from the star, the Alfvén radius R_A, where the magnetic and kinematic energy densities become equal. The zone around the star at distances $r < R_A$ is called the magnetosphere. Inside the magnetosphere the motion of the accreted gas is completely governed by the magnetic field. A portion of the material is funneled from the disk toward the star along the closed magnetic force lines in the region near the magnetic poles, while other material is flowing outwards along the open force lines at lower magnetic latitudes. An example of magnetic field topology near the magnetosphere of a T Tauri star is presented in Figure 12 from [14].

While models of magnetically driven accretion and outflows successfully reproduce many observational properties of the classical T Tauri stars, the picture is unclear for the HAeBes, since we have poor knowledge of whether their magnetic fields are reasonably strong to maintain the MA scenario of the disk/star interaction. Thus, the detection of a magnetosphere in HAeBes can be an important indicator for significant magnetic fields in these objects.

According to the well-known equation connecting the magnetospheric radius of the star, R_A, with other stellar parameters such as the magnetic field strength B, the radius R, the mass M and the mass accretion rate $\dot{M}$ [15]:

$$R_A = \left(\mu^2 / \dot{M}(2GM)^{1/2} \right)^{2/7},\tag{1}$$

where $\mu = BR^3$ denotes the star's dipole moment.

Assuming that the observational signs of a magnetosphere can be revealed if $R_A/R > 1.5$ and adopting a mean value of the mass accretion rate $\dot{M}$ for HAeBes equal to $5 \times 10^{-7} \, M_\odot/\text{yr}$ [16], and the solar value for the ratio M/R, we can estimate a lower limit of $B > 500 \, \text{G}$ for $R = 2.5 \, R_\odot$ (Herbig Ae stars) and $B > 200 \, \text{G}$ for $R = 6 \, R_\odot$ (Herbig Be stars). These values of B appear quite reasonable for the HAeBes.

Provided the magnetic axis of the star is not coincident with the axis of stellar rotation (i.e., a magnetic oblique rotator), the geometrical configuration inside the magnetosphere is not axially symmetric relative to the rotation axis. As a result, a global azimuthal inhomogeneity is formed in the CS environment near the star. In this case, the rotation of the star with its magnetosphere modulates different observational parameters and stimulates their cyclic variability with the period $P = P_{rot}$ (rotation period) or $P_{rot}/2$, depending on whether one or two magnetic poles can be seen during one rotation cycle.

The majority of the magnetosphere diagnostics are based on a search for such cyclic variability. Two types of inhomogeneity responsible for the generation of this variability are: (a) an accretion stream inside the magnetosphere, asymmetric relative to the rotation axis and (b) a hot spot on the stellar surface in the region where the accreted material falls onto the star. The first of them can be discovered by spectroscopy in lines originating near the region of the disk/star interaction. The local hot spots on the stellar surface can be revealed by means of photometry and high resolution spectroscopy of high accuracy using atmospheric lines sensitive to the temperature.

In this work, we present four examples of a successful detection of magnetospheres around Herbig Ae/Be stars.

3. The Magnetic Herbig Ae Star HD 101412

This Herbig A2e star has been chosen to test our approach in searching for signatures of magnetospheres as this object is certain to have a magnetosphere: its surface magnetic field was estimated as $B \sim 3\,kG$, which is a rather large value among HAeBes [17]. The magnetic field configuration of HD 101412 has a unique orientation relative to the observer: the angle between the line of sight and the rotation axis $i = 80° \pm 7°$ and the angle between the rotation and magnetic axes has been estimated as $\beta = 84° \pm 13°$. This implies that the magnetic poles are close to the accretion disk plane and the areas around the magnetic poles, where the accreted matter falls onto the star, are seen twice per rotation period, which was estimated as P_{rot} or $P_m = 42\overset{d}{.}076 \pm 0\overset{d}{.}017$. An artist's impression of the MA-scenario in HD 101412 is presented in Figure 1.

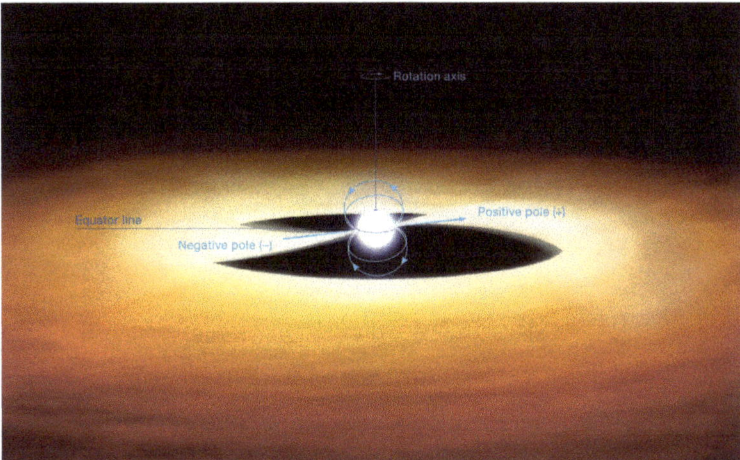

Figure 1. Artist's impression of the MA in HD 101412, looking at the magnetic equator. The exact size and shape of both the disk and the accretion streams are for illustration only, since we lack constraints on these parameters [18]. Credit: Schöller et al., A&A, 592, A50, 2016, reproduced with permission © ESO.

We obtained 30 spectra of HD 101412 between 2011 and 2014 with the spectrographs CRIRES and X-shooter, installed at the 8-m telescopes of the Very Large Telescope (VLT) at ESO (Chile). The spectra covered the near infrared (NIR) spectral range, including the lines He I λ10830 and Paγ. Both lines are formed in the CS environment close to the accretion region, and their parameters are expected to be modulated by the rotation of the star and its magnetosphere. Typical profiles of these two lines are shown in Figure 2.

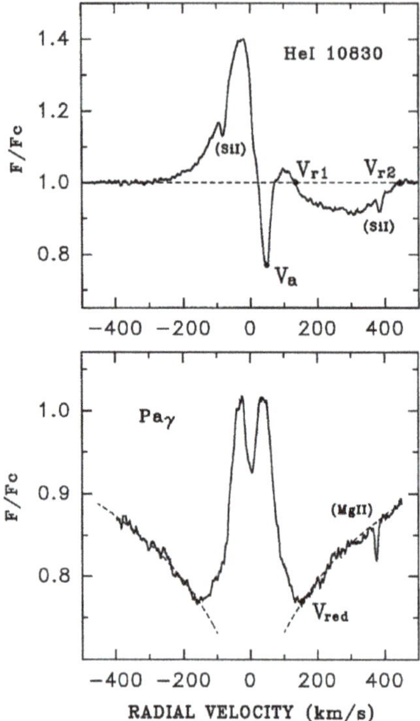

Figure 2. Typical profiles of the NIR CS lines He I λ10830 and Paγ in the spectrum of HD 101412 shown in [18]. Parameters of the profiles used in the quantitative analysis are indicated. Credit: Schöller et al., A&A, 592, A50, 2016, reproduced with permission © ESO.

The behavior of the lines confirms the described above structure of magnetospheric accretion in HD 101412. The central absorption of the profiles of both lines originates from the disk itself. The infalling streams where the broad redshifted absorption on the He I λ10830 line profile is forming screen the stellar disk near the regions of magnetic poles. The high-temperature region of the He I λ10830 formation is geometrically thinner in comparison to Paγ. The region of the Paγ line formation is much more extended, and the Paγ line profile includes emission not only from the accreted flow, but also from the inner disk outside the magnetosphere. Due to this, the redshifted absorption, which is clearly seen in the He I λ10830 line profile, is completely overlapped by a double-peaked emission from the inner disk in the case of the Paγ line profile.

Expecting that the variability of the He I λ10830 and Paγ line parameters could be of cyclic character, as a result of profile modulation by the rotation of the star with its magnetosphere, to detect this variability we have chosen the following line parameters: the equivalent width (EW) of both lines, the velocities of the blue (V_{r1}) and red (V_{r2}) edges of the redshifted absorption component of the He I λ10830 line, and of the red boundary of the Paγ emission profile (see Figure 2).

The periodogram analysis of these five parameters was carried out using the method of fitting phase dependencies of the observed data for each value of the trial period P with a sine for a range of P from 5 to 80 days. Parameters of the sinusoids were determined by the least-square method. The periodograms constructed for three parameters of the He I λ10830 line are shown in the left of Figure 3. We also calculated the noise periodogram in order to estimate the significance level of the separate peaks and to determine the window function (the details of the method can be found in [19,20]). The parameter A/σ denotes the amplitude of the sinusoid in the unites of standard deviation of the residuals of the sine function fit for a given P.

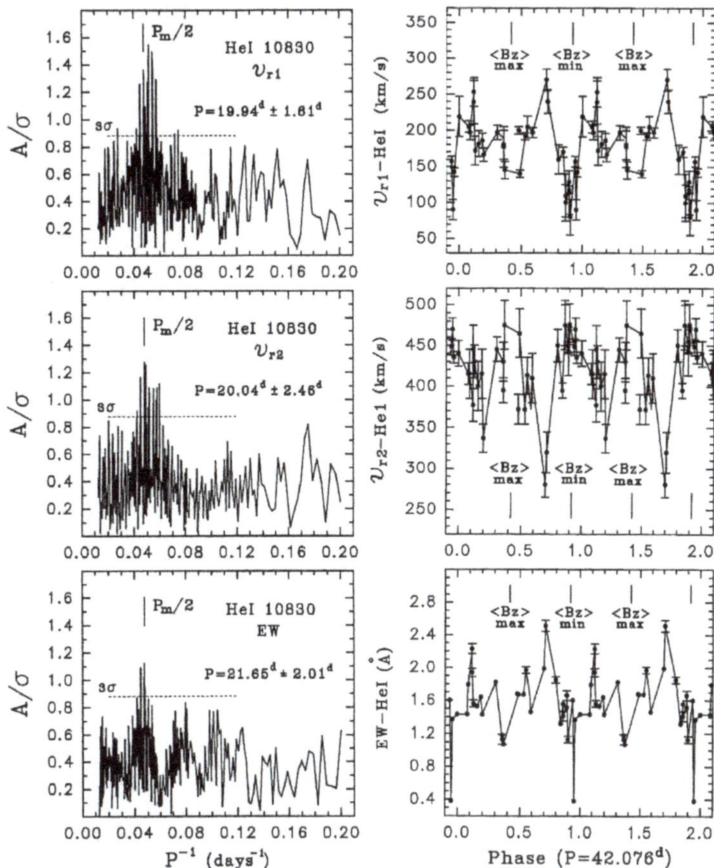

Figure 3. *Left*: A/σ periodograms for different parameters of the He I λ10830 line shown in Figure 2. The 3σ significance levels are marked by dashed lines. Short vertical lines indicate the value corresponding to half of the rotation period ($P_{rot}/2 = 21\overset{d}{.}04$). *Right*: Phase dependencies of different line parameters over the rotation period $P_m = 42\overset{d}{.}076$. The initial phase $\phi = 0$ corresponds to MJD 52797.4. Credit: Schöller et al., A&A, 592, A50, 2016, reproduced with permission © ESO.

The reappearance of the same period in all periodograms can be considered as a criterion of its validity. All periodograms contain the peak near $P_m/2$ at a rather high significance level. The discrete structure of these wide peaks which is caused by a relatively small number of observations makes it possible to determine the mean value P and the standard deviation of each peak (also presented on the left panels of Figure 3. The resulting period $P = 20\overset{d}{.}53 \pm 1\overset{d}{.}68$ is close to half of the magnetic rotation period $P_m/2 = 21\overset{d}{.}038$.

The right of Figure 3 illustrates the phase dependencies of the He I λ10830 line taken from [17,18]. As the MA model predicts, the V_{r2} parameter is at its maximum when $\langle B_z \rangle$ reaches its minimal or maximal value. In contrast to V_{r2}, the parameters V_{r1} and EW demonstrate minimum values at these phases (when each of two accretion flows screens the star).

We can conclude that our test delivers a positive result and that the behavior of the spectroscopic parameters for the lines originating close to the region of the disk/star interaction can be used for diagnostics of the magnetospheric character of the accretion from the disk onto the star.

4. Signatures of the Magnetosphere around the Early Type Herbig B6e Star HD 259431

HD 259431 is a Herbig B6e star associated with the reflection nebula NGC 2247. Its fundamental parameters are: $M = 6.6\,M_\odot$, $R = 6.63\,R_\odot$, $v\sin i = 90\,\mathrm{km/s}$ [21] and the accretion rate $\dot{M} = 7.8 \times 10^{-7}\,M_\odot/\mathrm{yr}$ [16]. No reliable magnetic field has been measured in this object [11,22].

Our spectroscopic observations of the object were carried out at four observatories: (a) the Crimean AO (2.6 m Shajn telescope, echelle spectrograph); (b) ESO (2.2 m telescope, FEROS spectrograph); (c) the OAN SPM observatory in Mexico (2.1 m telescope, echelle spectrograph); and (d) the Kourovka UFU AO (1.2 m telescope, echelle spectrograph). More than 250 high-resolution spectra were obtained from 2010 to 2019.

HD 259431 demonstrates a remarkable type of spectral line variability. During three days, the double-peaked emission line Hγ with a depression in the red wing transformed into a line with a P Cyg-type profile (top of Figure 4). The same picture is observed in the Balmer lines from Hβ up to Hϵ. At epochs where we see a depression in the red wings of the emission Balmer lines, a broad absorption wing of He I λ5876 is observed up to +400 km/s (see Figure 4). Such profile variability can be expected in the case of magnetospheric disk accretion, if the magnetic axis is inclined to the rotation axis. In this situation, the line of sight intersects repeatedly the region of the accretion flow or the wind zone, located at different magnetic latitudes. The existence of a magnetosphere is also evident from the strong increase of the red absorption wing of the He I λ5876 line, as can be seen in Figure 4.

This picture could be validated if a search for periodic variations of the line parameters revealed a period close to the expected period of stellar rotation, P_{rot}. In our periodogram analysis, we used two spectroscopic parameters: $V_{\mathrm{bis}}(\mathrm{H}\beta)$—a bisector velocity of the emission Hβ line profile at the continuum level F_c—and $V_{\mathrm{red}}(\mathrm{He\ I})$—a velocity of the red boundary of the absorption wing of the He I λ5876 line. Figure 5 illustrates that both parameters are in strong anti-correlation, with the coefficient $r = -0.805 \pm 0.049$. We used the standard Lomb–Scargle method [23], and tried to detect periodicity in the range of 2–3 days, which is likely to contain the P_{rot} of the star, due to the previous spectrointerferometric estimation of i in the range from $40°$ to $60°$ and $R = 6.6\,R_\odot$ [21]. The significance level was determined by the method described in [24]. As one can see in the periodograms displayed in Figure 6, both parameters demonstrate a period $P = 2.839\,\mathrm{d}$. The phase dependencies constructed for these two line parameters are presented in Figure 7. The dependencies look rather "noisy", with a large amplitude of scatter. This can be explained by the presence of other types of spectral variability that are not connected with stellar rotation, on the same timescales. As can be seen in Figure 7, the cyclic variations of both parameters take place in opposite phases. Adopting our estimation of $P = 2.839\,\mathrm{d}$ as the most probable value of the rotation period, we can estimate the inclination angle i as $48° \pm 7°$.

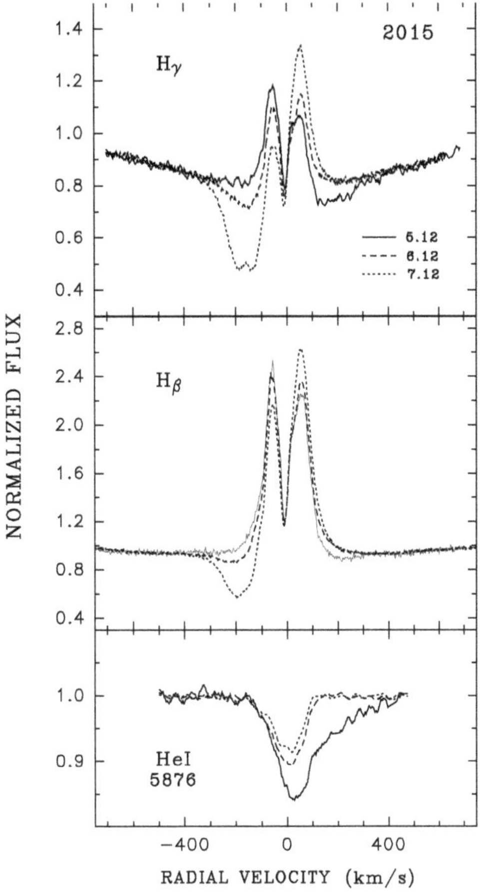

Figure 4. Variations of the line profiles for Hγ, Hβ, and He I λ5876 in the spectrum of HD 259431 during three epochs on 5–7 December 2015. X-scale is the same for all three plots.

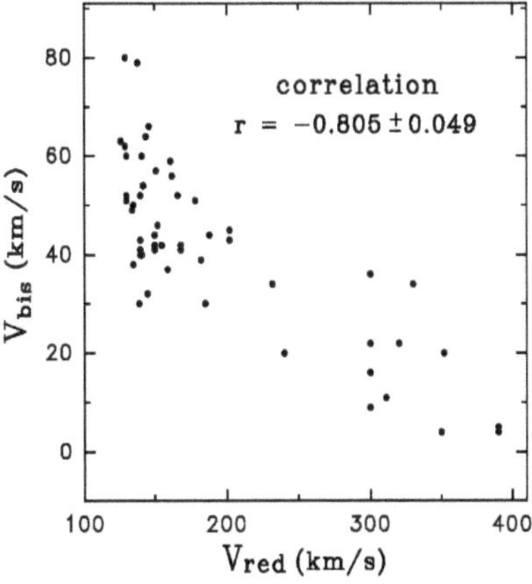

Figure 5. Dependence between the parameters V_{bis} (Hβ) and V_{red} (He I λ5876) in the spectra of HD 259431. r is the correlation coefficient.

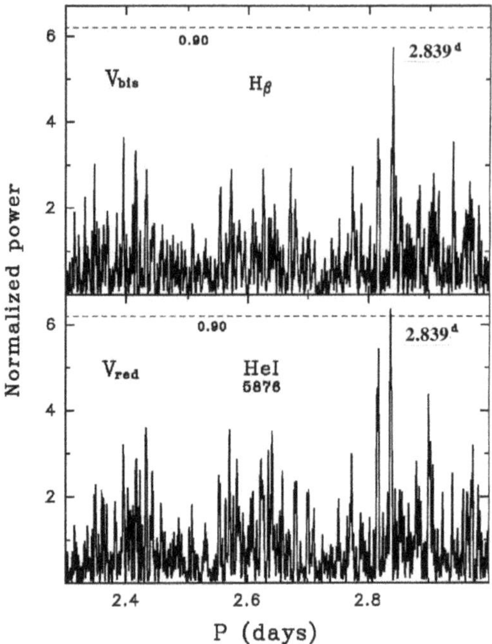

Figure 6. Lomb-Scargle periodogram for the parameters V_{bis} (Hβ) and V_{red} (He I λ5876) from the HD 259431 spectra. The same period, $P = 2.839$ d, is visible in both periodograms.

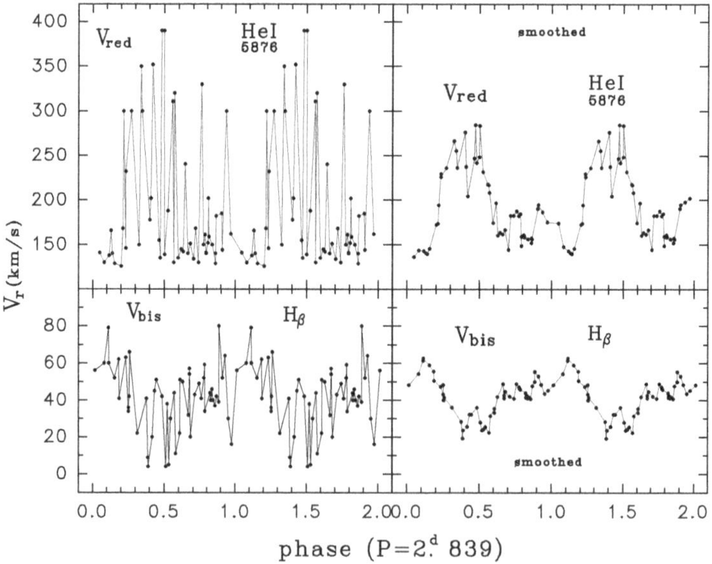

Figure 7. Phase dependencies of the parameters V_{bis} (Hβ) and V_{red} (He I λ5876) from the HD 259431 spectra, constructed for the period $P = 2.839$ d (**left**) and smoothed over five data points (**right**).

Using the time when the parameter V_{red} (He I λ5876) reaches its maximum as the initial phase $\phi = 0$, we can construct the phase variations of the profiles of some other CS lines in the spectrum of HD 259431. Figure 8 illustrates such variability for the lines Hδ, He I λ6678 and the OI $\lambda 7773$ triplet. All of them show the same behavior with the rotation phase as the lines Hβ and He I $\lambda 6678$. In general, the results of our spectroscopic investigation of HD 259431 confirm the presence of a magnetosphere. The estimated Alfvén radius of the object using Equation (1) is $R_A = 1.9\,R$ (for $B = 300\,\mathrm{G}$), and $R_A = 1.5\,R$ (for $B = 200\,\mathrm{G}$), which appear to be quite reasonable. The schematic picture of the disk/star interaction near the stellar surface is shown in Figure 9.

We would like to note that our result is the first detection of magnetospheric accretion among earlier-type Herbig Be stars (here: B6). This contradicts the conclusion of Cauley and Johns-Krull [25] that Herbig Be stars have no magnetospheres. For more details see Pogodin et al. [26].

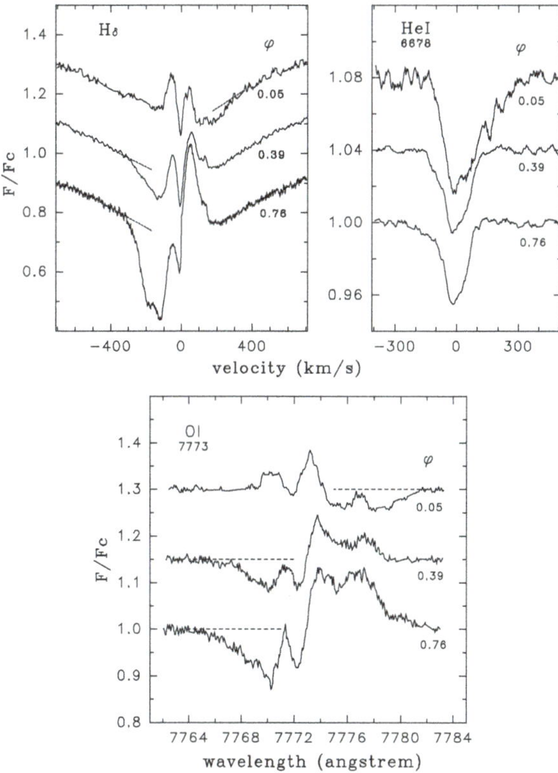

Figure 8. Profiles of the lines Hδ, He I $\lambda 6678$ and OI $\lambda 7773$ observed in the spectra of HD 259431 at the phases of 0.05, 0.39 and 0.76 of the period $P = 2.839\,\mathrm{d}$. The initial phase $\varphi = 0$ corresponds to the epoch when the parameter V_{red} (He I $\lambda 5876$) was maximal.

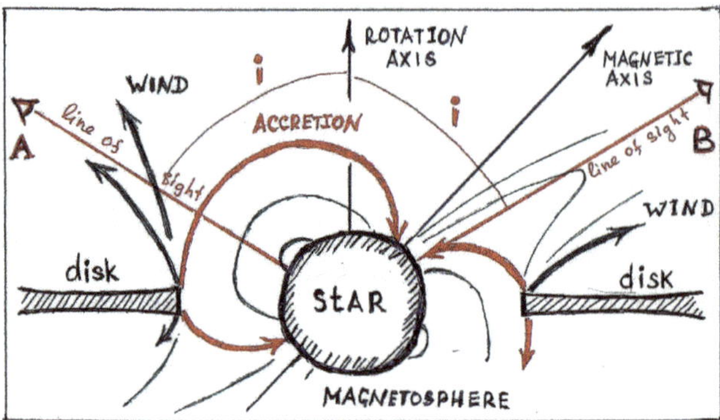

Figure 9. Schematic picture of the region of disk/star interaction near HD 259431, where the magnetic axis is inclined relative to the rotation axis. The star, disk, magnetosphere and wind zone are shown in the figure. The picture is presented in the reference frame of the star. Two directions of the line-of-sight are indicated. From direction A the line of sight intersects the wind zone at low magnetic latitudes, from direction B the line of sight intersects the accretion stream inside the magnetosphere at high magnetic latitudes.

5. The Young Magnetized Binary System HD 104237

HD 104237 is a young binary system containing a Herbig Ae primary ($M_1 = 2.2\,M_\odot$) and a T Tauri secondary of K2 spectral type ($M_2 = 1.4\,M_\odot$). Böhm et al. [27] derived the orbital elements and velocity curves for the components. They estimated $P_{\mathrm{orb}} = 19.859\,\mathrm{d}$, $i_{\mathrm{orb}} = 17°$, $e = 0.665$, and $\gamma = 13.9\,\mathrm{km/s}$, constructing the radial velocity curves for both components. The basic stellar parameters of the system components were determined in a number of studies [27–30]. The most recent analysis by Cowley, Castelli and Hubrig [30] indicated $T_{\mathrm{eff}} = 8\,250\,\mathrm{K}$, $\log g = 4.2$, $v \sin i = 8\,\mathrm{km/s}$ for the primary and $T_{\mathrm{eff}} = 4800\,\mathrm{K}$, $\log g = 3.7$ and $v \sin i = 12\,\mathrm{km/s}$ for the secondary.

Such low projected velocities are beneficial for the detection of the magnetic field by the spectropolarimetric method. A possible presence of a weak magnetic field in HD 104237 (of the order of 50 G) was announced over 20 years ago by Donati at al. [31]. Later Wade et al. [22] did not confirm this detection. Hubrig et al. [7] estimated $\langle B_z \rangle = 63 \pm 15\,\mathrm{G}$ for the primary from a high-resolution HARPSpol spectrum, and, more recently, Järvinen et al. [32] reported a definite detection of a magnetic field $\langle B_z \rangle = 129 \pm 12\,\mathrm{G}$ for the secondary T Tauri star and a marginal detection of $\langle B_z \rangle = 13\,\mathrm{G}$ for the primary.

The most recent HARPSpol spectropolarimetric analysis using the LSD and SVD methods allowed to measure the magnetic fields of the system's components more precisely. Eighty-eight spectra obtained in March 2015 were analyzed. The $\langle B_z \rangle$ value of the primary was found to be varying on different dates from $47 \pm 6\,\mathrm{G}$ to $72 \pm 6\,\mathrm{G}$ and that of the secondary from $609 \pm 27\,\mathrm{G}$ to $124 \pm 13\,\mathrm{G}$ [32].

The components of this binary system have a CS matter distribution in the region of the disk/star interaction that is different from expected for a single Herbig Ae/Be star with a magnetosphere. Dunhill et al. [33] constructed a spatial structure of the CS material distribution using their smoothed particle hydrodynamics (SPH) simulations of the circumbinary disk around the system HD 104237. The binary clears out a large cavity in the disk, which is not centered on the binary center of mass. Two streams of matter are accreting from the circumbinary disk onto the system components. The size of the two individual disks around the components is expected to be rather small due to tidal interaction. A connection of these accretion streams with the magnetic fields of both components is not clear, but in any case, local spots can be expected on the stellar surface in the stream impact regions. The rotation of such spots can modulate the photometric parameters of the object as well as the observed parameters of atmospheric lines, if the spectral data were obtained with sufficient accuracy. As a result, the method of diagnostics of local features on the stellar surface of the system components is the same as suggested for single objects that were tested for the presence of a magnetosphere.

The equivalent widths of the Stokes *I* LSD profiles of the metal lines in the spectrum of the primary component were used to search for periodicity corresponding to the expected rotation period P_{rot}. The search was carried out using a non-linear least-square fit to multiple harmonics using the Levenberg–Marquardt method [34]. The periodogram is presented in Figure 10 (top). We also made a statistical test to check the null hypothesis on the absence of periodicity, i.e., to check the statistical significance of the fit [35]. The window function is also shown in the figure. The contribution of the secondary to the EWs at the phases of conjunction was removed from the measured values. As a result, we obtained a best period solution at $P = 4.33717 \pm 0.00316\,\mathrm{d}$. Figure 10 (bottom) illustrates the phase dependence of the EWs constructed for the derived period $P = 4.33717\,\mathrm{d}$. The nature of this periodicity is likely to be connected with the existence of temperature spots on the stellar surface.

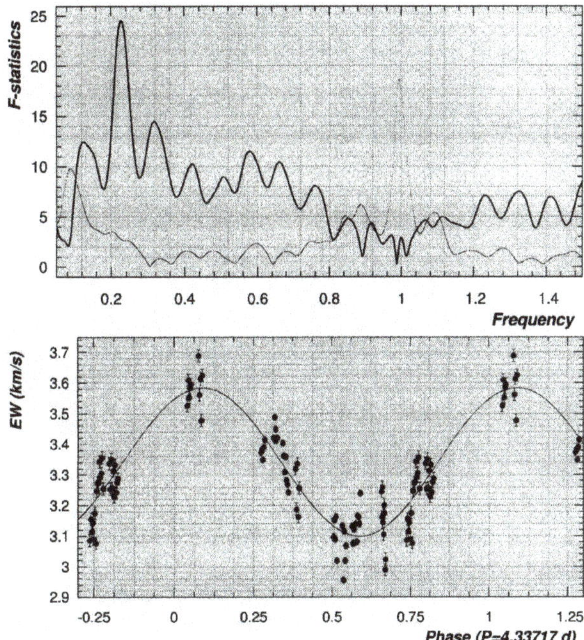

Figure 10. Top: Periodigram for the equivalent width measurements of the primary in HD 104237. The peak at $0.23057\,\mathrm{d}^{-1}$ corresponds to a period of $4.33717\,\mathrm{d}$. The window function is denoted by a dotted line. **Bottom**: EWs measured in the primary phased with the period of $4.33717\,\mathrm{d}$.

This estimate of the rotation period of the primary is in good agreement with the previous result by Böhm et al. [36], $P_{\mathrm{rot}} = 100 \pm 5\,\mathrm{h}$, which was derived from the analysis of CS Hα line variations. This favors the idea that these spots on the stellar surface can be connected with inhomogeneities rotating in the CS media around the star, in other words, with the accretion flows inside the magnetosphere.

6. The Herbig Ae/Be star HD 37806

Our study of the Herbig Ae/Be star HD 37806 was also devoted to a search for a magnetosphere, but in this case we applied a different method, which is not using the modulation of the observational parameters by a hypothetical rotating azimuthal inhomogeneity.

HD 37806 is an isolated Herbig star of A2–B8 type, associated with the Orion OB1b subgroup in the OB1 association. It has a $v \sin i = 120$ km/s and no magnetic field has been detected so far [11,22]. Our spectroscopic observations of the object were carried out in November 2012 with the Coudé spectrograph ASP-14 installed at the 2.6 m Shajn telescope of the Crimean AO. Twenty spectra near the He I λ5876 line were obtained during the two nights of 8/9 and 11/12 November. The analysis of the variability observed during each night has shown that the profile of this line originating in the high-temperature zone of the disk/star interaction demonstrates intensity variations in the form of standing waves in the region of the red absorption (see Figure 11, for the second observing night). These results are presented in more detail in Pogodin et al. [37].

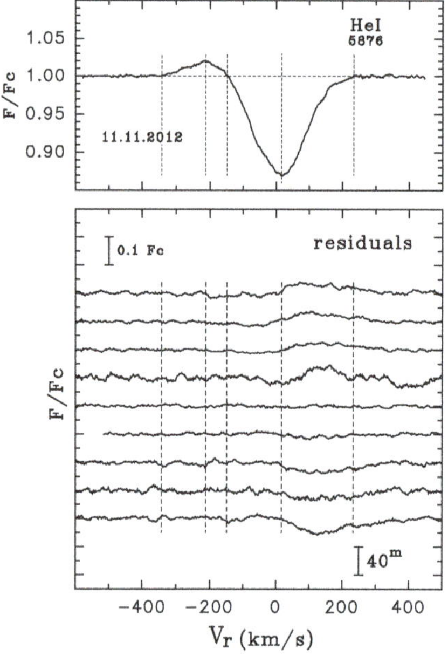

Figure 11. Top: Normalized profiles of the He I λ5876 line observed in the spectrum of HD 37806 on 11 November 2012. **Bottom:** Residual profiles constructed relative to the nightly mean. Both intensity and time scale are indicated. The time increases from top to bottom.

Extensive model calculations have shown that such variability can appear only if a rotating local stream intersects the line-of-sight, which is orthogonal to the Surfaces of Equal Radial Velocities (SERVs) of the moving gas. In the case of a purely Keplerian accretion disk (see left of Figure 12) or a disk with radial motion (see right of Figure 12), the SERVs are not orthogonal to the line-of-sight in the region between the star and the observer. In this case the short-term variations would appear as running waves on the residual plots. Figures 12 and 13 were first presented in slightly different form in the paper by Pogodin et al. [37].

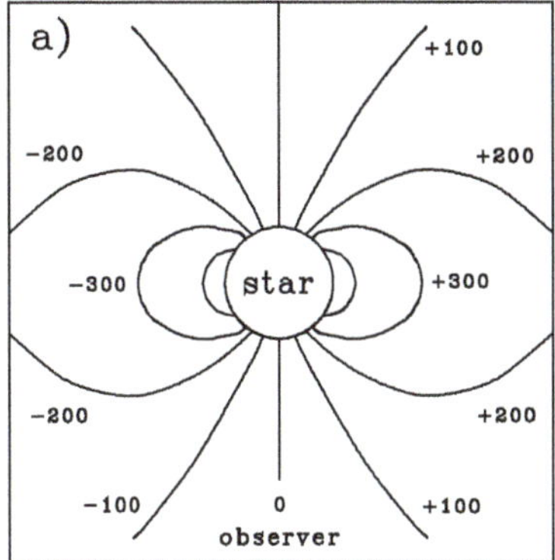
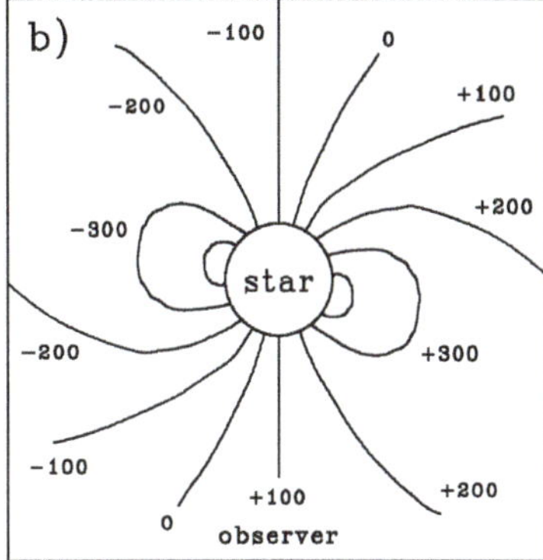

Figure 12. Picture of the SERVs distribution in the disk plane near the star. The case (**a**) is for the Keplerian disk; the case (**b**) is the same as (**a**) but with a radial velocity of the accretion onto the star of $V = 100\,\text{km/s}$. The corresponding values of the radial velocities are indicated near each SERV in the framework of the observer.

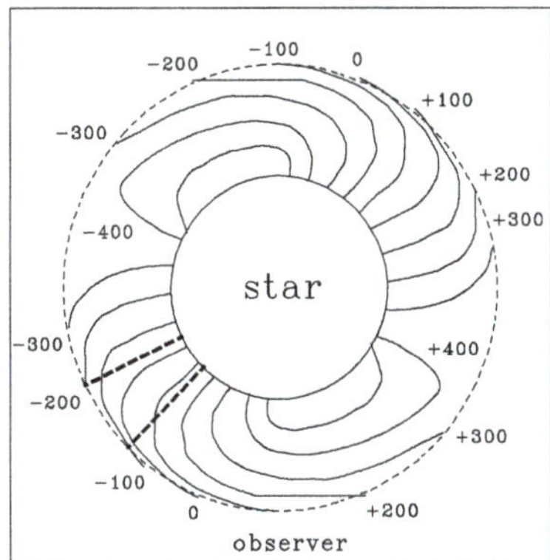

Figure 13. The same as Figure 12, but for the MA model. The parameters of the model are indicated in Section 6.

Only in the case of the MA scenario the SERVs intersecting the line-of-sight can become orthogonal to it. Figure 13 is constructed for the SERVs picture of the simplest geometric model ($i = 90°, \beta = 90°$) such as in the case of HD 101412 (see Section 3). The figure illustrates the SERVs' orientation corresponding to this model. Inside the magnetosphere, we assume the rigid rotation velocity law $V_{\text{rot}}(r) = U_0(r/R)$, and the radial velocity $V_{\text{r}} = V_0(R/r - R/R_{\text{A}})$ for the freefall regime, where $U_0 = 160\,\text{km/s}$ is the equatorial rotation velocity on the stellar surface, and V_0, the radial velocity at the outer magnetospheric boundary, where $R_{\text{A}} = 2\,R$, is equal to $635\,\text{km/s}$, the terminal velocity for the object. Furthermore, we calculated the He I $\lambda5876$ residuals for this model at different rotation phases, adopting the gas to be optically thick in this spectral line and the source function to be equal to 0.1 of the stellar intensity (see Figure 14).

In the general case, for any geometry and orientation of the magnetic field configuration the character of the SERVs is expected to be similar. It is determined by the specific kinematics of the gas inside of the magnetosphere: the radial velocity is minimal near the outer boundary of the magnetosphere and maximal near the stellar surface (with free-fall motion). Its rotation velocity is, on the contrary, maximal at the outer boundary of the magnetosphere and minimal near the stellar surface (rigidly rotating with the star).

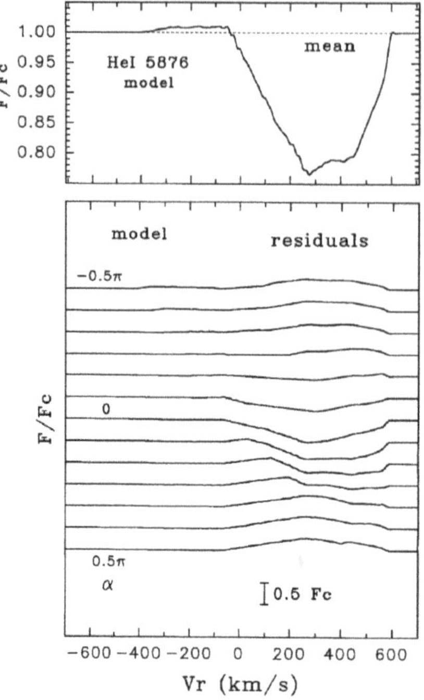

Figure 14. The same as Figure 11, but for the model described in detail in Section 6. α is an angle between the axis of the rotating stream and the line of sight.

We conclude that this type of short-term spectroscopic variability can be considered as an additional signature of the MA scenario. Besides that, the estimation of R_A using Equation (1) shows reasonable values: R_A = 1.7–2.2 R for B = 200–300 G, respectively. In these calculations, we used the following values: $M = 4\,M_\odot$, $R = 4.6\,R_\odot$, and a mass accretion rate $\dot{M} = 1.4 \times 10^{-7}\,M_\odot/\text{yr}$.

7. Conclusions

We consider several diagnostical methods for the magnetospheric accretion in Herbig Ae/Be stars, which are surrounded by accretion disks. The detection of a magnetosphere can be considered as a supplement to the widely used spectropolarimetric observations to measure stellar magnetic fields.

The magnetosphere is a CS region near the star where the disk accretion onto the star is completely governed by the magnetic field of the star. If the magnetic axis is inclined relative to the rotation axis, a global azimuthal inhomogeneity is likely to be formed. Its rotation can modulate different spectral or photometric parameters of the object under investigation. The registration of such a modulation in a form of cyclic parameter variations is the most common way to reveal the presence of the MA scenario of disk/star interaction. Besides that, the moving gas inside the magnetosphere has a specific kinematics, the radial velocity corresponds to the free-fall regime, and the gaseous streams rotate rigidly with the star.

These features of the magnetospheric accretion have been used in our work. Four cases of MA diagnostics have been considered in this paper. These are: (a) HD 101412—a slowly rotating Herbig Ae star with an anomalously strong magnetic field; (b) the early type Herbig B6e star HD 259431; (c) the young binary system HD 104237 consisting of a Herbig Ae star and a T Tauri K2-type star; and (d) the Herbig Ae/Be star HD 37806, where signs of the MA disk/star interaction were found from a particular type of short-term variability of the He I λ5876 line profile during one night rather than from a search for periodicities in the observed parameter variations. The results of our investigation was positive in each of the four cases. We have found that even the two rapidly rotating objects HD 259431 and HD 37806 demonstrate signatures of magnetospheric accretion, i.e., they should have significant magnetic fields. The commonly recognized spectropolarimetric method was ineffective to detect magnetic fields in these two stars.

We hope that the application of MA diagnostics considered in this paper to other Herbig Ae/Be stars will allow to significantly increase the percentage of stars that possess magnetic fields among these objects.

Author Contributions: M.P.—conceptualization, observations at the CrAO, data reduction and analysis and writing–original draft preparation; N.D.—reduction of the ESO data, synthetic spectra calculation and writing—editing; N.B.—observations at the OAN SPM and writing—LaTeX draft preparation, review and editing; S.H., M.S. and S.J.—observations at the ESO, acquisition of archive data, data reduction and analysis and writing—editing; S.P., O.K. and I.A.—observations at the CrAO and spectral data reduction. All authors have read and agreed to the published version of the manuscript.

Funding: M.A.P. and N.G.B. acknowledge the financial support of the Ministry of Science and Higher Education (grant no. 075-15-2020-780 "Exoplanets-4").

Data Availability Statement: The data underlying this article are available on request from the authors.

Acknowledgments: The authors are grateful to the reviewers for useful comments.

Conflicts of Interest: The authors declare no conflict of interest.

References

1. Herbig, G.H. The Spectra of Be- and Ae-Type Stars Associated with Nebulosity. *Astrophys. J. Suppl.* **1960**, *4*, 337. [CrossRef]
2. Finkenzeller, U.; Mundt, R. The Herbig Ae/Be stars associated with nebulosity. *Astron. Astrophys. Suppl.* **1984**, *55*, 109–141.
3. The, P.S.; de Winter, D.; Perez, M.R. A new catalogue of members and candidate members of the Herbig Ae/Be (HAEBE) stellar group. *Astron. Astrophys. Suppl.* **1994**, *104*, 315–339.
4. Gullbring, E.; Hartmann, L.; Briceño, C.; Calvet, N. Disk Accretion Rates for T Tauri Stars. *Astrophys. J.* **1998**, *492*, 323–341. [CrossRef]
5. Hubrig, S.; Schöller, M.; Yudin, R.V. Magnetic fields in Herbig Ae stars. *Astron. Astrophys.* **2004**, *428*, L1–L4. [CrossRef]
6. Hubrig, S.; Stelzer, B.; Schöller, M.; Grady, C.; Schütz, O.; Pogodin, M.A.; Curé, M.; Hamaguchi, K.; Yudin, R.V. Searching for a link between the magnetic nature and other observed properties of Herbig Ae/Be stars and stars with debris disks. *Astron. Astrophys.* **2009**, *502*, 283–301. [CrossRef]
7. Hubrig, S.; Ilyin, I.; Schöller, M.; Lo Curto, G. HARPS spectropolarimetry of Herbig Ae/Be stars. *Astron. Nachrichten* **2013**, *334*, 1093. [CrossRef]
8. Hubrig, S.; Carroll, T.A.; Scholler, M.; Ilyin, I. The prevalence of weak magnetic fields in Herbig Ae stars: The case of PDS 2. *Mon. Not. RAS* **2015**, *449*, L118–L122. [CrossRef]
9. Wade, G.A.; Drouin, D.; Bagnulo, S.; Landstreet, J.D.; Mason, E.; Silvester, J.; Alecian, E.; Böhm, T.; Bouret, J.C.; Catala, C.; et al. Discovery of the pre-main sequence progenitors of the magnetic Ap/Bp stars? *Astron. Astrophys.* **2005**, *442*, L31–L34. [CrossRef]
10. Alecian, E.; Catala, C.; Wade, G.A.; Donati, J.F.; Petit, P.; Landstreet, J.D.; Böhm, T.; Bouret, J.C.; Bagnulo, S.; Folsom, C.; et al. Characterization of the magnetic field of the Herbig Be star HD200775. *Mon. Not. RAS* **2008**, *385*, 391–403. [CrossRef]
11. Alecian, E.; Wade, G.A.; Catala, C.; Grunhut, J.H.; Landstreet, J.D.; Bagnulo, S.; Böhm, T.; Folsom, C.P.; Marsden, S.; Waite, I. A high-resolution spectropolarimetric survey of Herbig Ae/Be stars—I. Observations and measurements. *Mon. Not. RAS* **2013**, *429*, 1001–1026. [CrossRef]
12. Moss, D. The survival of fossil magnetic fields during pre-main sequence evolution. *Astron. Astrophys.* **2003**, *403*, 693–697. [CrossRef]
13. Hubrig, S.; Järvinen, S.P.; Schöller, M.; Carroll, T.A.; Ilyin, I.; Pogodin, M.A. Observations of Magnetic Fields in Herbig Ae/Be Stars. In *Physics of Magnetic Stars, Proceedings of a Conference Held at Special Astrophysical Observatory, Nizhny Arkhyz, Russia, 1–5 October 2018*; Kudryavtsev, D.O., Romanyuk, I.I., Yakunin, I.A., Eds., Astronomical Society of the Pasific: San Francisco, CA, USA, 2019; Volume 518, p. 18.
14. Camenzind, M. Magnetized Disk-Winds and the Origin of Bipolar Outflows. *Rev. Mod. Astron.* **1990**, *3*, 234–265. [CrossRef]
15. Wang, Y.M. Location of the Inner Radius of a Magnetically Threaded Accretion Disk. *Astrophys. J. Lett.* **1996**, *465*, L111–L113. [CrossRef]
16. Donehew, B.; Brittain, S. Measuring the Stellar Accretion Rates of Herbig Ae/Be Stars. *Astron. J.* **2011**, *141*, 46. [CrossRef]

the BTA-6m telescope and our relatively simple model. The methodology described in Silant'ev et al. [22] was taken as a basis for this work.

2. Basic Equations

2.1. Stokes Parameters

When considering radiation from an axially symmetric accretion disk with a magnetic field, its integral Stokes parameters can be written in the following form [22]:

$$\langle Q \rangle = Q(0,\mu)\frac{2}{\pi}\int_0^{\pi/2} d\Phi\, \frac{1+a^2+b^2\cos^2\Phi}{(1+a^2+b^2\cos^2\Phi)^2-(2ab\cos\Phi)^2},$$
$$\langle U \rangle = a\,Q(0,\mu)\frac{2}{\pi}\int_0^{\pi/2} d\Phi\, \frac{1+a^2-b^2\cos^2\Phi}{(1+a^2+b^2\cos^2\Phi)^2-(2ab\cos\Phi)^2}, \tag{1}$$

where $\mu = \cos i$, where i is the angle between the line of sight and the axis of the disk, $Q(0,\mu)$ is the value of the Stokes parameter without a magnetic field. The parameters a and b are expressed, in turn, as follows:

$$a = 0.8\mu\lambda^2(\mu m)B_{\parallel}(G),$$
$$b = 0.8\sqrt{1-\mu^2}\lambda^2(\mu m)B_{\perp}(G), \tag{2}$$

where λ is the wavelength, $B_{\parallel}$ and $B_{\perp}$ are, respectively, the component of the magnetic field in the disk parallel and perpendicular to the disk axis (see Figure 1 from Silant'ev et al. [22]). Taking into account that in the Milne problem (multiple scattering of light in optically thick flattened atmospheres [23,24]) without a magnetic field, the Stokes parameter $U(0,\mu) \equiv 0$, we obtained the following value of the relative polarization and positional angle χ:

$$P_{\text{rel}} = P(B,\mu)/P(0,\mu) = \sqrt{\langle Q\rangle^2 + \langle U\rangle^2}/Q(0,\mu),$$
$$\chi = 0.5\arctan(\langle U\rangle/\langle Q\rangle). \tag{3}$$

Note that P_{rel} depends only on the dimensionless parameters a and b and does not depend on $Q(0,\mu)$.

The polarization value $P(0,\mu)$ without a magnetic field was previously calculated by us numerically using the Sobolev–Chandrasekhar model [23,24] and is tabulated in Gnedin et al. [25]. In addition, we note that the polarization has a small effect on the radiation intensity [23]. Thus, we have the opportunity to accurately calculate the polarization value $P = P_{\text{rel}}P(0,\mu)$, using numerical integration for the parameter P_{rel} and tabular values for $P(0,\mu)$.

2.2. Magnetic Field

When considering the magnetic field in the accretion disk, it is usually assumed (see, for example, Pariev et al. [18]) that its dependence on the radius has a power-law form:

$$B(R) = B_{\text{H}}(R_{\text{H}}/R)^s, \tag{4}$$

where B_{H} is the value of the magnetic field intensity at the event horizon of SMBH in AGN, $R_{\text{H}} = GM_{\text{BH}}(1+\sqrt{1-a_*^2})/c^2$ is the radius of the event horizon, M_{BH} is the mass of the SMBH, G is the gravitational constant, c is the speed of light, $a_* = cJ/GM_{\text{BH}}^2$ is the dimensionless spin of the SMBH, J is the angular momentum of the SMBH rotation. As for the s parameter, there are models with different values of this parameter [18], but for the Shakura–Sunyaev disk, the most often adopted value is $s = 5/4$ [19]. In our work, we decided to investigate in more detail the influence of this parameter on the model and therefore we tried a number of values within $0.5 < s < 2$.

2.3. Dependence of the Polarization Degree on the Wavelength

Since the polarization degree depends on the magnetic field, and the magnetic field depends on the radius, then in order to obtain the dependence of the polarization on the wavelength, we need the dependence of the radius on the wavelength. For the Shakura–Sunyaev accretion disk, we have [26]:

$$R_\lambda(cm) = 0.97 \times 10^{10} \lambda [\mu m]^{4/3} \left(\frac{M_{BH}}{M_\odot} \right)^{2/3} \left(\frac{l_E}{\varepsilon} \right)^{1/3}. \tag{5}$$

where R_λ is the distance in the accretion disk, which corresponds to wavelength λ, $l_E = L_{bol}/L_{Edd}$ is the Eddington ratio, L_{bol} is the bolometric luminosity, $L_{Edd} = 1.3 \times 10^{38} M_{BH}/M_\odot \, erg/s$ is the Eddington luminosity, $\varepsilon = L_{bol}/(\dot{M}c^2)$ is the radiative efficiency, $\dot{M}$ is the accretion rate.

3. Results of Theoretical Calculations

We estimated dependencies of radius R_λ, magnetic field strength B, polarization degree P and position angle χ on wavelength λ (in optical range) for $M_{BH}/M_\odot = 10^8$, $B_H = 10^4$ G, $a_* = 0.9$, $\varepsilon = 0.155$, $l_E = 0.2$ and various values of s. Results are presented in Table 1. In Inoue and Doi [27], the authors obtained $B \approx 10$ G at $R \approx 40R_H$ for AGNs with $M_{BH} \sim 10^8 M_\odot$. These data are in good agreement with our model, in which, for $B_H = 10^4$ G and $s = 1.85$, $B(40R_H) \approx 10.7$ G.

Table 1. Dependencies of radius R_λ, magnetic field strength B, polarization degree P and position angle χ on wavelength λ for $M_{BH}/M_\odot = 10^8$, $B_H = 10^4$ G, $a_* = 0.9$, $\varepsilon = 0.155$, $l_E = 0.2$ and various values of s.

		$s = 1.25$			$s = 1.5$			$s = 1.75$			$s = 2.0$		
λ [μm]	R_λ/R_H	B [G]	P [%]	χ [deg]	B [G]	P [%]	χ [deg]	B [G]	P [%]	χ [deg]	B [G]	P [%]	χ [deg]
0.38	29.4	145.2	0.27	41.5	62.3	0.61	37.0	26.8	1.23	28.3	11.5	1.87	16.7
0.40	31.5	133.3	0.27	41.6	56.3	0.61	37.0	23.8	1.24	28.1	10.0	1.89	16.2
0.42	33.6	122.9	0.26	41.6	51.0	0.61	37.0	21.2	1.26	27.8	08.8	1.91	15.8
0.44	35.8	113.7	0.26	41.7	46.5	0.61	37.0	19.0	1.27	27.6	07.8	1.93	15.4
0.46	38.0	105.6	0.25	41.7	42.5	0.61	37.0	17.1	1.28	27.4	06.9	1.94	15.1
0.48	40.2	98.4	0.25	41.8	39.1	0.61	37.0	15.5	1.29	27.2	06.2	1.95	14.7
0.50	42.4	91.9	0.25	41.8	36.0	0.61	37.0	14.1	1.31	27.1	05.5	1.97	14.4
0.52	44.7	86.1	0.24	41.8	33.3	0.61	37.0	12.9	1.32	26.9	05.0	1.98	14.1
0.54	47.0	80.8	0.24	41.9	30.9	0.61	37.0	11.8	1.33	26.7	04.5	1.99	13.8
0.56	49.3	76.1	0.24	41.9	28.7	0.61	37.0	10.8	1.34	26.6	04.1	2.00	13.5
0.58	51.7	71.8	0.24	42.0	26.8	0.61	37.0	10.0	1.35	26.4	03.7	2.01	13.2
0.60	54.1	67.8	0.23	42.0	25.0	0.61	37.0	9.2	1.36	26.2	03.4	2.02	13.0
0.62	56.5	64.2	0.23	42.0	23.4	0.61	37.0	8.5	1.37	26.1	03.1	2.03	12.7
0.64	59.0	60.9	0.23	42.1	22.0	0.61	37.0	7.9	1.38	25.9	02.9	2.04	12.5
0.66	61.4	57.9	0.23	42.1	20.7	0.61	37.0	7.4	1.39	25.8	02.6	2.04	12.3
0.68	63.9	55.1	0.22	42.1	19.5	0.61	37.0	6.9	1.39	25.7	02.4	2.05	12.1
0.70	66.4	52.5	0.22	42.1	18.4	0.61	37.0	6.4	1.40	25.5	02.3	2.06	11.9
0.72	69.0	50.1	0.22	42.2	17.4	0.61	37.0	6.0	1.41	25.4	02.1	2.06	11.7
0.74	71.6	47.8	0.22	42.2	16.4	0.61	37.0	5.7	1.42	25.3	01.9	2.07	11.5
0.76	74.1	45.7	0.22	42.2	15.6	0.61	37.0	5.3	1.43	25.2	01.8	2.07	11.3
0.78	76.8	43.8	0.21	42.2	14.8	0.61	37.0	5.0	1.43	25.0	01.7	2.08	11.1

We calculated the value of the polarization and the positional angle as a function of the value of the magnetic field intensity at the event horizon B_H and the parameter s in the visible range. For this purpose we adopted the following parameter values characteristic of Seyfert type 1 AGN: $M_{BH} = 10^8 M_\odot$, $a_* = 0.9$ [17,28], radiative efficiency $\varepsilon = 0.155$, Eddington ratio $l_E = 0.2$.

As for the angle i, since it is rather difficult to determine this angle from the observables, research usually adopts $i \approx 45°$ in the calculations. In our work, we used a more complex and arguably a more accurate method. Initially, the values were calculated for all angles from 0 to 90 degrees, and then the resulting data were convolved with a Gaussian centered at 45 degrees. If an angle other than 45 degrees was required for calculations, then a Gaussian with a center at this angle was taken.

Based on the generally assumed dipole nature of the magnetic field, and also based, for example, on the conclusions of Piotrovich et al. [29], the ratio between the components of the magnetic field parallel and perpendicular to the disk axis was taken as $B_\perp = 0.1 B_\parallel$.

Figures 1 and 2 show the results of these calculations in the form of three-dimensional graphs. In Figure 1, one can see that the average value of the polarization in the visible range depends rather strongly on the parameters B_H and s. The rapid decrease in polarization with increasing field and decreasing s is due to Faraday depolarization (see Equation (2) from Silant'ev et al. [22]). It can be seen in Figure 2 that the gradient of the positional angle depends on the parameters in a rather complex way, which theoretically can make it possible to accurately determine the parameters from the observed gradient. However, the amplitude of this dependence, unfortunately, is rather small.

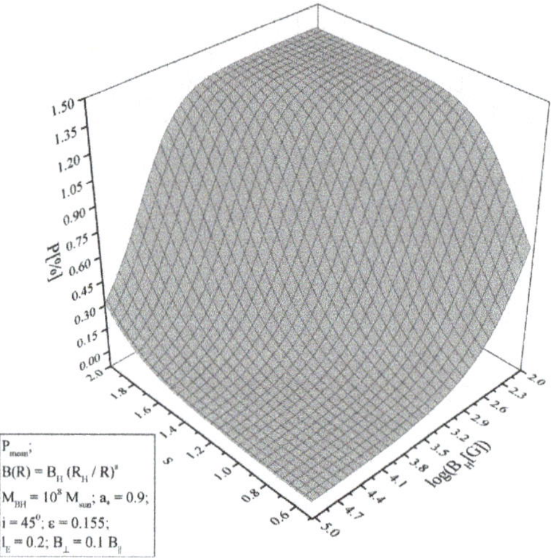

Figure 1. Wavelength-averaged degree of polarization P depending on the value of the magnetic field intensity at the event horizon B_H and the exponent of the power-law dependence s of the magnetic field on the radius in the disk.

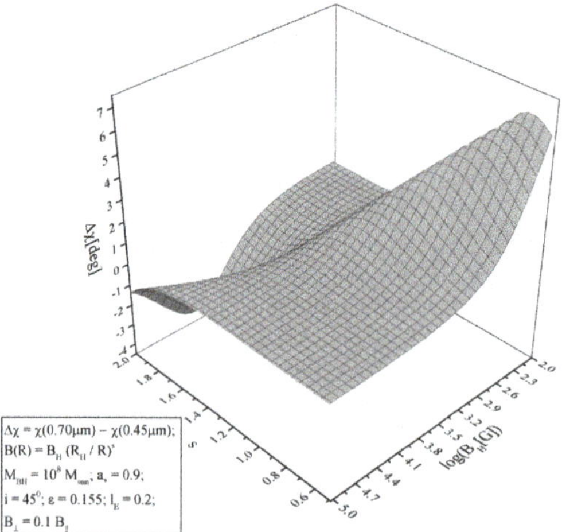

Figure 2. The difference between the positional angles of polarized radiation $\Delta\chi$ at a wavelength of 0.70 μm and 0.45 μm depending on the value of the magnetic field intensity at the event horizon B_H and the exponent of the power-law dependence s of the magnetic field on the radius in the disk.

In addition, we also plotted the dependencies of the degree of polarization and the positional angle on the wavelength for the above parameter values for different values of B_H and s (see Figures 3–6).

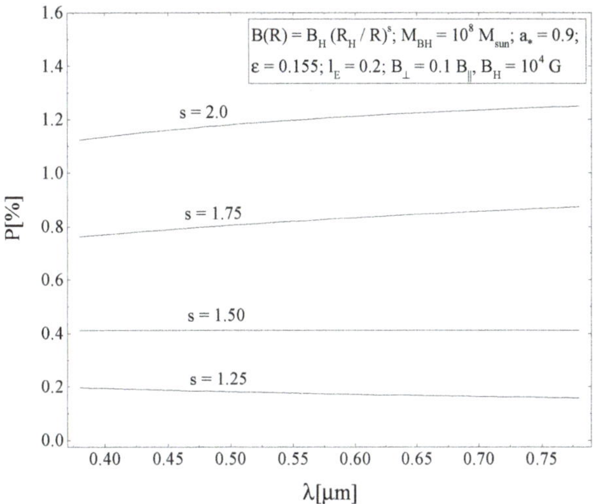

Figure 3. Dependence of the degree of polarization P on the wavelength λ for different values of the parameter s.

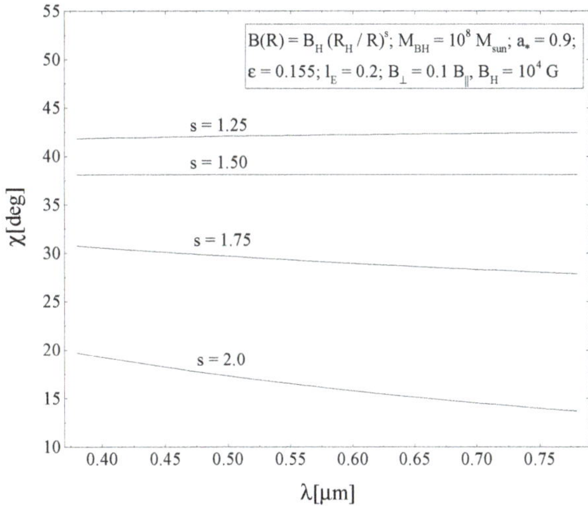

Figure 4. Dependence of the positional angle χ on the wavelength λ for different values of the parameter s.

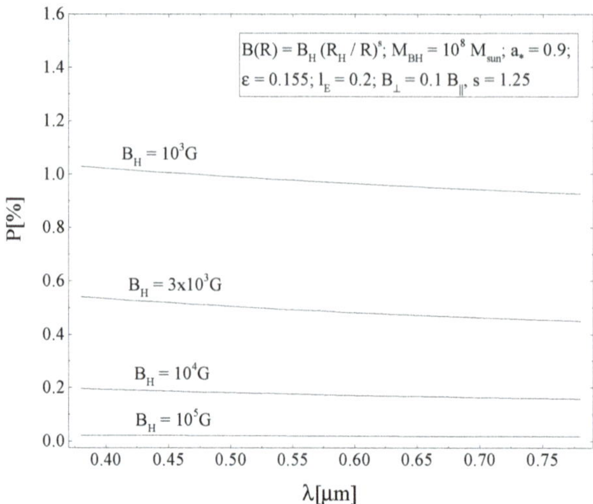

Figure 5. Dependence of the degree of polarization P on the wavelength λ for different values of the parameter B_{H}.

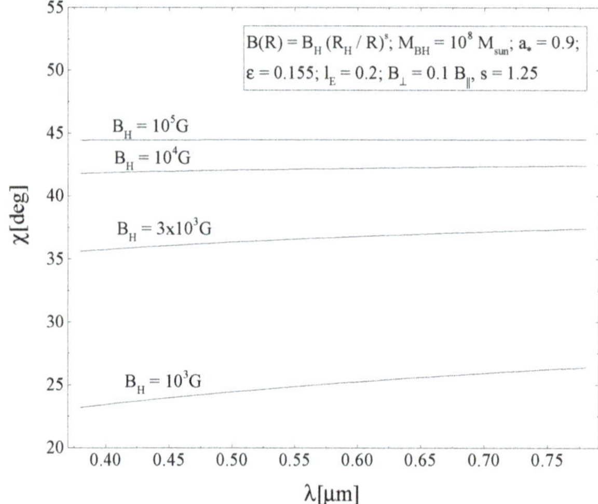

Figure 6. Dependence of the positional angle χ on the wavelength λ for different values of the parameter B_{H}.

It should be noted that the dependence of the degree of polarization and positional angle on the wavelength in our model was found to be relatively weak. It is practically impossible to reliably measure such a gradient of polarization and position angle in real astronomical observations with the existing signal-to-noise level. Therefore, in further calculations, we used the average value of the polarization, making sure that the dependence of the observed polarization on the wavelength was not too strong and had a monotonic form. As for the observed gradient of the position angle, it was evaluated only qualitatively; in particular, objects with a pronounced non-monotonic dependence of the position angle on the wavelength were discarded, for example, in the presence of the so-called "S-shaped" feature, which appears to be due to other physical mechanisms (for example, scattering by a spherical optically thin envelope with a magnetic field [30]).

4. Estimations of B_H and s Based on Optical Spectropolarimetry of AGNs

In our work, we used the already published data of spectropolarimetric observations of a sample of 33 AGN in type 1 Seyfert galaxies, carried out on the BTA-6m telescope with the participation of the authors [16,17,31]. First round of observations were performed in 2008–2009. The observations were carried out with the SCORPIO focal reducer in the spectropolarimetric mode mounted at the prime focus. We used an EEV42-40 2048 × 2048 pixel CCD array with a pixel size of 13.5 × 13.5 μm as the detector and a VPHG550g volume holographic phase grating from the SCORPIO kit operating in the range 3500–7200 Å as the dispersing element. The reciprocal linear dispersion in the detector plane was 1.8 Å/pixel. In the spectrograph, we used a set of five circular diaphragms 4".5 in diameter arranged in the form of a pseudoslit with a step of 9.7-arcsec. A Savart plate placed behind the diaphragms was used as the polarization analyzer. We used the central diaphragm to take the spectra of an object in perpendicular polarization planes and the remaining diaphragms to take the night-sky spectra. The actual spectral resolution of our data was determined by the monochromatic image of the diaphragms and was 40–42 Å. The seeing in all sets of observations was at least 2". The technique of polarization observations and calculations was described by Afanasiev and Moiseev [32]. To calibrate the wavelengths and the relative transmission of the diaphragms, we used an Ar–Ne–He filled line-spectrum lamp and a quartz lamp. To calibrate the spectropolarimetric channel of the spectrograph, we observed standards from Turnshek et al. [33]. Second round of observations were performed in 2012–2016. The observations were carried out with the SCORPIO-2 spectrograph [34]. The spectra were taken in two ranges: 4200–7500 Å for redshifts z < 0.1 (VPHG940 grating) and 5700–9500 Å for z > 0.1 (VPHG940 grating). The spectral resolution for a working 2" slit was 14 and 12 Å, respectively. For objects at Galactic latitudes < 30° we took into account the interstellar polarization that was determined from the observations of stars around the object. The technique of observations and data reduction is described in detail in Afanasiev and Amirkhanyan [35].

We have formed a sample of 33 sources with published mass estimates for their central SMBHs. Since our model assumes a geometrically thin, optically thick disk [19], we considered only objects with Eddington ratio in the $0.01 < l_\mathrm{E} < 0.3$ range [36].

Note that in this work we used the average values of the parameters M_BH, a_* and l_E, neglecting the errors. This is explained by the following arguments. The parameters a_* and l_E themselves have little effect on the polarization value. The M_BH parameter has a more noticeable effect, but this effect can be neglected in comparison with the error of the observational spectropolarimetric data.

For each object, the dependence of the polarization and the gradient of the positional angle on B_H and s was constructed as it was shown in Section 3. Then, this dependence was compared with observational data. The result is a set of values for B_H and s that satisfies these conditions. After that, the B_H values were additionally subject to the condition that they must fall within the limits obtained for these objects by independent methods [37,38]. In the paper Daly [38], errors in the determination of B_H are not indicated, so we took them as ± 0.3 in a logarithmic scale. For those objects for which the magnetic field strength was not previously estimated, we adopted the value $\log B_H[G] = 4.0 \pm 1.0$, since the results of [37,38] estimates of the magnetic field for type 1 Seyfert nuclei give this characteristic range. The values of B_H and s were averaged to obtain the average value and its associated dispersion.

For the objects PG 1545+210 and 2MASX J06021107+2828382, the measured polarization value was found to be greater than the maximum possible value with this calculation method. Since the polarization increases with the inclination angle, the angle value $i = 60°$ was used for these objects. It is believed that for objects of the type under study (Seyfert type 1 galaxies) this angle usually lies within $20° \leq i \leq 60°$ (see, for example, Wu and Han [39]).

Our results are presented in Table 2.

Table 2. Results of determination of the magnetic field intensity at the event horizon $B_{\rm H}$ and parameter s for our objects. P_{obs} is the observed polarization. $M_{\rm BH}$, a and $l_{\rm E}$ are mass, spin and Eddington ratio of SMBH. $B_{\rm H,pr}$ is magnetic field at the event horizon estimated by the method described in Piotrovich et al. [37], except for objects Mkn 509, NGC 3227, NGC 5548 and Mkn 590 whose magnetic fields were estimated in Daly [38]. For objects for which the magnetic field was not previously estimated, we took the value 4.0 ± 1.0.

Object	P_{obs}	$\log\left(\frac{M_{\rm BH}}{M_\odot}\right)$	a_*	$l_{\rm E}$	$\log(B_{\rm H,pr}[{\rm G}])$	$\log(B_{\rm H}[{\rm G}])$	s
2MASS J02093740+5226396 [5]	1.47 ± 0.46	8.53 [5]	0.970 [5]	0.045 [5]	$4.00^{+1.00}_{-1.00}$	$3.33^{+0.17}_{-0.30}$	1.81 ± 0.15
2MASX J02421465+0530361 [5]	0.89 ± 0.43	8.33 [5]	0.930 [5]	0.100 [5]	$4.00^{+1.00}_{-1.00}$	$3.84^{+0.29}_{-0.84}$	1.57 ± 0.27
2MASX J06021107+2828382 [5]	2.34 ± 0.40	8.15 [5]	0.960 [5]	0.009 [5]	$4.00^{+1.00}_{-1.00}$	$3.39^{+0.17}_{-0.28}$	1.73 ± 0.17
3C 390.3 [12]	0.65 ± 0.36	9.12 [6]	0.998 [6]	0.004 [6]	$3.41^{+0.28}_{-0.37}$	$3.18^{+0.08}_{-0.10}$	1.87 ± 0.11
MCG 08-11-011 [5]	1.46 ± 0.90	8.12 [7]	0.950 [5]	0.060 [5]	$4.00^{+1.00}_{-1.00}$	$3.67^{+0.30}_{-0.67}$	1.68 ± 0.24
Mrk 79 [7]	1.31 ± 0.38	7.69 [7]	0.995 [8]	0.040 [3]	$3.87^{+0.15}_{-0.20}$	$3.86^{+0.09}_{-0.12}$	1.81 ± 0.11
Mrk 352 [5]	1.05 ± 0.44	7.19 [5]	0.750 [5]	0.033 [5]	$4.00^{+1.00}_{-1.00}$	$4.04^{+0.30}_{-1.04}$	1.55 ± 0.28
Mrk 509 [7]	1.21 ± 0.43	8.16 [2]	0.840 [9]	0.160 [9]	$4.69^{+0.30}_{-0.30}$	$4.44^{+0.03}_{-0.04}$	1.99 ± 0.02
Mrk 590 [14]	0.61 ± 0.25	7.20 [3]	0.815 [10]	0.214 [3]	$5.05^{+0.30}_{-0.30}$	$4.89^{+0.07}_{-0.08}$	1.72 ± 0.08
Mrk 1095 [7]	0.48 ± 0.24	8.27 [7]	0.930 [8]	0.069 [11]	$3.74^{+0.26}_{-0.19}$	$3.80^{+0.12}_{-0.16}$	1.46 ± 0.15
Mrk 1146 [5]	0.59 ± 0.40	7.41 [6]	0.997 [6]	0.130 [6]	$4.60^{+0.21}_{-0.30}$	$4.58^{+0.12}_{-0.17}$	1.59 ± 0.16
Mrk 1506 [7]	1.06 ± 0.41	7.74 [7]	0.950 [8]	0.050 [9]	$4.43^{+0.18}_{-0.11}$	$4.43^{+0.07}_{-0.08}$	1.96 ± 0.05
NGC 3227 [7]	1.18 ± 0.31	7.22 [7]	0.998 [10]	0.040 [13]	$4.47^{+0.30}_{-0.30}$	$4.42^{+0.13}_{-0.19}$	1.89 ± 0.08
NGC 4051 [7]	0.55 ± 0.54	6.77 [7]	0.970 [8]	0.209 [4]	$5.08^{+0.18}_{-0.17}$	$4.96^{+0.03}_{-0.03}$	1.28 ± 0.31
NGC 4593 [7]	1.06 ± 0.50	6.73 [2]	0.905 [10]	0.120 [3]	$4.00^{+1.00}_{-1.00}$	$4.34^{+0.30}_{-1.34}$	1.39 ± 0.30
NGC 5548 [7]	0.55 ± 0.23	7.68 [7]	0.970 [10]	0.050 [3]	$4.48^{+0.30}_{-0.30}$	$4.50^{+0.14}_{-0.21}$	1.80 ± 0.12
PG 0003+199 [5]	0.52 ± 0.35	7.42 [6]	0.990 [6]	0.308 [6]	$4.68^{+0.17}_{-0.17}$	$4.69^{+0.09}_{-0.11}$	1.52 ± 0.13
PG 0007+106 [5]	0.48 ± 0.30	8.14 [6]	0.993 [6]	0.100 [6]	$4.21^{+0.12}_{-0.12}$	$4.22^{+0.06}_{-0.08}$	1.60 ± 0.14
PG 0026+129 [12]	0.71 ± 0.23	8.09 [6]	0.806 [6]	0.311 [6]	$4.42^{+0.29}_{-0.24}$	$4.48^{+0.13}_{-0.18}$	1.78 ± 0.11
PG 0049+171 [5]	0.81 ± 0.27	8.35 [6]	0.998 [6]	0.025 [6]	$3.92^{+0.14}_{-0.18}$	$3.88^{+0.08}_{-0.10}$	1.91 ± 0.07
PG 0050+124 [5]	1.95 ± 1.70	7.44 [1]	0.997 [5]	0.155 [5]	$4.00^{+1.00}_{-1.00}$	$4.15^{+0.30}_{-1.15}$	1.52 ± 0.34
PG 0054+144 [5]	1.15 ± 0.53	8.97 [6]	0.996 [6]	0.017 [6]	$3.64^{+0.22}_{-0.29}$	$3.42^{+0.05}_{-0.05}$	1.96 ± 0.05
PG 0804+761 [12]	0.27 ± 0.18	8.22 [6]	0.912 [6]	0.157 [6]	$4.25^{+0.22}_{-0.20}$	$4.28^{+0.11}_{-0.14}$	1.42 ± 0.16
PG 0923+129 [5]	0.21 ± 0.11	7.25 [6]	0.998 [6]	0.220 [6]	$4.67^{+0.15}_{-0.18}$	$4.67^{+0.09}_{-0.11}$	1.31 ± 0.10
PG 0923+201 [5]	0.74 ± 0.47	9.02 [6]	0.996 [6]	0.027 [6]	$3.66^{+0.17}_{-0.24}$	$3.61^{+0.10}_{-0.13}$	1.81 ± 0.14
PG 1022+519 [12]	0.50 ± 0.43	7.15 [1]	0.650 [5]	0.275 [5]	$4.98^{+0.20}_{-0.20}$	$4.91^{+0.06}_{-0.07}$	1.51 ± 0.19
PG 1309+355 [5]	1.41 ± 0.73	9.06 [6]	0.991 [6]	0.024 [6]	$3.63^{+0.30}_{-0.36}$	$3.36^{+0.06}_{-0.07}$	1.94 ± 0.06
PG 1501+106 [5]	0.93 ± 0.73	8.53 [6]	0.998 [6]	0.025 [6]	$3.88^{+0.17}_{-0.25}$	$3.83^{+0.11}_{-0.14}$	1.74 ± 0.17
PG 1545+210 [5]	2.03 ± 0.44	9.32 [6]	0.916 [6]	0.021 [6]	$3.52^{+0.42}_{-0.31}$	$3.27^{+0.04}_{-0.04}$	1.96 ± 0.05
PG 1613+658 [5]	0.75 ± 0.43	9.18 [6]	0.998 [6]	0.006 [6]	$3.40^{+0.10}_{-0.13}$	$3.32^{+0.04}_{-0.04}$	1.96 ± 0.05
PG 2112+059 [12]	0.72 ± 0.20	8.69 [6]	0.583 [6]	0.299 [6]	$4.31^{+1.12}_{-0.42}$	$4.19^{+0.14}_{-0.22}$	1.84 ± 0.11
PG 2214+139 [5]	1.23 ± 0.24	8.55 [6]	0.988 [6]	0.055 [6]	$3.96^{+0.23}_{-0.26}$	$3.77^{+0.04}_{-0.05}$	1.98 ± 0.03
PG 2233+134 [12]	0.55 ± 0.40	8.04 [1]	0.500 [5]	0.270 [5]	$4.00^{+1.00}_{-1.00}$	$4.30^{+0.30}_{-1.30}$	1.34 ± 0.36

Sources: (1) Vestergaard and Peterson [40]; (2) Peterson et al. [41]; (3) Satyapal et al. [42]; (4) Gnedin et al. [43]; (5) Afanasiev et al. [17]; (6) Piotrovich et al. [37]; (7) Afanasiev et al. [31]; (8) Piotrovich et al. [44]; (9) Piotrovich et al. [45]; (10) Our estimations based on the method from Piotrovich et al. [37]; (11) Marin [46]; (12) Afanasiev et al. [16]; (13) Devereux [47]; (14) Savić et al. [48].

5. Analysis of the Estimated Parameters of the Magnetic Field

As mentioned earlier, the value of the spin a_* has a rather weak effect on the results, especially taking into account the fact that this value itself varies within rather narrow boundaries. Therefore, the dependence of $B_{\rm H}$ and s on a is negligible.

Figure 7 shows the obtained values of the magnetic field at the event horizon $B_{\rm H}$ and the exponent of the power-law dependence s in graphical form. No pronounced dependence among these parameters on each other is observed. However, one can notice that the values are concentrated in the right side of the graph. It should be noted that all the values of s we obtained are greater than the $5/4$ value usually adopted for accretion disks in type 1 Seyfert nuclei [19]. Figure 8 presents the dependence of the magnetic field strength at the event horizon of the SMBH on its mass. There is a linear dependence of the form $\log B_{\rm H}[{\rm G}] \approx (-0.69 \pm 0.04)\log(M_{\rm BH}/M_\odot) + (9.76 \pm 0.35)$, which in a close agreement with a similar relation obtained by us in Piotrovich et al. [37]. Figure 9 depicts the dependence of the exponent of the power-law index s on the SMBH mass $M_{\rm BH}$. We find a linear dependence $s \approx (0.19 \pm 0.04)\log M_{\rm BH}/M_\odot + (0.18 \pm 0.34)$. It should be noted here that the accuracy of this linear approximation is lower than that obtained for B_{H}. Figure 10 displays the dependence of the magnetic field at the event horizon of the SMBH on its Eddington ratio l_{E}. We derive a linear dependence of the form $\log B_{\mathrm{H}}[G] \approx (1.05 \pm 0.09) \log l_{\mathrm{E}} + (5.38 \pm 0.11)$ in agreement with the results obtained by Piotrovich et al. [37]. Figure 11 gives the dependence of the exponent of the power-law dependence of s on the Eddington ratio l_{E}. A linear dependence of the form $s \approx (-0.25 \pm 0.06) \log l_{\mathrm{E}} + (1.41 \pm 0.08)$ is visible, which, however, also has a lower accuracy than for B_{H}.

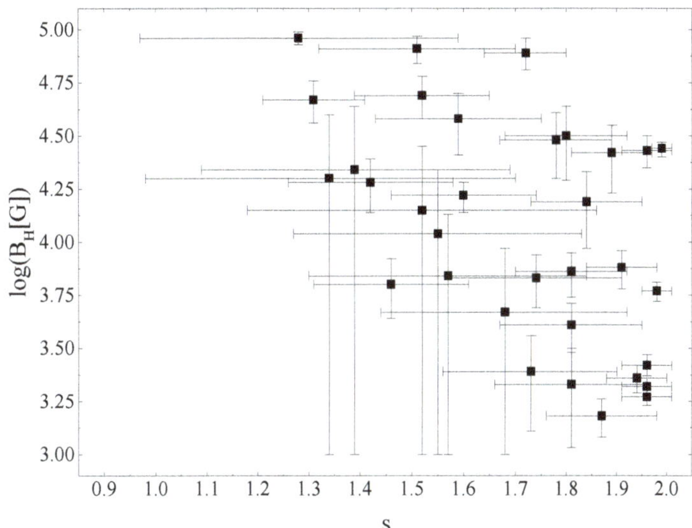

Figure 7. The obtained values of the magnetic field intensity B_{H} and the exponent of the power-law dependence s.

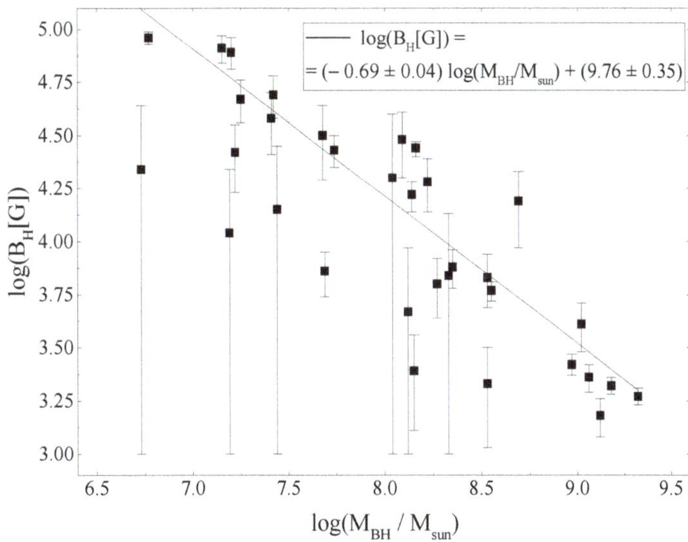

Figure 8. Dependence of the magnetic field intensity B_{H} on the mass of the SMBH M_{BH}.

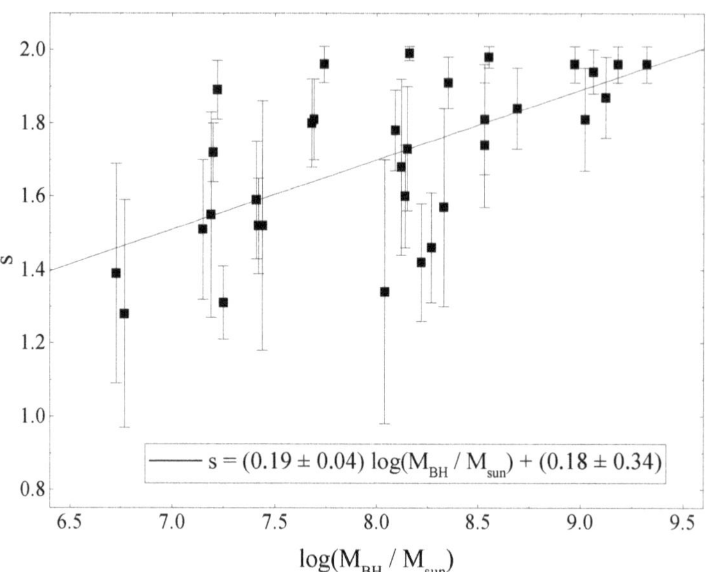

Figure 9. Dependence of the exponent of the power-law dependence s on the SMBH mass M_{BH}.

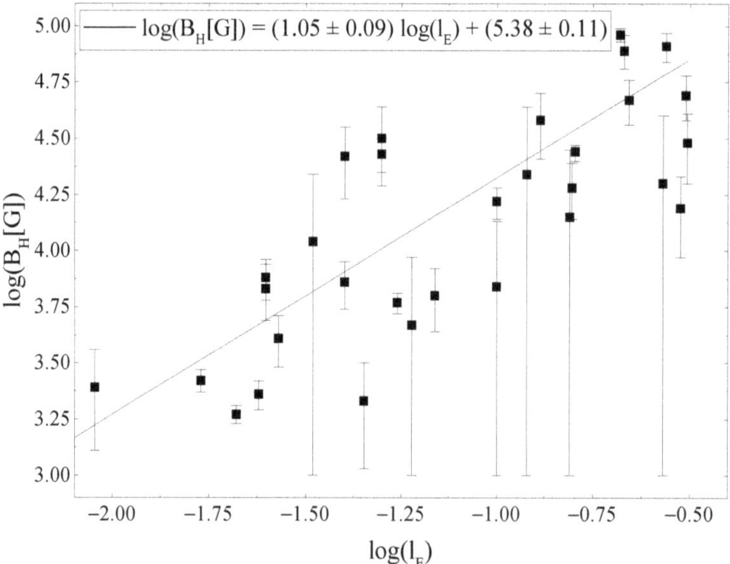

Figure 10. Dependence of the magnetic field intensity B_H on the Eddington ratio l_E.

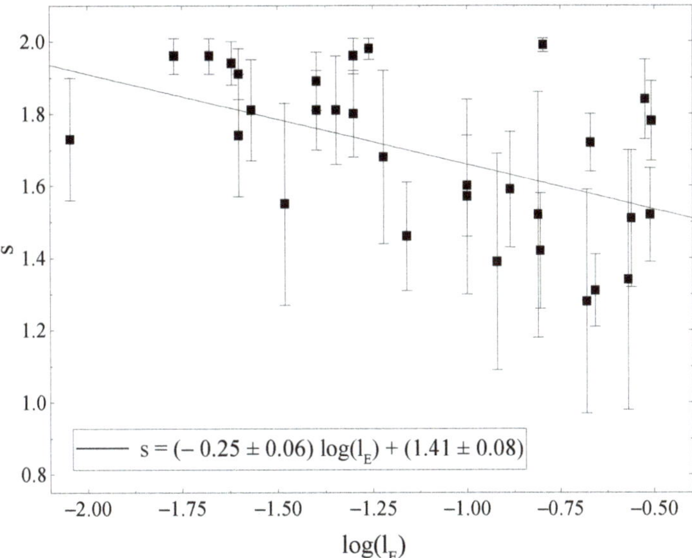

Figure 11. Dependence of the exponent of the power-law dependence s on the Eddington ratio l_E.

The histograms in Figures 12 and 13 show the distributions of objects by the values B_H and s. It should be noted that these histograms cannot be regarded as a source of completely accurate statistical data due to the limited number of objects. Our results are only estimates. It can be seen that, for the magnetic field, the peak of the distribution falls on the region $4.0 < \log B_H[G] < 4.5$. For comparison, we can mention that in our work Piotrovich et al. [37] the peak of the distribution was in the region of $3.5 < \log B_H[G] < 4.0$. This statistical difference is most likely due to the low statistics in current study. In general, these results are consistent with results from Daly [38].

As for the parameter s, the peak of the distribution is in the region of $1.85 < s < 2.00$, which is quite interesting, given that, as mentioned earlier, the standard value of s for accretion disks of the Shakura–Sunyaev type is usually considered to be $5/4$ [19].

Table 3 presents the main statistical properties of the parameters of our model and the results of our calculations. The statistical properties of B_H were found to be close to the values from Piotrovich et al. [37] and Daly [38].

Table 3. Basic statistical properties of the parameters. "Mean" indicates the arithmetic mean, "Median" the median value, "SD" the standard deviation.

Parameter	Mean	Median	SD
$\log M_{BH}/M_\odot$	8.05	8.14	0.72
$\log l_E$	−1.19	−1.22	0.51
$\log B_H[G]$	4.06	4.15	0.52
s	1.70	1.74	0.22

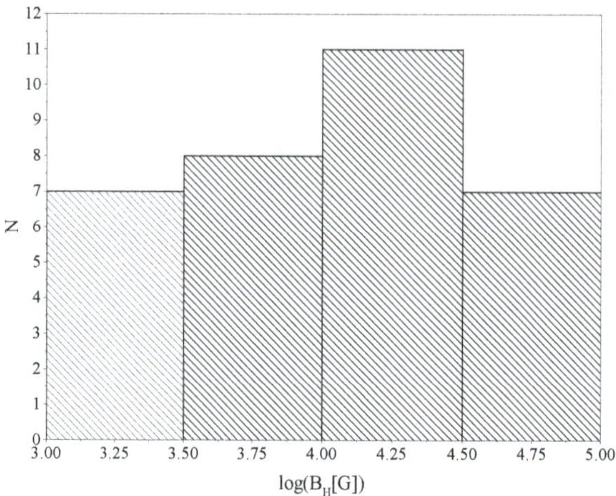

Figure 12. A histogram showing the number of objects with a certain B_H value.

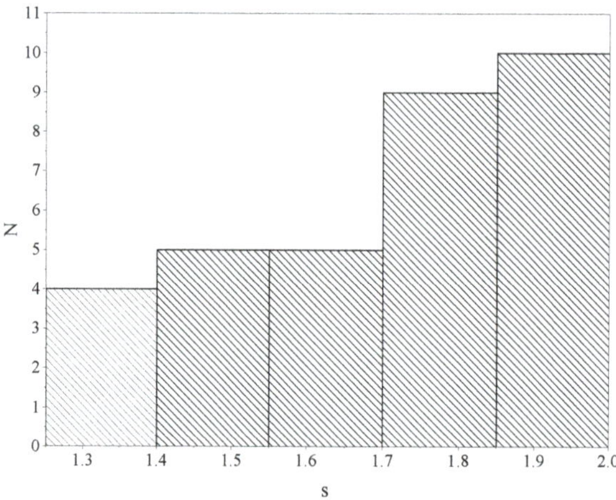

Figure 13. A histogram showing the number of objects with a certain s value.

6. Conclusions

Based on the spectropolarimetric data of 33 Seyfert type 1 galaxies obtained with the BTA-6m telescope of the Special Astrophysical Observatory, estimates of the values of magnetic fields at the event horizon of the SMBHs B_H and the values of the exponents of the power-law dependence s of the magnetic field on the radius $B(R) = B_\mathrm{H}(R_\mathrm{H}/R)^s$, where R_H is the radius of the event horizon.

The average value of $\log B_\mathrm{H}[\mathrm{G}]$ was found to be ~ 4, which is in good agreement with the results obtained by other methods [37,38], in which the magnetic field strength was estimated using the physical parameters of the relativistic jets. It was possible to reveal the dependence of the magnetic field on the SMBHs mass and the Eddington ratio of the form $\log B_\mathrm{H}[\mathrm{G}] \approx (-0.69 \pm 0.04)\log(M_\mathrm{BH}/M_\odot) + (9.76 \pm 0.35)$ and $\log B_\mathrm{H}[\mathrm{G}] \approx (1.05 \pm 0.09)\log l_\mathrm{E} + (5.38 \pm 0.11)$, which agree well with the results of Piotrovich et al. [37].

The average value of s is $s \approx 1.7$, and the maximum distribution over s is within $1.85 < s < 2.0$. This is a rather interesting result, since $s = 5/4$ is usually taken in calculations for accretion disks in type 1 Seyfert nuclei. We also managed to estimate the dependence of s on the SMBHs mass and the Eddington ratio of the form $s \approx (0.19 \pm 0.04)\log M_\mathrm{BH}/M_\odot + (0.18 \pm 0.34)$ and $s \approx (-0.25 \pm 0.06)\log l_\mathrm{E} + (1.41 \pm 0.08)$.

In addition, although these approximations have a larger error than in the case of B_{H}, they are still of interest. In particular, it may indicate that the more complex disk models are required than the Shakura–Sunyaev model. This problem undoubtedly requires further study.

In addition, for two objects PG 1545+210 and 2MASX J06021107+2828382, the measured polarization value was found to be greater than the maximum possible value at the inclination angle between the line of sight and the axis of the accretion disk $i = 45°$. Since the polarization increases with the angle, it was concluded that for these objects the angle should be closer to $i = 60°$.

Author Contributions: Conceptualization, M.P.; methodology, M.P.; validation, M.P.; formal analysis, M.P., S.B. and T.N.; investigation, S.B. and T.N.; resources, S.B. and T.N.; data curation, S.B. and T.N.; writing—original draft preparation, M.P.; writing—review and editing, M.P.; visualization, M.P.; supervision, M.P.; project administration, M.P.; funding acquisition, M.P. and S.B. All authors have read and agreed to the published version of the manuscript.

Funding: This research was supported by the grant of Russian Science Foundation project number 20-12-00030 "Investigation of geometry and kinematics of ionized gas in active galactic nuclei by polarimetry methods". Observations with the SAO RAS telescope are supported by the Ministry of Science and Higher Education of the Russian Federation (including agreement No.05.619.21.0016, project IDRFMEFI61919X0016).

Data Availability Statement: The data underlying this article are available in the article.

Acknowledgments: The authors are grateful to the employees of the Special Astrophysical Observatory V.L. Afanasiev (deceased 21 December 2020) and E.S. Shablovinskaya for help in carrying out observations and processing the results and the employee of the Central Astronomical Observatory at Pulkovo N.A. Silant'ev for useful comments and advice. The authors are also grateful to the reviewers for useful comments.

Conflicts of Interest: The authors declare no conflict of interest.

References

1. Blaes, O.M. Course 3: Physics Fundamentals of Luminous Accretion Disks around Black Holes. In *Accretion Discs, Jets and High Energy Phenomena in Astrophysics*; Beskin, V., Henri, G., Menard, F., Pelletier, G., Dalibard, J., Eds.; Springer: Berlin/Heidelberg, Germany, 2004; Volume 78, pp. 137–185.
2. Moran, J.M. The Black-Hole Accretion Disk in NGC 4258: One of Nature's Most Beautiful Dynamical Systems. In *Frontiers of Astrophysics: A Celebration of NRAO's 50th Anniversary*; Bridle, A.H., Condon, J.J., Hunt, G.C., Eds.; Astronomical Society of the Pacific Conference Series; National Radio Astronomy Observatory: Charlottesville, VA, USA, 2008; Volume 395, p. 87.
3. Moderski, R.; Sikora, M.; Lasota, J.P. On Black Hole Spins and Dichotomy of Quasars. In *Relativistic Jets in AGNs*; Ostrowski, M., Sikora, M., Madejski, G., Begelman, M., Eds.; Jagiellonian University: Krakow, Poland, 1997; pp. 110–116.
4. Li, L.X. Accretion Disk Torqued by a Black Hole. *Astrophys. J.* **2002**, *567*, 463–476. [CrossRef]
5. Wang, D.X.; Xiao, K.; Lei, W.H. Evolution characteristics of the central black hole of a magnetized accretion disc. *Mon. Not. R. Astron. Soc.* **2002**, *335*, 655–664. [CrossRef]
6. Wang, D.X.; Ma, R.Y.; Lei, W.H.; Yao, G.Z. Magnetic Coupling of a Rotating Black Hole with Its Surrounding Accretion Disk. *Astrophys. J.* **2003**, *595*, 109–119. [CrossRef]
7. Zhang, W.M.; Lu, Y.; Zhang, S.N. The Black Hole Mass and Magnetic Field Correlation in Active Galactic Nuclei. *Chin. J. Astron. Astrophys. Suppl.* **2005**, *5*, 347–352. [CrossRef]
8. Ma, R.Y.; Yuan, F.; Wang, D.X. Influence of the Magnetic Coupling Process on Advection-dominated Accretion Flows around Black Holes. *Astrophys. J.* **2007**, *671*, 1981–1989. [CrossRef]
9. Martin, P.G.; Thompson, I.B.; Maza, J.; Angel, J.R.P. The polarization of Seyfert galaxies. *Astrophys. J.* **1983**, *266*, 470–478. [CrossRef]
10. Webb, W.; Malkan, M.; Schmidt, G.; Impey, C. The Wavelength Dependence of Polarization of Active Galaxies and Quasars. *Astrophys. J.* **1993**, *419*, 494. [CrossRef]
11. Impey, C.D.; Malkan, M.A.; Webb, W.; Petry, C.E. Ultraviolet Spectropolarimetry of High-Redshift Quasars with the Hubble Space Telescope. *Astrophys. J.* **1995**, *440*, 80. [CrossRef]
12. Wilkes, B.J.; Schmidt, G.D.; Smith, P.S.; Mathur, S.; McLeod, K.K. Optical Detection of the Hidden Nuclear Engine in NGC 4258. *Astrophys. J. Lett.* **1995**, *455*, L13. [CrossRef]
13. Barth, A.J.; Tran, H.D.; Brotherton, M.S.; Filippenko, A.V.; Ho, L.C.; van Breugel, W.; Antonucci, R.; Goodrich, R.W. Polarized Narrow-Line Emission from the Nucleus of NGC 4258. *Astron. J.* **1999**, *118*, 1609–1617. [CrossRef]

14. Smith, J.E.; Young, S.; Robinson, A.; Corbett, E.A.; Giannuzzo, M.E.; Axon, D.J.; Hough, J.H. A spectropolarimetric atlas of Seyfert 1 galaxies. *Mon. Not. R. Astron. Soc.* **2002**, *335*, 773–798. [CrossRef]

15. Modjaz, M.; Moran, J.M.; Kondratko, P.T.; Greenhill, L.J. Probing the Magnetic Field at Subparsec Radii in the Accretion Disk of NGC 4258. *Astrophys. J.* **2005**, *626*, 104–119. [CrossRef]

16. Afanasiev, V.L.; Borisov, N.V.; Gnedin, Y.N.; Natsvlishvili, T.M.; Piotrovich, M.Y.; Buliga, S.D. Spectropolarimetric observations of active galactic nuclei with the 6-m BTA telescope. *Astron. Lett.* **2011**, *37*, 302–310. [CrossRef]

17. Afanasiev, V.L.; Gnedin, Y.N.; Piotrovich, M.Y.; Natsvlishvili, T.M.; Buliga, S.D. Determination of Supermassive Black Hole Spins Based on the Standard Shakura–Sunyaev Accretion Disk Model and Polarimetric Observations. *Astron. Lett.* **2018**, *44*, 362–369. [CrossRef]

18. Pariev, V.I.; Blackman, E.G.; Boldyrev, S.A. Extending the Shakura–Sunyaev approach to a strongly magnetized accretion disc model. *Astron. Astrophys.* **2003**, *407*, 403–421. [CrossRef]

19. Shakura, N.I.; Sunyaev, R.A. Black holes in binary systems. Observational appearance. *Astron. Astrophys.* **1973**, *24*, 337–355.

20. Gnedin, Y.N.; Buliga, S.D.; Silant'ev, N.A.; Natsvlishvili, T.M.; Piotrovich, M.Y. Topology of magnetic field and polarization in accretion discs of AGN. *Astrophys. Space Sci.* **2012**, *342*, 137–145. [CrossRef]

21. Baczko, A.K.; Schulz, R.; Kadler, M.; Ros, E.; Perucho, M.; Krichbaum, T.P.; Böck, M.; Bremer, M.; Grossberger, C.; Lindqvist, M.; et al. A highly magnetized twin-jet base pinpoints a supermassive black hole. *Astron. Astrophys.* **2016**, *593*, A47. [CrossRef]

22. Silant'ev, N.A.; Piotrovich, M.Y.; Gnedin, Y.N.; Natsvlishvili, T.M. Magnetic fields of AGNs and standard accretion disk model: testing by optical polarimetry. *Astron. Astrophys.* **2009**, *507*, 171–182. [CrossRef]

23. Chandrasekhar, S. *Radiative Transfer*; Clarendon Press: Oxford, UK, 1950.

24. Sobolev, V.V. *A Treatise on Radiative Transfer*; Van Nostrand: Princeton, NJ, USA, 1963.

25. Gnedin, Y.N.; Piotrovich, M.Y.; Silant'ev, N.A.; Natsvlishvili, T.M.; Buliga, S.D. Polarization of Radiation and Basic Parameters of the Circumnuclear Region of Active Galactic Nuclei. *Astrophysics* **2015**, *58*, 443–452. [CrossRef]

26. Poindexter, S.; Morgan, N.; Kochanek, C.S. The Spatial Structure of an Accretion Disk. *Astrophys. J.* **2008**, *673*, 34–38. [CrossRef]

27. Inoue, Y.; Doi, A. Detection of Coronal Magnetic Activity in nearby Active Supermassive Black Holes. *Astrophys. J.* **2018**, *869*, 114. [CrossRef]

28. Trakhtenbrot, B. The Most Massive Active Black Holes at z ~ 1.5–3.5 have High Spins and Radiative Efficiencies. *Astrophys. J. Lett.* **2014**, *789*, L9. [CrossRef]

29. Piotrovich, M.Y.; Silant'ev, N.A.; Gnedin, Y.N.; Natsvlishvili, T.M.; Buliga, S.D. The magnetic-field structure in a stationary accretion disk. *Astron. Rep.* **2016**, *60*, 486–497. [CrossRef]

30. Gnedin, Y.N.; Silant'Ev, N.A.; Piotrovich, M.Y.; Pogodin, M.A. Polarization Effects in the Radiation of Magnetized Envelopes and Extended Accretion Structures. *Astron. Rep.* **2005**, *49*, 179–189. [CrossRef]

31. Afanasiev, V.L.; Popović, L.Č.; Shapovalova, A.I. Spectropolarimetry of Seyfert 1 galaxies with equatorial scattering: Black hole masses and broad-line region characteristics. *Mon. Not. R. Astron. Soc.* **2019**, *482*, 4985–4999. [CrossRef]

32. Afanasiev, V.L.; Moiseev, A.V. The SCORPIO Universal Focal Reducer of the 6-m Telescope. *Astron. Lett.* **2005**, *31*, 194–204. [CrossRef]

33. Turnshek, D.A.; Bohlin, R.C.; Williamson, R.L.I.; Lupie, O.L.; Koornneef, J.; Morgan, D.H. An Atlas of Hubble Space Telescope Photometric, Spectrophotometric, and Polarimetric Calibration Objects. *Astron. J.* **1990**, *99*, 1243. [CrossRef]

34. Afanasiev, V.L.; Moiseev, A.V. Scorpio on the 6 m Telescope: Current State and Perspectives for Spectroscopy of Galactic and Extragalactic Objects. *Balt. Astron.* **2011**, *20*, 363–370. [CrossRef]

35. Afanasiev, V.L.; Amirkhanyan, V.R. Technique of polarimetric observations of faint objects at the 6-m BTA telescope. *Astrophys. Bull.* **2012**, *67*, 438–452. [CrossRef]

36. Netzer, H.; Trakhtenbrot, B. Bolometric luminosity black hole growth time and slim accretion discs in active galactic nuclei. *Mon. Not. R. Astron. Soc.* **2014**, *438*, 672–679. [CrossRef]

37. Piotrovich, M.Y.; Mikhailov, A.G.; Buliga, S.D.; Natsvlishvili, T.M. Determination of magnetic field strength on the event horizon of supermassive black holes in active galactic nuclei. *Mon. Not. R. Astron. Soc.* **2020**, *495*, 614–620. [CrossRef]

38. Daly, R.A. Black Hole Spin and Accretion Disk Magnetic Field Strength Estimates for More Than 750 Active Galactic Nuclei and Multiple Galactic Black Holes. *Astrophys. J.* **2019**, *886*, 37. [CrossRef]

39. Wu, X.B.; Han, J.L. Inclinations and Black Hole Masses of Seyfert 1 Galaxies. *Astrophys. J. Lett.* **2001**, *561*, L59–L62. [CrossRef]

40. Vestergaard, M.; Peterson, B.M. Determining Central Black Hole Masses in Distant Active Galaxies and Quasars. II. Improved Optical and UV Scaling Relationships. *Astrophys. J.* **2006**, *641*, 689–709. [CrossRef]

41. Peterson, B.M.; Ferrarese, L.; Gilbert, K.M.; Kaspi, S.; Malkan, M.A.; Maoz, D.; Merritt, D.; Netzer, H.; Onken, C.A.; Pogge, R.W.; et al. Central Masses and Broad-Line Region Sizes of Active Galactic Nuclei. II. A Homogeneous Analysis of a Large Reverberation-Mapping Database. *Astrophys. J.* **2004**, *613*, 682–699. [CrossRef]

42. Satyapal, S.; Dudik, R.P.; O'Halloran, B.; Gliozzi, M. The Link between Star Formation and Accretion in LINERs: A Comparison with Other Active Galactic Nucleus Subclasses. *Astrophys. J.* **2005**, *633*, 86–104. [CrossRef]

43. Gnedin, Y.N.; Globina, V.N.; Piotrovich, M.Y.; Buliga, S.D.; Natsvlishvili, T.M. Spins of Supermassive Black Holes and the Magnetic Fields of Accretion Disks in Active Galactic Nuclei with Maser Emission. *Astrophysics* **2014**, *57*, 163–175. [CrossRef]

44. Piotrovich, M.; Gnedin, Y.; Natsvlishvili, T.; Buliga, S. Estimates of supermassive black hole (SMBH) spins for the standard accretion disk model: Comparison with relativistic fitting of SMBH spectra. *New Astron.* **2018**, *65*, 25–28. [CrossRef]

45. Piotrovich, M.Y.; Gnedin, Y.N.; Natsvlishvili, T.M.; Buliga, S.D. Constraints on spin of a supermassive black hole in quasars with big blue bump. *Astrophys. Space Sci.* **2017**, *362*, 231. [CrossRef]
46. Marin, F. Are there reliable methods to estimate the nuclear orientation of Seyfert galaxies? *Mon. Not. R. Astron. Soc.* **2016**, *460*, 3679–3705. [CrossRef]
47. Devereux, N. The dynamics of the broad-line region in NGC 3227. *Mon. Not. R. Astron. Soc.* **2021**, *500*, 786–794. [CrossRef]
48. Savić, D.; Goosmann, R.; Popović, L.Č.; Marin, F.; Afanasiev, V.L. AGN black hole mass estimates using polarization in broad emission lines. *Astron. Astrophys.* **2018**, *614*, A120. [CrossRef]

Review

Scattered Radiation of Protoplanetary Disks

Vladimir P. Grinin [1,2,*] **and Larisa V. Tambovtseva** [1]

[1] Pulkovo Observatory of the Russian Academy of Sciences, 196140 Saint Petersburg, Russia; lvtamb@mail.ru
[2] V.V. Sobolev Astronomical Institute, Saint Petersburg State University, 198504 Saint Petersburg, Russia
* Correspondence: grinin@gao.spb.ru

Abstract: Scattered radiation of circumstellar (CS) dust plays an important role in the physics of young stars. Its observational manifestations are various but more often they are connected with the appearance of intrinsic polarization in young stars and their CS disks. In our brief review we consider two classes of astrophysical objects in which the participation of scattered radiation is key for understanding their nature. First of all, these are irregular variables (UX Ori type stars). The modern idea of their nature and the mechanism of their variability has been formed thanks to synchronous observations of their linear polarization and brightness. The second class of objects is the CS disks themselves. Their detailed investigation became possible due to observations in polarized light using a coronographic technique and large telescopes.

Keywords: protoplanetary disk; scattered radiation; linear polarization; UX Ori stars; RW Aur

Citation: Grinin, V.P.; Tambovtseva, L.V. Scattered Radiation of Protoplanetary Disks. *Universe* 2022, *8*, 224. https://doi.org/10.3390/universe8040224

Academic Editors: Galina L. Klimchitskaya, Vladimir M. Mostepanenko and Nazar R. Ikhsanov

Received: 14 February 2022
Accepted: 30 March 2022
Published: 2 April 2022

Publisher's Note: MDPI stays neutral with regard to jurisdictional claims in published maps and institutional affiliations.

1. Introduction

Scattered radiation of circumstellar disks, as a rule, makes a small contribution to the optical radiation of young stars. The exception is a subclass of irregular variable stars with UX Ori as a prototype [1], and a small number of highly embedded stars and stars with edge-on disks. The family of UX Ori stars includes mainly the stars of the Ae spectral type. For a long time broadband photometric observations were the only method for their investigation. However, such observations could not unambiguously determine the mechanism of their unusual variability, representing a sequence of stochastic brightness weakening with an amplitude up to $2\text{--}3^m$ and duration from a few days to a few weeks. The same observations can often be explained in completely different ways. For example, the reddening of the star with a decrease in its brightness has been equally well interpreted with both an increase in the CS extinction [2], and an appearance of magnetic spots on the star [3].

The important role of scattered radiation in understanding the nature of the variability of these objects was first pointed out by one of the authors of this paper [4]. During the deep brightness minima caused by screening the star with the CS gas and dust clouds, the direct radiation of the star is blocked by the screen. At such moments, the scattered radiation of the CS dust dominates the observed radiation. This explains a range of properties of these objects, including the unusual behavior of color indices at brightness fading, restriction of the brightness amplitudes and increases in the linear polarization in the brightness minima. Based on these facts, Grinin et al. [5] determined the evolutionary status of the stars from this subclass: the UX Ori stars (or UXOrs) are usually young stars, namely, intermediate-mass Herbig Ae stars surrounded by protoplanetary disks, and they differ from the usual photometrically inactive Herbig stars only with a small inclination of their CS disks to the line of sight. This conclusion has been supported by further investigations [6,7], including interferometric observations in the near-infrared region of the spectrum [8,9].

It should be noted that an addition of the photometrically active UX Ori type stars to the photometrically inactive ("classical") Herbig AeBe stars had a strong influence on the further development of our ideas about all classes of Herbig stars. It turned out that these stars are not surrounded by spherical gas and dust envelopes, as previously assumed (see, e.g., [10]), but by circumstellar disks. It was also found that the Herbig stars demonstrate not only spectral signs of the matter outflow but also signs of accretion. It all depends on the angle between the disk plane and the line of sight [11,12].

2. Coronographic Effect

During the long-standing photopolarimetric observations of the UX Ori type stars, it turned out that the observed changing in the linear polarization parameters is well described with the model suggested in [4]. It claims that the CS dust clouds obscure the star from the observer but do not influence the optical properties of the scattered radiation of the CS disk. The latter means that the shadow zones on the disk created by the clouds are much smaller in comparison with its size. Herewith the dust cloud themselves also do not influence the polarization of the stellar radiation that passes through them. The latter means that the dust grains in the clouds are not lined up. In this condition, changes in the intensity of the observed radiation during the eclipse I_{obs} are described with the following simple relationship:

$$I_{obs} = I_* e^{-\tau_\lambda} + I_{sc},\tag{1}$$

where I_* is the intensity of the stellar radiation out of the eclipse, τ_λ is the optical depth of the cloud screening the star at the wavelength λ, and I_{sc} is the intensity of the scattered radiation, which is considered unchanged during the eclipse, as was mentioned above.

The dependence of τ on λ is assumed as $\tau_\lambda = \tau f(\lambda)$, which suggests that within the stellar disk the screen is homogeneous (this permits us to treat the star as a point source of light), and dependence of its optical depth on the wavelength does not change during the eclipse. The latter assumption is based on a very important observational fact: the existence of the straight section on the color–magnitude diagram that is used to determine the CS extinction law (see, e.g., [13]).

From Formula (1) one can directly obtain a link between the intensity of the scattered radiation of the disk and the possible maximum amplitude of the stellar brightness reduction:

$$(\Delta m)_{max} = 2.5 \log(1 + I_*/I_{sc}),\tag{2}$$

Knowing the magnitudes $(\Delta m)_{max}$ from the photometric observations, one can immediately find for each star the contribution of the scattered radiation of the CS disk to the optical radiation of the star out of the eclipse. On average about 10% [4] is confirmed when modeling the interferometric observations [9].

This model explains changes in the color indices observed in UX Ori type stars during the brightness minima (Figure 1), and describes well the observational link between the brightness variations and parameters of the linear polarization of these stars. It permits us to distinguish with high accuracy between the intrinsic polarization of the star caused by the scattered radiation of the CS disk and the interstellar one:

$$\mathbf{P}_{obs} = \mathbf{P}_{IS} + \mathbf{P}_{in}(\Delta m),\tag{3}$$

Here $\mathbf{P}_{obs}$, $\mathbf{P}_{IS}$ and $\mathbf{P}_{in}$ are pseudo-vectors of the observed, interstellar and intrinsic polarization of the investigated star, and Δm is the amplitude of the diminution of the star brightness counted from its brightest state.

$$\mathbf{P}_{in}(\Delta m) = \mathbf{P}_{in}(0)\, e^{-0.4\,\Delta m}.\tag{4}$$

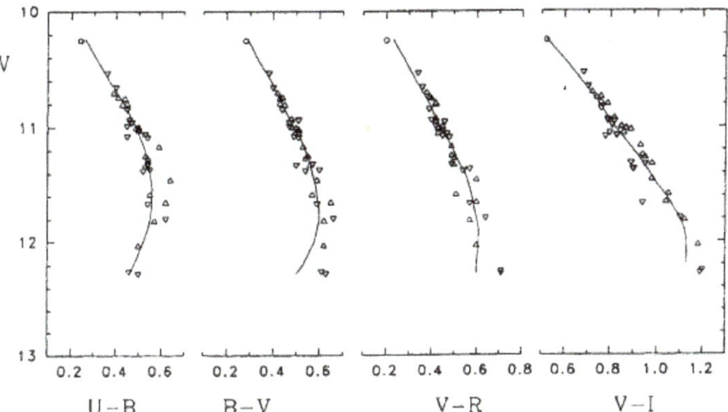

Figure 1. Color–magnitude diagram of the UX Ori type star WW Vul from observations of the deep brightness minimum in 1997 [14]. Open circles mark the ascending part of the minimum; triangles mark its descending part. Lines show theoretical dependencies calculated in [15] on the base of Equation (1).

After making up such equations for each *i*-th observation, we obtain a redundant system of equations for each photometric band. The solution of such a system with the least square method allows us to find two unknown quantities, $\mathbf{P}_{IS}$ and $\mathbf{P}_{in}(0)$. Each of them is a pseudo-vector that gives the degree of the interstellar and intrinsic polarization and their position angles. Thus, for each photometric band the model solution is obtained on the base of the observed photometry and polarimetry of the object investigated. Then, this information is used for modeling the physical parameters of the CS dust (including the chemical composition and the particle size distribution), and what is more important, parameters of the protoplanetary disks (see, e.g., [16–18]). The most important results obtained in this field are as follows: (1) the circumstellar dust in the surface layers of the protoplanetary disks is close to the dust in the interstellar medium with its chemical composition and differs from it in minimum grain size, which is about an order of magnitude larger than that in the interstellar medium; and (2) the best agreement with observations is provided by the model of the protoplanetary disk with thickening in the dust evaporation zone [18]. In particular, this model explains a non-trivial observational fact: the hopping change in the position angle of the linear polarization in some young stars with changing in the wavelength of the radiation [19].

As an example, in Figure 2 it is shown that the linear polarization degrees depend on the brightness of the UX Ori type star WW Vul in the *U* band from [14]. One can see that it corresponds rather well with the model curve calculated on the base of Equations (2) and (3). Generally, the results of the synchronous observations of the linear polarization and the brightness of the UX Ori type stars show that, like in T Tauri stars [20], the main source of the linear polarization in the Herbig Ae stars in the visible spectrum is the scattered radiation of the circumstellar dust. The role of the optical dichroism in the formation of intrinsic polarization in the young stars is negligible. The circumstellar clouds intersecting the line of sight play the role of a natural coronograph: in blocking the direct radiation of the star, they permit us to observe the weak scattered radiation of the disk. This radiation turned out highly polarized (5–8%) [5], leading to the conclusion about the small inclinations of the disks in UX Ori type stars to the line of sight.

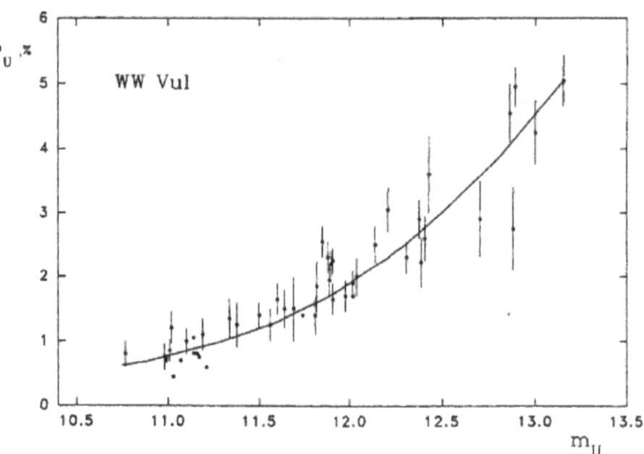

Figure 2. Dependence of the linear polarization degree on the brightness of the UX Ori type star WW Vul in the U band from [14]. Lines show theoretical dependencies calculated using Equations (2) and (3).

2.1. Simulations of Long-Lasting Eclipses

Cases are known in which eclipses of the UX Ori type stars lasted several months. During such events an interesting phenomenon was observed: the change in the position angle of polarization was out of sync with the change in the star brightness [14,21]. Simulation of such events showed [22] that the reason of these anomalies was an obscuration of a noticeable part of the disk with the dust screen. This led to the appearance of vast shadows on the disk whose movement on the disk behind the screen generated changes in the parameters of the disk intrinsic polarization. Of course, in such cases the coronographic effect mentioned above cannot be realized completely.

Very seldom are deep eclipses lasting more than a year observed in young stars [23–27]. Their physics apparently differs from the simple model of the CS dust cloud transit across the stellar disk accepted for the UX Ori type stars. Such eclipses indicate the appearance of a large amount of matter in the nearest vicinity of the star, for example, as a result of the fall of massive gas and dust blobs from the remnant of the protostellar cloud onto the CS disk. Such a type of cloudy accretion onto the young objects was discussed in the literature according to FUORs outbursts [28]. Certain hopes were placed on this mechanism in connection with the discussion of the nature of the eclipses of UX Ori type stars [29]. However, this hypothesis did not receive support because of the significantly modest scale of the eclipses observed in these stars. This can be due to density fluctuations in the dusty atmosphere of the disk, or in the disk wind.

In the case of eclipses lasting years, the cloudy accretion may well be a source of matter into which the young object is temporarily immersed. In favor of this suggestion two observational facts testify: (1) an increase in infrared fluxes at the wavelengths ≥ 2 μm observed during the optical minima in some T Tauri stars [30–32] (this means that during such events the CS dust blocks the large amount of stellar radiation and that this dust is fairly close to the star); (2) an increase in the linear polarization of one of them (RW Aur) up to 30% in the I band [33]. Such high polarization is typical for very young stars immersed in gas and dust cocoons. Its source is the radiation of the star scattered in the polar (optically semi-transparent) regions of the cocoon [34]. The position angle (P.A.) of linear polarization in such objects is parallel to the disk plane (see, e.g., [33]).

2.2. Scattering by Moving Dust

As is known, the main part of the thermal radiation of the protoplanetary disks in the near-infrared spectrum region originates near their inner boundary of the dust sublimation zone [35,36]. In the vicinity of this region, the main part of the scattered radiation is also formed. In T Tauri stars this region is at a distance of 5–10 stellar radii and rotates with velocities of about 150–200 km/s. Therefore, in the scattering of stellar radiation by dust particles the frequency of the scattered radiation will change due to the Doppler effect [37,38]. For the broadband observations this effect has no meaning. However, when studying the spectra of the UX Ori type stars and relative objects it has to be taken into account. Such a problem was first solved in [39]. An example of the photospheric line transformation in the spectrum of the T Tauri type star during the deep brightness minimum is presented in Figure 3. It is seen that at first in the absorption line wide wings appear due to an increase in the contribution of the scattered radiation. In the deep minimum the photospheric line transforms into a shallow but wide absorption band. Keeping in mind that the spectrum of T Tauri type star is rich with photospheric lines, one should expect that in its spectrum of the scattered radiation wide bands will overlap because of blending and form a quasi-continuum [40]. This case is illustrated by the fragment of the synthetic spectrum of the typical T Tauri star in the vicinity of Ca I 6103 and 6122 Ålines shown in Figure 4 in the bright state and in the deep minimum. Namely, the same spectrum transformation was recently observed at the deep brightness minimum for RW Aur [41,42].

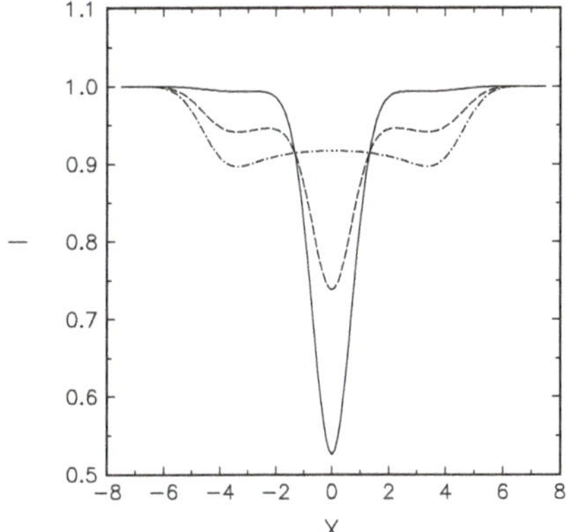

Figure 3. Photospheric line broadening in the T Tauri type star (solid line) during an eclipse due to scattering at the inner boundary of the protoplanetary disk. The dot-dashed line corresponds to the total eclipse.

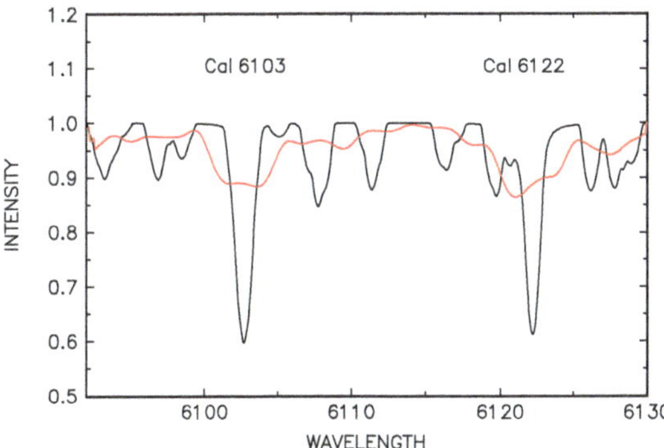

Figure 4. A part of the synthetic spectrum of RW Aur in the vicinity of the Ca I 6103 and 6122 Å lines in the bright state (black) and the deep minimum (red).

The other source of the intrinsic polarization of young stars can be the dust component of the disk wind [18]. In this case the moving dust has two velocity components: tangential and poloidal ones. Therefore, the scattering of stellar radiation by the disk wind can lead to more complex transformation of the photospheric lines: to the line broadening due to rotation and redshift due to the poloidal motion. This issue is briefly discussed in [38,39] and deserves a more detailed quantitative analysis.

In UX Ori type stars the effect considered above is revealed in a significantly weaker form [38]. This is caused by two reasons: (1) in most of these stars the photospheric lines are broadened by the strong stellar rotation; (2) the dust sublimation zone is located much farther from the star due to the higher luminosity. Therefore, the Keplerian velocities of the disk in this region are significantly less compared to those in T Tauri stars. Figure 5 demonstrates the part of the synthetic spectrum of the UX Ori type star CQ Tau in the vicinity of the Fe II 5018 line in the bright state and the deep minimum. It is seen that in this case the spectrum of the scattered radiation differs little from the photospheric one.

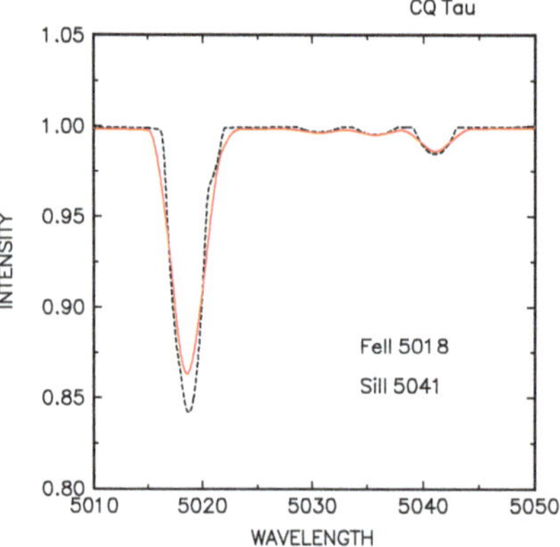

Figure 5. A part of the synthetic spectrum of CQ Tau in the vicinity of the Fe II 5018 Å line in the bright state (black) and the deep minimum (red).

The same occurs in the emission lines in the spectra of young stars. Nevertheless, the influence of the scattered radiation on the emission lines is well known from the observations of the linear polarization (see [43] and cited papers therein). When traversing the Hα line profile, both the value and position angle change, which reflects the contribution of various parts of the CS disk to the polarization of the different parts of the line profile. As shown in [43], this effect is sensitive to the parameters of the inner CS disk regions, and may shed additional light on their structure.

3. Images of Circumstellar Disks in the Scattered Light

The coronographic method has been very successful in observations of the scattered radiation of the CS debris disks. In particular, with its help Smith and Terrile [44] first observed the circumstellar disk of β Pictoris. This disk is seen nearly edge-on on the coronographic image. Therefore, its radiation (as well as the radiation of the UX Ori type stars) was found to be strongly polarized [45]. The coronographic observations with HST revealed that the inner region of the β Pic disk was slightly curved relative to its outer part [46]. The authors of the paper quoted above suggested that this curvature was caused by a disturbing body (a massive planet) whose orbit was slightly inclined relative to the plain of the outer disk. Fourteen years later this planet has been found [47], also with the coronagraphic technique. An important contribution to such observations has been made with the Hubble Space Telescope (see, e.g., [48,49] and references there).

At the present time the technique of observations of weak objects in the vicinity of stars based on the coronographic method is widely used in the different astrophysical fields, among them the study of the fine structure of the circumstellar disks in polarized light (see, e.g., [50–53] and the references therein). Of particular interest are the first and rather successful attempts to monitor the protoplanetary disk images in the polarized light [51]. They showed that on the images of the CS disk of HD 135344B (observed nearly pole-on), the shadows were caused by absorption of the stellar radiation by local perturbations in the inner disk. These shadows are manifested as narrow radial bands of the variable brightness, and also as the wide bands caused by absorption by large scale structures. These observations point to the direct physical connection between eclipses of the UX Ori type stars and shadow formation on the circumstellar disks.

Another example of successful monitoring of a circumstellar disk in the polarized light is the long-term observations of CQ Tau [54]. This star is one of the most active members of the UXOrs family [55], having a very complex light curve [56]. According to interferometric observations in the near-IR [57] and millimeter wavelengths [58,59], the inner and outer parts of the CS disk of CQ Tau have different inclinations: $i = 48.5 \pm 5°$ and $i = 35°$, correspondingly. In such cases, a narrow shadow from the inner disk can be observed on the outer one (see, e.g., [60–62]). The observations of CQ Tau have demonstrated that such a shadow exists on the peripheral region of its CS disk [54].

The aforementioned interferometric observations of CQ Tau support the point of view according to which the extinction events in UX Ori stars take place in the innermost part of the CS disk where the NIR radiation is formed. Continuation of such observations is of undeniable interest for understanding the nature of the perturbations in the inner regions of CS disks, which can be caused by different reasons such as an azimuthal heterogeneity of the dusty disk wind, collisions of planetesimals or hydrodynamical fluctuations in the dust evaporation zone of the disk.

If the extinction events are driven by the disk wind, the UXOrs activity will depend not only on the disk inclination but also on the mass loss rate and the dusty wind loading area. The latter depends on the magnetic field in the disk and the star luminosity. The mass loss and accretion rates are closely connected. Therefore, in this case the UXOrs activity will be sensitive to the disk inclination, as well as the stellar luminosity and mass accretion rate.

The Edge-On Disks

Young stars with edge-on CS disks are also observed through scattered light. The prototype of such objects is the well-known T Tauri star HH 30. The first image of this object was obtained with high resolution by Burrows et al. [46] with the Hubble Space Telescope. It revealed a flared CS disk (in accordance with the prediction of the Shakura and Sunyaev [63] model (see Kenyon and Hartmann [64]) and the highly asymmetric jet. In addition to the jet, a conical molecular outflow is also observed in the CO lines [65]. Photometric and polarimetric observations have shown that HH 30 is a variable object [66–68]. It demonstrates the periodic variability of brightness and linear polarization with a period of 7.49 days [67]. The physical model of such a periodicity is not clear. It could be a hot spot on the rotating star or periodic variations of the CS extinction in the star's vicinity.

Variable illumination of the disk leads to changes in its shape [67]. These changes are one of the most interesting manifestations of the circumstellar activity of young stars reminiscent of the moving shadows on the disk images [50]. The other non-trivial special property of HH 30 is its almost one-side jet and molecular outflow [46,64,69]. The origin of such asymmetric outflows is discussed in [70]. An opposite case of the edge-on disk with a well-developed and almost symmetrical jet is HH 212 [71].

To date, more than ten young stars with edge-on disks are known (see, e.g., [72] and the references there). Most of them are T Tauri stars. Observations and modeling of such objects permit us to study in detail the internal structure of CS disks (see, e.g., [73,74]). It is obvious that the spectra of such objects are strongly distorted by the scattered radiation [37].

4. Intrinsic Polarization of Young Stars and Orientation of Their CS Disks

The position angle of linear polarization depends on the geometry of the scattering medium. Calculations show [75] that in models with the classical accretion disk (with a flared surface) the P.A. is orthogonal to the disk plane. In the stars at earlier evolutionary stages surrounded by accreting envelopes the P.A. is parallel to the disk plane [76]. The average age of the UX Ori type stars is several million years [77]. Therefore, most of them belong to the first group of young stars. Keeping in mind what was mentioned above, one can use the position angle of the intrinsic polarization in order to determine the orientation of protoplanetary disks in the projection on the sky plane. Such a possibility was confirmed by results of interferometric observations of UX Ori in the near-infrared spectrum region: according to [9] the position angle of the symmetry axis of its circumstellar disk is equal to $127.5 \pm 24.5°$. Polarimetric observations of UX Ori in the deep minima give P.A. = $125°–129°$ [21,78]. Orientation of the circumstellar disk of another UX Ori type star VV Ser is determined with the position of the shadow formed by the disk on the reflective nebula behind the star. According to [79], the P.A. of the shadow is equal to $15 \pm 5°$, which corresponds to the P.A. of the disk symmetry axis $105 \pm 5°$. Observations of the intrinsic polarization of VV Ser gives the position angle in the close range: P.A. = $88°–100°$ [80].

These two examples testify that the linear polarization of most UX Ori type stars in the deep brightness minima in fact characterizes the position of the symmetry axis of the circumstellar disk on the sky plane, and this position can be compared with the direction of the interstellar magnetic field determined with the help of polarization of the neighboring stars. Unfortunately, this is not always possible because of the complex structure of the interstellar magnetic fields in star formation regions. Figure 6 shows an example of the polarization map of the BM And, as well as the vicinity in which the polarization pseudo-vectors of BM And itself are shown in the bright stage and deep brightness minimum. It is seen that the interstellar magnetic field in the stellar vicinity is fairly homogeneous in direction, and the position angle of the stellar intrinsic polarization coincides with this direction with high accuracy [81]. This implies that the interstellar magnetic filed controlled the star formation process from the protostellar cloud, and that the circumstellar disk of the star "remembered" the direction of the magnetic lines.

Similar results have also been obtained during photopolarimetric observations for three other UX Ori type stars: WW Vul [14], BF Ori [82] and VV Ser [79,83,84]. Furthermore, the same method was used in [85] for a large group of Herbig AeBe stars. It was shown that subsamples of the more polarized stars from their list present a statistically significant tendency toward intrinsic polarization aligned with the interstellar magnetic field.

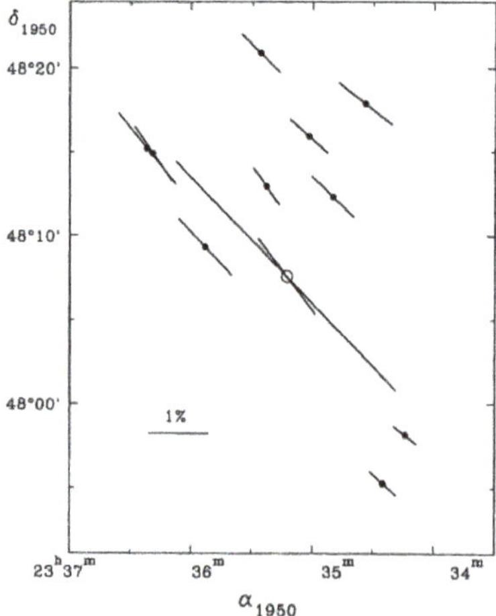

Figure 6. Polarization map of the BM and as well as the neighborhood from [81,84]. Two values of the star polarization are given: at the bright state and at the deep minimum. Credit: Grinin et al. 1995, A&AS, 112, 457, reproduced with permission © ESO.

5. Conclusions

Thus, although the scattered radiation of protoplanetary disks makes up only a very small part of the radiation of young stars, its existence provides us very important information about young stars and their circumstellar environment. Of great interest is the possibility to study the orientation of CS disks not resolved in a telescope with the help of polarimetric observations. Polarized radiation makes it possible to see fine details on the disk images and study their structure and variability. Photopolarimetric observations of UXOrs and their modeling permit us to investigate the perturbations in the innermost regions of the disks where planetary systems are formed.

Author Contributions: V.P.G.—conceptualization, data analysis and writing-editing; L.V.T.—numerical modelling and writing—editing; All authors have read and agreed to the published version of the manuscript.

Funding: We acknowledge the financial support of the Ministry of Science and Higher Education of the Russian Federation (grant no. 075-15-2020-780).

Institutional Review Board Statement: Not applicable.

Informed Consent Statement: Not applicable.

Data Availability Statement: All necessary data are contained in this paper.

Acknowledgments: The authors are grateful to the reviewers for useful comments.

Conflicts of Interest: The authors declare no conflict of interest.

References

1. Herbst, W.; Herbst, D.K.; Grossman, E.J.; Weinstein, D. Catalogue of UBVRI Photometry of T Tauri Stars and Analysis of the Causes of Their Variability. *Astrophys. J.* **1994**, *108*, 1906–1923. [CrossRef]
2. Wenzel, W. Extremely young stars. In *Non-Periodic Phenomena in Variable Stars*; Detre, L., Eds.; Academic Press: Budapest, Hungary, 1969; pp. 61–73.
3. Herbst, W. T Tauri Variables. *Publ. Astron. Soc. Pac.* **1986**, *98*, 1088–1094. [CrossRef]
4. Grinin, V.P. On the Blue Emission Visible during Deep Minima of Young Irregular Variables. *Sov. Astron. Lett.* **1988**, *14*, 27–28.
5. Grinin, V.P.; Kiselev, N.N.; Minikulov, N.K.; Chernova, G.P.; Voshchinnikov, N.V. The investigations of 'zodiacal light' of isolated Ae-Herbig stars with non-periodic Algol-type minima. *Astrophys. Space Sci.* **1991**, *186*, 283–298 [CrossRef]
6. Natta, A.; Grinin, V.P.; Mannings, V.; Ungerechts, H. The Evolutionary Status of UX Orionis-Type Stars. *Astrophys. J.* **1997**, *491*, 885–890. [CrossRef]
7. Pontoppidan, K.M.; Dullemond, C.P.; Blake, G.A.; Boogert, A.C.A.; van Dishoeck, E.F.; Evans, N.J., II; Kessler-Silacci, J.; Lahuis, F. Modeling Spitzer Observations of VV Ser. I. The Circumstellar Disk of a UX Orionis Star. *Astrophys. J.* **2007**, *656*, 980–990 [CrossRef]
8. Kreplin, A.; Weigelt, G.; Kraus, S.; Grinin, V.; Hofmann, K.-H.; Kishimoto, M.; Schertl, D.; Tambovtseva, L.; Clausse, J.-M.; Massi, F.; et al. Revealing the inclined circumstellar disk in the UX Ori system KK Ophiuchi. *Astron. Astrophys.* **2013**, *551*, A21–A27. [CrossRef]
9. Kreplin, A.; Madlener, D.; Chen, L.; Weigelt, G.; Kraus, S.; Grinin, V.; Tambovtseva, L.; Kishimoto, M. Resolving the inner disk of UX Orionis. *Astron. Astrophys.* **2016**, *590*, A96–A101. [CrossRef]
10. Miroshnichenko, A.; Ivezic, Z.; Elitzur, M. On Protostellar Disks in Herbig Ae/Be Stars. *Astrophys. J.* **1997**, *475*, L41–L44. [CrossRef]
11. Grinin, V.P.; Rostopchina, A.N. Orientation of circumstellar disks and the statistics of H_α profiles of Ae/Be Herbig stars. *Astron. Rep.* **1996**, *40*, 171–178.
12. Vioque, M.; Oudmaijer, R.D.; Baines, D.; Mendigutia, I.; Perez-Martinez, R. Gaia DR2 study of Herbig Ae/Be stars. *Astron. Astrophys.* **2018**, *620*, A128–A145. [CrossRef]
13. Pugach, A.F. Optical Properties of the Circumstellar Dust around Stars with Aperiodic Fadings. *Astron. Rep.* **2004**, *48*, 470–475. [CrossRef]
14. Grinin, V.P.; Kiselev, N.N.; Minikulov, N.K.; Chernova, G.P. Linear Polarization in Deep Minima of WW-Vulpeculae. *Sov. Astron. Lett.* **1988**, *14*, 219–223.
15. Voshchinnikov, N.V.; Grinin, V.P. Dust around Young Stars—Model of Envelope of the Ae Herbig Star WW-Vulpeculae. *Astrophysics* **1991**, *34*, 84–95. [CrossRef]
16. Voshchinnikov, N.V.; Grinin, V.P.; Karyukin, V.V. Monte Carlo simulation of light scattering in the envelopes of young stars. *Astron. Astrophys.* **1995**, *294*, 547–554.

17. Natta, A.; Whitney, B.A. Models of scattered light in UXORs. *Astron. Astrophys.* **2000**, *364*, 633–640.
18. Shulman, S.G.; Grinin, V.P. Influence of the Disk Wind on the Intrinsic Polarization of Young Stars. *Astron. Lett.* **2019**, *45*, 384–395. [CrossRef]
19. Pereyra, A.; Girart, J.M.; Magalhaes, A.M.; Rodrigues, C.V.; de Araujo, F.X. Near infrared polarimetry of a sample of YSOs. *Astron. Astrophys.* **2009**, *501*, 595–607. [CrossRef]
20. Bastien, P.; Landstreet, J.D. Polarization observations of the T Tauri stars RY Tauri, T Tauri, and V866 Scorpii. *Astrophys. J.* **1979**, *229*, L137–L140. [CrossRef]
21. Grinin, V.P.; The, P.S.; de Winter, D.; Giampapa, M.; Rostopchina, A.N.; Tambovtseva, L.V.; van den Ancker, M.E. The β Pictoris phenomenon among young stars. I. The case of the Herbig Ae star UX Orionis. *Astron. Astrophys.* **1994**, *292*, 165–174.
22. Shulman, S.G.; Grinin, V.P. Influence of Large-Scale Perturbations in Circumstellar Disks on the Linear Polarization Parameters of UX Ori Stars. *Astron. Lett.* **2019**, *45*, 664–676. [CrossRef]
23. Bouvier, J.; Grankin, K.; Ellerbroek, L.E.; Bouy, H.; Barrado, D. AA Tauri's sudden and long-lasting deepening: Enhanced extinction by its circumstellar disk. *Astron. Astrophys.* **2013**, *557*, A77–A86. [CrossRef]
24. Rodriguez, J.E.; Pepper, J.; Stassun, K.G.; Siverd, R.J.; Cargile, P.; Beatty, T.G.; Gaudi, B.S. Occultation of the T Tauri Star RW Aurigae A by its Tidally Disrupted Disk. *Astrophys. J.* **2013**, *146*, 112–122. [CrossRef]
25. Semkov, E.H.; Peneva, S.P.; Ibryamov, S.I. The pre-main-sequence star V1184 Tauri (CB 34V) at the end of prolonged eclipse. *Astron. Astrophys.* **2015**, *582*, A113–A119. [CrossRef]
26. Grinin, V.P.; Barsunova, O.Y.; Sergeev, S.G.; Shugarov, S.Y.; Fedorova, E.I. Unusual Eclipse of the UX Ori Type Star V719 Per. *Astron. Rep.* **2021**, *65*, 864–868. [CrossRef]
27. Bozhinova, I.; Scholz, A.; Costigan, G.; Lux, O.; Davis, C.J.; Ray, T.; Boardman, N.F.; Hay, K.L.; Hewlett, T.; Hodosan, G.; et al. The disappearing act: A dusty wind eclipsing RW Aur. *Mon. Not. R. Astron. Soc.* **2016**, *463*, 4459–4468. [CrossRef]
28. Hartmann, L.; Kenyon, S.J. The FU Orionis Phenomenon. *Annu. Rev. Astron. Astrophys.* **1996**, *34*, 207–240. [CrossRef]
29. Graham, J.A. Clumpy Accretion onto Pre-Main Sequence Stars. *Publ. Astron. Soc. Pac.* **1992**, *104*, 479–488. [CrossRef]
30. Grinin, V.P.; Arkharov, A.A.; Barsunova, O.Y.; Sergeev, S.G.; Tambovtseva, L.V. Photometric activity of the UX Ori star V1184 Tau in the optical and near-infrared spectral ranges. *Astron. Lett.* **2009**, *35*, 114–120. [CrossRef]
31. Shenavrin, V.I.; Petrov, P.P.; Grankin, K.N. Hot Dust Revealed during the Dimming of the T Tauri Star RW Aur A. *Inf. Bull. Var. Stars.* **2015**, *6143*, 1.
32. Covey, K.R.; Larson, K.A.; Herczeg, G.J.; Manara, C.F. A Differential Measurement of Circumstellar Extinction for AA Tau's 2011 Dimming Event. *Astrophys. J.* **2021**, *161*, 61–77. [CrossRef]
33. Dodin, A.; Grankin, K.; Lamzin, S.; Nadjip, A.; Safonov, B.; Shakhovskoi, D.; Shenavrin, V.; Tatarnikov, A.; Vozyakova, O. Analysis of colour and polarimetric variability of RW Aur A in 2010–2018. *Mon. Not. R. Astron. Soc.* **2019**, *482*, 5524–5541. [CrossRef]
34. Elsasser, H.; Staude, H.J. On the Polarization of Young Stellar Objects. *Astron. Astrophys.* **1978**, *70*, L3–L6.
35. Natta, A.; Prusti, T.; Neri, R.; Wooden, D.; Grinin, V.P.; Mannings, V. A reconsideration of disk properties in Herbig Ae stars. *Astron. Astrophys.* **2001**, *371*, 186–197.
36. Dullemond, C.P.; Dominik, C.; Natta, A. Passive Irradiated Circumstellar Disks with an Inner Hole. *Astrophys. J.* **2001**, *560*, 957–969. [CrossRef]
37. Appenzeller, I.; Bertout, C.; Stahl, O. Edge-on T Tauri stars. *Astron. Astrophys.* **2005**, *434*, 1005–1019. [CrossRef]
38. Grinin, V.P.; Tambovtseva, L.V.; Weigelt, G. Spectral line profiles changed by dust scattering in heavily obscured young stellar objects. *Astron. Astrophys.* **2012**, *544*, A45–A50.
39. Grinin, V.P.; Mitskevich, A.S.; Tambovtseva, L.V. Light scattering by moving dust grains in the immediate vicinity of young stars. *Astron. Lett.* **2006**, *32*, 110–119. [CrossRef]
40. Grinin, V.P.; Tambovtseva, L.V.; Dmitriev, L.V. On the spectrum of the scattered radiation of protoplanetary disks. *Astron. Rep.* **2022**, *66*, 314–320.
41. Takami, M.; Wei, Y.-J.; Chou, M.-Y.; Karr, J.L.; Beck, T.L.; Manset, N.; Chen, W.-P.; Kurosawa, R.; Fukagawa, M.; White, M.; et al. Stable and Unstable Regimes of Mass Accretion onto RW Aur A. *Astrophys. J.* **2016**, *820*, 139–152. [CrossRef]
42. Facchini, S.; Manara, C.F.; Schneider, P.C.; Clarke, C.J.; Bouvier, J.; Rosotti, G.; Booth, R.; Haworth, T.J. Violent environment of the inner disk of RW Aurigae A probed by the 2010 and 2015 dimming events. *Astron. Astrophys.* **2016**, *596*, A38–A48. [CrossRef]
43. Vink, J.S.; Drew, J.E.; Harries, T.J.; Oudmaijer, R.D.; Unruh, Y. Probing the circumstellar structures of T Tauri stars and their relationship to those of Herbig stars. *Mon. Not. R. Astron. Soc.* **2005**, *359*, 1049–1064. [CrossRef]
44. Smith, B.A.; Terrile, R.J. A Circumstellar Disk Around β Pictoris. *Science* **1984**, *226*, 1421–1424. [CrossRef] [PubMed]
45. Gledhill, T.M.; Scarrott, S.M.; Wolstencroft, R.D. Optical polarization in the disc around β Pictoris. *Mon. Not. R. Astron. Soc.* **1991**, *252*, 50–54. [CrossRef]
46. Burrows, C.J.; Stapelfeldt, K.R.; Watson, A.M.; Krist, J.E.; Ballester, G.E.; Clarke, J.T.; Crisp, D.; Gallagher, J.S., III; Griffiths, R.E.; Hester, J.J. Hubble Space Telescope Observations of the Disk and Jet of HH 30. *Astrophys. J.* **1996**, *473*, 437–451. [CrossRef]
47. Lagrange, A.-M.; Gratadour, D.; Chauvin, G.; Fusco, T.; Ehrenreich, D.; Mouillet, D.; Rousset, G.; Rouan, D.; Allard, F.; Gendron, E.; et al. A probable giant planet imaged in the β Pictoris disk. VLT/NaCo deep L'-band imaging. *Astron. Astrophys.* **2009**, *493*, L21–L25. [CrossRef]
48. Grady, C.A.; Schneider, G.; Hamaguchi, K.; Sitko, M.L.; Carpenter, W.J.; Hines, D.; Collins, K.A.; Williger, G.M.; Woodgate, B.E.; Henning, T.; et al. The Disk and Environment of a Young Vega Analog: HD 169142. *Astrophys. J.* **2007**, *665*, 1391–1406. [CrossRef]

49. Hornbeck, J.B.; Swearingen, J.R.; Grady, C.; Williger, G.M.; Brown, A.; Sitko, M.L.; Wisniewski, J.P.; Perrin, M.D.; Lauroesch, J.T.; Schneider, G.; et al. Panchromatic Imaging of a Transitional Disk: The Disk of GM Aur in Optical and FUV Scattered Light. *Astrophys. J.* **2016**, *829*, 65–81. [CrossRef]

50. Garufi, A.; Quanz, S.P.; Avenhaus, H.; Buenzli, E.; Dominik, C.; Meru, F.; Meyer, M.R.; Pinilla, P.; Schmid, H.M.; Wolf, S. Small vs. large dust grains in transitional disks: Do different cavity sizes indicate a planet?. SAO 206462 (HD 135344B) in polarized light with VLT/NACO. *Astron. Astrophys.* **2013**, *560*, A105–A115. [CrossRef]

51. Stolker, T.; Sitko, M.; Lazareff, B.; Benisty, M.; Dominik, C.; Waters, R.; Min, M.; Perez, S.; Milli, J.; Garufi, A.; et al. Variable Dynamics in the Inner Disk of HD 135344B Revealed with Multi-epoch Scattered Light Imaging. *Astrophys. J.* **2017**, *849*, 143–158. [CrossRef]

52. van Holstein, R.G.; Stolker, T.; Jensen-Clem, R.; Ginski, C.; Milli, J.; de Boer, J.; Girard, J.H.; Wahhaj, Z.; Bohn, A.J.; Millar-Blanchaer, M.A.; et al. A survey of the linear polarization of directly imaged exoplanets and brown dwarf companions with SPHERE-IRDIS. First polarimetric detections revealing disks around DH Tau B and GSC 6214–210 B. *Astron. Astrophys.* **2021**, *647*, A21–A49. [CrossRef]

53. Garufi, A.; Dominik, C.; Ginski, C.; Benisty, M.; van Holstein, R.G.; Henning, T.; Pawellek, N.; Pinte, C.; Avenhaus, H.; Facchini, S.; et al. A SPHERE survey of self-shadowed planet-forming disks. *Astron. Astrophys.* **2022**, *658*, A137–A152.

54. Safonov, B.S.; Strakhov, I.A.; Goliguzova, M.V.; Voziakova, O.V. Apparent Motion of the Circumstellar Envelope of CQ Tau in Scattered Light. *Astrophys. J.* **2022** *163*, 31–42. [CrossRef]

55. Berdyugin, A.V.; Berdyugina, S.V.; Grinin, V.P.; Minikulov, N.K. Discovery of High Linear Polarization at Brightness Minima of CQ-Tauri. *Sov. Astron.* **1990**, *34*, 408–416.

56. Shakhovskoj, D.N.; Grinin, V.P.; Rostopchina, A.N. Analysis of the Historical Light Curve of the UX Ori Star CQ Tau. *Astrophysics* **2005**, *48*, 135–142 [CrossRef]

57. Eisner, J.A.; Lane, B.F.; Hillenbrand, L.A.; Akeson, R.L.; Sargent, A.I. Resolved Inner Disks around Herbig Ae/Be Stars. *Astrophys. J.* **2004**, *613*, 1049–1071. [CrossRef]

58. Chapillon, E.; Guilloteau, S.; Dutrey, A.; Piétu, V. Disks around CQ Tau and MWC 758: Dense PDRs or gas dispersal? *Astron. Astrophys.* **2008**, *488*, 565–578. [CrossRef]

59. Ubeira Gabellini, M.G.; Miotello, A.; Facchini, S.; Ragusa, E.; Lodato, G.; Testi, L.; Benisty, M.; Bruderer, S.; Kurtovic, N.T.; Andrews, S.; et al. A dust and gas cavity in the disc around CQ Tau revealed by ALMA. *Mon. Not. R. Astron. Soc.* **2019**, *486*, 4638–4654. [CrossRef]

60. Benisty, M.; Stolker, T.; Pohl, A.; de Boer, J.; Lesur, G.; Dominik, C.; Dullemond, C.P.; Langlois, M.; Min, M.; Wagner, K.; et al. Shadows and spirals in the protoplanetary disk HD 100453. *Astron. Astrophys.* **2017**, *597*, A42–A53. [CrossRef]

61. Pinilla, P.; Benisty, M.; de Boer, J.; Manara, C.F.; Bouvier, J.; Dominik, C.; Ginski, C.; Loomis, R.A.; Sicilia Aguilar, A. Variable Outer Disk Shadowing around the Dipper Star RXJ1604.3-2130. *Astrophys. J.* **2018**, *868*, 85–99. [CrossRef]

62. Casassus, S.; Avenhaus, H.; Pérez, S.; Navarro, V.; Carcamo, M.; Marino, S.; Cieza, L.; Quanz, S.P.; Alarcon, F.; Zurlo, A.; et al. An inner warp in the DoAr 44 T Tauri transition disc. *Mon. Not. R. Astron. Soc.* **2018**, *477*, 5104–5114. [CrossRef]

63. Shakura, N.I.; Sunyaev, R.A. Black holes in binary systems. Observational appearance. *Astron. Astrophys.* **1973**, *24*, 337–355.

64. Kenyon, S.J.; Hartmann, L. Spectral Energy Distributions of T Tauri Stars: Disk Flaring and Limits on Accretion. *Astrophys. J.* **1987**, *323*, 714–733. [CrossRef]

65. Pety, J.; Gueth, F.; Guilloteau, S.; Dutrey, A. Plateau de Bure interferometer observations of the disk and outflow of HH 30. *Astron. Astrophys.* **2006**, *458*, 841–854. [CrossRef]

66. Wood, K.; Wolk, S.J.; Stanek, K.Z.; Leussis, G.; Stassun, K.; Whitney, B.; Wolff, M.; Whitney, B. Optical Variability of the T Tauri Star HH 30 IRS. *Astrophys. J.* **2000**, *542*, L21–L24. [CrossRef]

67. Durán-Rojas, M.C.; Watson, A.M.; Stapelfeldt, K.R.; Hiriart, D. The Polarimetric and Photometric Variability of HH 30. *Astrophys. J.* **2009**, *137*, 4330–4338. [CrossRef]

68. Stapelfeldt, K.R.; Watson, A.M.; Krist, J.E. A Variable Asymmetry in the Circumstellar Disk of HH 30. *Astrophys. J.* **1999**, *516*, L95–L98. [CrossRef]

69. Hartigan, P.; Morse, J. Collimation, Proper Motions, and Physical Conditions in the HH 30 Jet from Hubble Space Telescope Slitless Spectroscopy. *Astrophys. J.* **2007**, *660*, 426–440. [CrossRef]

70. Dyda, S.; Lovelace, R.V.E.; Ustyugova, G.V.; Lii, P.S.; Romanova, M.M.; Koldoba, A.V. Asymmetric MHD outflows/jets from accreting T Tauri stars. *Mon. Not. R. Astron. Soc.* **2015**, *450*, 481–493. [CrossRef]

71. Lee, C.-F.; Ho, P.T.P.; Li, Z.-Y.; Hirano, N.; Zhang, Q.; Shang, H. A rotating protostellar jet launched from the innermost disk of HH 212. *Nat. Astron.* **2017**, *1*, 152L–175L. [CrossRef]

72. Luhman, K.L.; Adame, L.; D'Alessio, P.; Calvet, N.; McLeod, K.K.; Bohac, C.J.; Forrest, W.J.; Hartmann, L.; Sargent, B.; Watson, D.M. Hubble and Spitzer Observations of an Edge-on Circumstellar Disk around a Brown Dwarf. *Astrophys. J.* **2007**, *666*, 1219–1225. [CrossRef]

73. Madlener, D.; Wolf, S.; Dutrey, A.; Guilloteau, S. The circumstellar disk of HH 30. Searching for signs of disk evolution with multi-wavelength modeling. *Astron. Astrophys.* **2012**, *543*, A81–A95. [CrossRef]

74. Louvet, F.; Dougados, C.; Cabrit, S.; Mardones, D.; Ménard, F.; Tabone, B.; Pinte, C.; Dent, W.R.F. The HH30 edge-on T Tauri star. A rotating and precessing monopolar outflow scrutinized by ALMA. *Astron. Astrophys.* **2018**, *618*, A120–A148. [CrossRef]

75. Whitney, B.A.; Hartmann, L. Model Scattering Envelopes of Young Stellar Objects. I. Method and Application to Circumstellar Disks. *Astrophys. J.* **1992**, *395*, 529–539. [CrossRef]
76. Whitney, B.A.; Hartmann, L. Model Scattering Envelopes of Young Stellar Objects. II. Infalling Envelopes. *Astrophys. J.* **1993**, *402*, 605–622. [CrossRef]
77. Rostopchina, A.N. The location of UX Ori stars on the Hertzsprung-Russell diagram. *Astron. Rep.* **1999**, *43*, 113–118.
78. Voshchinnikov, N.V.; Grinin, V.P.; Kiselev, N.N.; Minikulov, N.K. Dust around Young Stars: Observations of the Polarization of UX Orionis in Deep Minima. *Astrophysics* **1988**, *28*, 182–193. [CrossRef]
79. Pontoppidan, K.M.; Dullemond, C.P.; Blake, G.A.; Evans, N.J., II; Geers, V.C.; Harvey, P.M.; Spiesman, W. Modeling Spitzer Observations of VV Ser. II. An Extended Quantum-heated Nebula and a Disk Shadow. *Astrophys. J.* **2007**, *656*, 991–1000. [CrossRef]
80. Rostopchina, A.N.; Grinin, V.P.; Shakhovskoi, D.N. Photometry and Polarimetry of the Classical Herbig Ae Star VV Ser. *Astron. Rep.* **2001**, *45*, 51–59. [CrossRef]
81. Grinin, V.P.; Kolotilov, E.A.; Rostopchina, A.N. Dust around young stars. Photopolarimetric observations of the T Tauri star BM Andromedae. *Astron. Astrophys.* **1995**, *112*, 457–473.
82. Grinin, V.P.; Kiselev, N.N.; Minikulov, N.K. Observations of Zodiacal Light around the Isolated Herbig Ae-Star BF-Orionis. *Sov. Astron. Lett.* **1989**, *15*, 448–452.
83. Shakhovskoi, D.N.; Rostopchina, A.N. Photometry of the Herbig Ae Star VV Ser B in an Anomalously Deep Brightness Minimum. *Astrophysics* **2000**, *43*, 487–489. [CrossRef]
84. Rostopchina, A.N.; Shakhovskoi, D.N. Interstellar polarization in a molecular cloud in Serpens. *Astrophysics* **2000**, *43*, 289–294. [CrossRef]
85. Rodrigues, C.V.; Sartori, M.J.; Gregorio-Hetem, J.; Magalhaes, A.M. The Alignment of the Polarization of Herbig Ae/Be Stars with the Interstellar Magnetic Field. *Astrophys. J.* **2009**, *698*, 2031–2035. [CrossRef]

Communication

Cosmological Model with Interconnection between Dark Energy and Matter

Gennady S. Bisnovatyi-Kogan

1   Space Research Institute of Russian Academy of Sciences, Profsoyuznaya 84/32, 117997 Moscow, Russia; gkogan@iki.rssi.ru

2   National Research Nuclear University MEPhI (Moscow Engineering Physics Institute), Kashirskoe Shosse 31, 115409 Moscow, Russia

Abstract: It is accepted in the present cosmology model that the scalar field, which is responsible for the inflation stage in the early universe, transforms completely into matter, and the accelerated universe expansion is presently governed by dark energy (DE), whose origin is not connected with the inflationary scalar field. We suppose here that dark matter (DM) has a common origin with a small variable component of dark energy (DEV). We suggest that DE may presently have two components, one of which is the Einstein constant Λ, and another, smaller component DEV (Λ_V) comes from the remnants of the scalar field responsible for inflation, which gave birth to the origin of presently existing matter. In this note we consider only the stages of the universe expansion after recombination, $z \simeq 1100$, when DM was the most abundant component of the matter, therefore we suggest for simplicity that a connection exists between DM and DEV so that the ratio of their densities remains constant over all the stages after recombination, $\rho_{DM} = \alpha \rho_{DEV}$, with a constant α. One of the problems revealed recently in cosmology is a so-called Hubble tension (HT), which is the difference between values of the present Hubble constant, measured by observation of the universe at redshift $z \lesssim 1$, and by observations of a distant universe with CMB fluctuations originated at $z \sim 1100$. In this paper we suggest that this discrepancy may be explained by deviation of the cosmological expansion from a standard Lambda-CDM model of a flat universe, due to the action of an additional variable component DEV. Taking into account the influence of DEV on the universe's expansion, we find the value of α that could remove the HT problem. In order to maintain the almost constant DEV/DM energy density ratio during the time interval at $z < 1100$, we suggest the existence of a wide mass DM particle distribution.

Keywords: dark energy; dark matter; Hubble constant

Citation: Bisnovatyi-Kogan, G.S. Cosmological Model with Interconnection between Dark Energy and Matter. *Universe* **2021**, *7*, 412. https://doi.org/10.3390/universe7110412

Academic Editors: Galina L. Klimchitskaya, Vladimir M. Mostepanenko and Nazar R. Ikhsanov

Received: 11 October 2021
Accepted: 27 October 2021
Published: 29 October 2021

Publisher's Note: MDPI stays neutral with regard to jurisdictional claims in published maps and institutional affiliations.

1. Introduction

During his long and successful scientific career, Yu. N. Gnedin worked also on problems on the border of physics and cosmology, namely on the possibility of direct searching, in astronomical observations, for axions, belonging to the family of Goldston bosons. These tiny particles, introduced by theoretical physicists, have been proposed as candidates for dark matter, whose presence was necessary for interpretation of different astronomical observations. During the years 1992–2009, Yu. N. Gnedin published at least six papers on this intriguing topic [1–6]. The cosmological model considered in this paper demands the presence of very light particles in DM, and axions could be the best candidates for that.

Here we consider a cosmological model that differs slightly from the widely accepted ΛCDM model by having two components of the dark energy instead of the usually considered one component represented by Λ. The model considered here is constructed in order to explain the so called "Hubble Tension", which is the observed discrepancy between different averaged "local" measurements of the Hubble constant from one side, and distant measurements based on the analysis of the CMB fluctuations, connected with the recombination epoch at redshift $z \sim 1100$. In order to compare these two measurements, it was necessary to specify the cosmological model for determining the present value of the Hubble constant from distant measurements. We accept that the origin of HT is connected with the not fully appropriate ΛCDM model, and suggest its modification removing this discrepancy.

While the present value of the Hubble constant (HC) is one of the most important cosmological parameters, its measurements for many years have been performed by different astronomical groups ("local measurements"). These measurements have been based on several steps in order to have a possibility for independent distance measurements of more and more distant objects with observed runaway velocities. The use of different steps in this sequence by different groups has resulted in finding different values for the HC, from ~ 50 km/s/Mps by the Sandage–Tamman group to ~ 100 km/s/Mps by the de Vacouleurs group [7]. Over time the local HC measurements at redshift $z \lesssim 1$ have been substantially improved due to construction of big telescopes and measurements by the Hubble mission, which permitted the narrowing of this interval of the HC to 72–75 km/s/Mps.

The measurements of the CMB fluctuations by instruments in the satellites WMAP and PANCK gave the possibility for fully independent HC measurements at the recombination epoch. HT appears when you compare the present values of the HC obtained from CMB using a simple ΛCDM model with local measurements. It is claimed that the discrepancy in these two values is statistically significant in the range 4.5σ to 6.3σ [8]; see, nevertheless [9].

It is accepted in the present cosmology model that the scalar field, which is responsible for the inflation stage in the early universe, transforms completely into matter, and the accelerated universe expansion is presently governed by dark energy (DE), whose origin is not connected with the inflationary scalar field. We suppose here that dark matter (DM) has a common origin with a small variable component of dark energy (DEV). We suggest that DE presently may have two components, one of which is the Einstein constant Λ, and another, smaller component DEV (Λ_V) comes from the remnants of the scalar field responsible for inflation, which gave birth to the origin of presently existing matter. In this note we consider only the stages of the universe's expansion after recombination, $z \simeq 1100$, when DM is the most abundant component of the matter; therefore, we suggest for simplicity that a connection exists between DM and DEV so that the ratio of their densities remains constant over all the stages after recombination $\rho_{DM} = \alpha\rho_{DEV}$, with a constant α.

In this paper we suggest that this discrepancy may be explained by the deviation of the cosmological expansion from a standard Lambda-CDM model of a flat universe, due to the action of an additional variable component DEV. Taking into account the influence of DEV on the universe's expansion, we find the value of α that could remove the HT problem. In order to maintain the almost constant DEV/DM energy density ratio during the time interval at $z < 1100$, we suggest the existence of a wide mass DM particle distribution.

2. Universe with Common Origin of DM and DE

The scalar field with the potential $V(\phi)$, where ϕ is the intensity of the scalar field, is considered the main reason for the inflation [10,11], but see [12]. The equation for the scalar field in the expanding universe is written as [13]

$$\ddot{\phi} + 3\frac{\dot{a}}{a}\dot{\phi} = -\frac{dV}{d\phi}.$$ (1)

Here a is a scale factor in the flat expanding universe [7]. The density ρ_V and pressure P_V of the scalar field [1] are defined as [13]

$$\rho_V = \frac{\dot{\phi}^2}{2} + V, \quad P_V = \frac{\dot{\phi}^2}{2} - V.$$ (2)

Consider the universe with the initial scalar field, at initial intensity ϕ_{in} and initial potential V_{in}, and at zero derivative $\dot{\phi}_{in} = 0$. The derivative of the scalar field intensity is growing on the initial stage of inflation.

Let us suggest that after reaching the relation

$$\dot{\phi}^2 = 2\alpha V, \tag{3}$$

it is preserved during further expansion. The kinetic part of the scalar field is transforming into matter, presumably dark matter, and the constant α determines the ratio of the the the dark energy density, represented by V, to the matter density, represented by the kinetic term. As follows from observations, the main part of DE is represented presently by DE, which may be considered as the Einstein constant Λ. At earlier times the input of constant Λ is smaller than the input of Λ_V, for a wide interval of constant α values.

Let us consider an expanding flat universe, described by the Friedmann equation [7]

$$\frac{\dot{a}^2}{a^2} = \frac{8\pi G}{3}\rho + \frac{\Lambda}{3}. \tag{4}$$

Introduce

$$\rho_\phi = V, \quad P_\phi = -V, \quad \rho_m = \frac{\dot{\phi}^2}{2}, \quad P_m = \beta\frac{\dot{\phi}^2}{2}, \quad \text{with} \quad \rho_m = \alpha\rho_\phi. \tag{5}$$

We suggest that only part β of the kinetic term makes the input into the pressure of the matter, so it follows from (3) and (5)

$$\rho = \rho_\phi + \rho_m = (1+\alpha)V, \quad P = P_\phi + P_m = -(1-\alpha\beta)V. \tag{6}$$

The adiabatic condition

$$\frac{d\rho}{\rho + P} = -\frac{dV}{V} = -3\frac{da}{a}, \quad V \text{ is the volume,} \tag{7}$$

may be written as

$$\dot{\rho} = -3\frac{\dot{a}}{a}(\rho_\phi + \rho_m + P_\phi + P_m) = -3\frac{\dot{a}}{a}(\rho_m + P_m) = -3\alpha\frac{1+\beta}{1+\alpha}\frac{\dot{a}}{a}\rho. \tag{8}$$

$$\frac{\rho}{\rho_*} = \left(\frac{a}{a_*}\right)^{\frac{3\alpha}{2(1+\alpha)}} \quad \text{for} \quad \beta = 0. \tag{9}$$

2.1. A Universe with $\Lambda = 0$

Suggest first that cosmological constant $\Lambda = 0$, and DE is created only by the part of the scalar field Λ_V, represented by V. The expressions for the total density ρ, scaling factor a, and Hubble "constant" H follow from (4)–(8) as

$$\frac{a}{a_*} = (6\pi G\rho_* t^2)^{\frac{1+\alpha}{3\alpha(1+\beta)}}\left(\frac{\alpha(1+\beta)}{1+\alpha}\right)^{\frac{2(1+\alpha)}{3\alpha(1+\beta)}} = \left(\frac{\rho_*}{\rho}\right)^{\frac{1+\alpha}{3\alpha(1+\beta)}} = \left(\frac{t}{t_*}\right)^{\frac{2(1+\alpha)}{3\alpha(1+\beta)}},$$

$$\rho = \left(\frac{1+\alpha}{\alpha(1+\beta)}\right)^2\frac{1}{6\pi G t^2}, \quad H = \frac{\dot{a}}{a} = \frac{2(1+\alpha)}{3\alpha(1+\beta)t}. \tag{10}$$

Here $\rho_* = \rho(t_*)$, $a_* = a(t_*)$, t_* is an arbitrary time moment. Write the expressions for particular cases. For $\beta = 1/3$ (radiation dominated universe) it follows from (10)

$$\frac{a}{a_*} = (6\pi G\rho_* t^2)^{\frac{1+\alpha}{4\alpha}} \left[\frac{4\alpha}{3(1+\alpha)}\right]^{\frac{1+\alpha}{2\alpha}} = \left(\frac{\rho_*}{\rho}\right)^{\frac{1+\alpha}{4\alpha}} = \left(\frac{t}{t_*}\right)^{\frac{1+\alpha}{2\alpha}},$$

$$\rho = \left(\frac{3(1+\alpha)}{4\alpha}\right)^2 \frac{1}{6\pi G t^2}, \quad H = \frac{\dot{a}}{a} = \frac{1+\alpha}{2\alpha t}. \tag{11}$$

For the value of $\beta = 0$ (dusty universe, $z < 1100$) we have

$$\frac{a}{a_*} = (6\pi G\rho_* t^2)^{\frac{1+\alpha}{3\alpha}} \left[\frac{\alpha}{1+\alpha}\right]^{\frac{2(1+\alpha)}{3\alpha}} = \left(\frac{\rho_*}{\rho}\right)^{\frac{1+\alpha}{3\alpha}} = \left(\frac{t}{t_*}\right)^{\frac{2(1+\alpha)}{3\alpha}},$$

$$\rho = \left(\frac{1+\alpha}{\alpha}\right)^2 \frac{1}{6\pi G t^2}, \quad H = \frac{\dot{a}}{a} = \frac{2(1+\alpha)}{3\alpha t}. \tag{12}$$

2.2. A Universe in the Presence of the Cosmological Constant Λ

Equations (5)–(8) are valid in the presence of Λ. The solution of Equation (4) with nonzero Λ is written in the form

$$\left(\frac{a}{a_*}\right)^{\frac{3\alpha(1+\beta)}{2(1+\alpha)}} = \sqrt{\frac{8\pi G\rho_*}{\Lambda c^2}} \sinh\left(\sqrt{\frac{\Lambda}{3}}\frac{3\alpha(1+\beta)}{2(1+\alpha)}ct\right) = \sqrt{\frac{\rho_*}{\rho}}; \tag{13}$$

$$\sqrt{\frac{\Lambda c^2}{8\pi G\rho}} = \sinh\left(\sqrt{\frac{\Lambda}{3}}\frac{3\alpha(1+\beta)}{2(1+\alpha)}ct\right), \quad H = \frac{\dot{a}}{a} = \sqrt{\frac{\Lambda c^2}{3}} \coth\left(\sqrt{\frac{\Lambda}{3}}\frac{3\alpha(1+\beta)}{2(1+\alpha)}ct\right). \tag{14}$$

For the dusty universe ($\beta = 0$), after recombination at $z < 1100$ we have

$$\left(\frac{a}{a_*}\right)^{\frac{3\alpha}{2(1+\alpha)}} = \sqrt{\frac{8\pi G\rho_*}{\Lambda c^2}} \sinh\left(\sqrt{\frac{\Lambda}{3}}\frac{3\alpha}{2(1+\alpha)}ct\right) = \sqrt{\frac{\rho_*}{\rho}}; \tag{15}$$

$$\sqrt{\frac{\Lambda c^2}{8\pi G\rho}} = \sinh\left(\sqrt{\frac{\Lambda}{3}}\frac{3\alpha}{2(1+\alpha)}ct\right), \quad H = \frac{\dot{a}}{a} = \sqrt{\frac{\Lambda c^2}{3}} \coth\left(\sqrt{\frac{\Lambda}{3}}\frac{3\alpha}{2(1+\alpha)}ct\right). \tag{16}$$

The dusty universe without DEV is described by relations following from (15) and (16) at $\alpha \to \infty$, giving

$$\left(\frac{a}{a_*}\right)^{\frac{3}{2}} = \sqrt{\frac{8\pi G\rho_*}{\Lambda c^2}} \sinh\left(\sqrt{\frac{\Lambda}{3}}\frac{3}{2}ct\right) = \sqrt{\frac{\rho_*}{\rho}}; \tag{17}$$

$$\sqrt{\frac{\Lambda c^2}{8\pi G\rho}} = \sinh\left(\sqrt{\frac{\Lambda}{3}}\frac{3}{2}ct\right), \quad H = \frac{\dot{a}}{a} = \sqrt{\frac{\Lambda c^2}{3}} \coth\left(\sqrt{\frac{\Lambda}{3}}\frac{3}{2}ct\right). \tag{18}$$

3. Hubble Tension

Recently, a challenge in cosmology was formulated because of different values, obtained from different experiments, of the Hubble constant at the present epoch. There is a significant discrepancy (tension) between the *Planck* measurement from cosmic microwave background (CMB) anisotropy, where the best-fit model gives [14–16],

$$H_0^{\text{P18}} = 67.36 \pm 0.54 \ \text{km}\,\text{s}^{-1}\,\text{Mpc}^{-1}, \tag{19}$$

and measurements using type Ia supernovae (SNIa) calibrated with Cepheid distances [17–21],

$$H_0^{\text{R19}} = 74.03 \pm 1.42 \ \text{km}\,\text{s}^{-1}\,\text{Mpc}^{-1}. \tag{20}$$

Measurements using time delays from lensed quasars [22] gave the value $H_0 = 73.3^{+1.7}_{-1.8} \, \text{km}\,\text{s}^{-1}\,\text{Mpc}^{-1}$, while in [23] it was found that

$H_0 = 72.4 \pm 1.9 \, \text{km} \, \text{s}^{-1} \, \text{Mpc}^{-1}$ using the tip of the red giant branch applied to SNIa, which is independent of the Cepheid distance scale. Analysis of a compilation of these and other recent high- and low-redshift measurements shows [24] that the discrepancy between *Planck* [16] and any three independent late-Universe measurements is between 4σ and 6σ. Different sophisticated explanations for the appearance of HT have been proposed [25–30] (see also [31–34]) and new experiments have been proposed for checking the reliability of this tension [35] (see also review [36]).

Dark matter (DM) and dark energy (DE) represent about 96% of the universe constituents [14,17,18], but their origin is still not clear. The present value of DE density may be represented by the Einstein cosmological constant Λ [37], but may also be a result of the action of the Higgs-type scalar fields, which are supposed to be the reason for the inflation in the early universe [10] (see also [11,12,38]). The value of the induced Λ, suggested for the inflation, is many orders of magnitude larger than its present value, and no attempts have been made to find a connection between them. The origin of DM is even more vague. There are numerous suggestions for its origin [39–41], but none of these possibilities has been experimentally or observationally confirmed, while many of them have been disproved.

To explain the origin of the Hubble Tension, we introduce a variable part of the cosmological "constant" Λ_V, proportional to the matter density $\rho_{DM} = \alpha \rho_{DEV}$. This part of Λ_V influences the cosmological expansion at large redshifts, where the influence of the real Einstein constant Λ is negligible. The value of Λ_V is represented by a small component of DE, which we define as DEV. We suppose here, without knowledge of the physical properties of DM particles, that there is a wide spectrum of DM particle, which could be produced by DEV until present time.

This seems necessary because at decreasing of the DEV field strength in the expanding universe it would be able to make mutual transformations only with DM particles of decreasing mass. The existence of particles with a very low rest mass (axions [42]) is considered often as a candidate for DM. Note that in the paper of Yu N. Gnedin [5], mutual transformations between axions and electromagnetic photons have been considered, instead of the hypothetical "scalar field" in our model.

4. Removing the Hubble Tension

We consider a model of the expanding universe after recombination, at $z \leq 1100$, with a fixed ratio α of energy densities between DEV, connected with a scalar field, cosmological constant Λ, and DM. If the mass spectrum of DM particles prolongs to very small masses, then we may expect an almost constant DEV–DM ratio. In the inflation model of the universe, only a scalar field was born at the very beginning, and matter was created in the process of expansion from the dynamic part of the scalar field density.

Here we show that in presence of DEV the Hubble value H is decreasing with time slower than without it. This creates a larger present value of H_0, removing the Hubble tension at $\alpha \sim 140$. The main idea of removing the tension is the following. The CMB measurements give the value of the Hubble constant H_r, at the redshift $z \sim 1100$, close to the moment of recombination. This value is used for calculation of the present value of H_0.

For analysis of the Hubble tension it is more convenient to use logarithmic variables, so that from (10), (19) and (20) we have

$$\log H_0^{P18} = \log 67.36 = 1.83; \quad \log H_0^{R19} = \log 74.03 = 1.87;$$

$$\log \frac{H_0^{R19}}{H_0^{P18}} = \Delta \log H_0 = 0.04. \tag{21}$$

The Planck value H_r^P was measured at the moment of recombination $z_r \approx 1100$, and extrapolated to present time using a dusty flat Friedmann model with a very small input from the cosmological constant Λ, when the values z, ω, a are connected as [7]

$$z + 1 = \frac{\omega}{\omega_0} = \frac{a_0}{a}. \tag{22}$$

In the case of an equipartition universe the extrapolation should be performed using Equation (16). Numerical modeling of large-scale structure formation gives preference to the cold dark matter model, corresponding to $P_m \approx 0$, $\beta = 0$. In our interpretation the Hubble tension is connected with incorrect extrapolation by Equation (16) without DEV. From the value of Hubble tension in Equation (21) we may estimate α from the condition that both measurements are correct, but the reported value H_0^{P18} is coming from the incorrect extrapolation, and the actual present epoch value of the Hubble constant is determined by H_0^{R19}.

From Equations (16) and (22) we obtain for a dusty universe, at $\beta = 0$, the following connection between the recombination redshift $z_r \approx 1100$, the present age of the universe t_0 and the age of the universe t_r, corresponding to the recombination, in the form

$$z_{r\alpha} + 1 = \frac{a_*}{a_{r\alpha}} = \left[\sqrt{\frac{8\pi G \rho_*}{\Lambda c^2}} \sinh\left(\sqrt{\frac{\Lambda}{3}} \frac{3\alpha}{2(1+\alpha)} ct_{r\alpha} \right) \right]^{-\frac{2(1+\alpha)}{3\alpha}} = \left(\frac{\rho_*}{\rho_{r\alpha}} \right)^{-\frac{(1+\alpha)}{3\alpha}}, \tag{23}$$

$$H = \frac{\dot{a}}{a} = \sqrt{\frac{\Lambda c^2}{3}} \coth\left(\sqrt{\frac{\Lambda}{3}} \frac{3\alpha}{2(1+\alpha)} ct_{r\alpha} \right). \tag{24}$$

Here indices with α indicate the values in the universe with DEV. The index "0" is related to present time at $z_0 = 0$. The index "r" is related to the moment of recombination. The values in the universe without DEV do not contain α in the indices, and are written as

$$z_r + 1 = \frac{a_*}{a_r} = \left[\sqrt{\frac{8\pi G \rho_*}{\Lambda c^2}} \sinh\left(\sqrt{\frac{\Lambda}{3}} \frac{3}{2} ct_r \right) \right]^{-\frac{2}{3}} = \left(\frac{\rho_*}{\rho_r} \right)^{-\frac{1}{3}}, \tag{25}$$

$$H = \frac{\dot{a}}{a} = \sqrt{\frac{\Lambda c^2}{3}} \coth\left(\sqrt{\frac{\Lambda}{3}} \frac{3}{2} ct_r \right). \tag{26}$$

The observational data are connected with a redshift, so we find the expression for the time as a function of the redshift in the form

$$ct_{r\alpha} = \frac{2(1+\alpha)}{3\alpha} \sqrt{\frac{3}{\Lambda}} \sinh^{-1}\left[\sqrt{\frac{\Lambda c^2}{8\pi G \rho_*}} (z_{r\alpha} + 1)^{\frac{3\alpha}{2(1+\alpha)}} \right] \tag{27}$$

$$\approx \frac{\alpha + 1}{\alpha \sqrt{6\pi G \rho_*}} (z_{r\alpha} + 1)^{\frac{3\alpha}{2(1+\alpha)}},$$

$$ct_{0\alpha} = \frac{2(1+\alpha)}{3\alpha} \sqrt{\frac{3}{\Lambda}} \sinh^{-1}\left(\sqrt{\frac{\Lambda c^2}{8\pi G \rho_*}} \right). \tag{28}$$

At $t = t_{r\alpha}$ the argument of "sinh" is very small, so only the first term in the expansion remains. Corresponding values for the universe without DEV are written as

$$ct_r = \frac{2}{3} \sqrt{\frac{3}{\Lambda}} \sinh^{-1}\left[\sqrt{\frac{\Lambda c^2}{8\pi G \rho_*}} (z_r + 1)^{\frac{3}{2}} \right] \approx \frac{1}{\sqrt{6\pi G \rho_*}} (z_r + 1)^{\frac{3}{2}}, \tag{29}$$

$$ct_{0\alpha} = \frac{2(1+\alpha)}{3\alpha} \sqrt{\frac{3}{\Lambda}} \sinh^{-1}\left(\sqrt{\frac{\Lambda c^2}{8\pi G \rho_*}} \right). \tag{30}$$

For $z_{r\alpha} = z_r \equiv z_{rec}$ we obtain a connection between $t_{r\alpha}$ and t_r as

$$t_{r\alpha} = \frac{\alpha + 1}{\alpha} (z_{rec} + 1)^{\frac{3}{2(1+\alpha)}} t_r. \tag{31}$$

The ratio of Hubble constants at present time t_0 to its value at the recombination time $t_{r\alpha}$, in the presence of DEV, using (16), is written as

$$\frac{H_{0\alpha}}{H_{r\alpha}} = \frac{\coth\left(\sqrt{\frac{\Lambda}{3}}\frac{3\alpha}{2(1+\alpha)}ct_0\right)}{\coth\left(\sqrt{\frac{\Lambda}{3}}\frac{3\alpha ct_{r\alpha}}{2(1+\alpha)}\right)} = \frac{\tanh\left(\sqrt{\frac{\Lambda}{3}}\frac{3\alpha ct_{r\alpha}}{2(1+\alpha)}\right)}{\tanh\left(\sqrt{\frac{\Lambda}{3}}\frac{3\alpha}{2(1+\alpha)}ct_0\right)} = \frac{\sqrt{\frac{\Lambda}{3}}\frac{3\alpha}{2(1+\alpha)}ct_{r\alpha}}{\tanh\left(\sqrt{\frac{\Lambda}{3}}\frac{3\alpha}{2(1+\alpha)}ct_0\right)}. \tag{32}$$

The corresponding values for the case without DEV, at $\alpha \to \infty$, with recombination time t_r are written as

$$\frac{H_0}{H_r} = \frac{\coth\left(\sqrt{\frac{\Lambda}{3}}\frac{3}{2}ct_0\right)}{\coth\left(\sqrt{\frac{\Lambda}{3}}\frac{3ct_r}{2}\right)} = \frac{\tanh\left(\sqrt{\frac{\Lambda}{3}}\frac{3ct_r}{2}\right)}{\tanh\left(\sqrt{\frac{\Lambda}{3}}\frac{3}{2}ct_0\right)} = \frac{\sqrt{\frac{\Lambda}{3}}\frac{3}{2}ct_r}{\tanh\left(\sqrt{\frac{\Lambda}{3}}\frac{3}{2}ct_0\right)}. \tag{33}$$

Using Equations (31)–(33), we obtain the ratio between the correct value $H_{0\alpha}$, identified with the local measurement of H, and the value H_0, obtained by calculations without account of DEV, in the form

$$\frac{H_{0\alpha}}{H_0} = \frac{\tanh\left(\sqrt{\frac{\Lambda}{3}}\frac{3}{2}ct_0\right)}{\tanh\left(\sqrt{\frac{\Lambda}{3}}\frac{3\alpha}{2(1+\alpha)}ct_0\right)}(z_{rec}+1)^{\frac{3}{2(1+\alpha)}}. \tag{34}$$

Numerical Estimations

Let us consider the following commonly accepted, averaged parameters of the universe, used in relation to the HT explanation.

Local Hubble constant $H_l = 73$ km/s/Mpc.

Distantly measured Hubble constant $H_d = 67.5$ km/s/Mpc.

Total density of the flat universe $\rho_{tot} = 2 \cdot 10^{-29}h^2 = 1.066 \cdot 10^{-29}$ g/cm^3, where h = H/100 km/s/Mpc, with $h = 0.73$.

Distantly measured densities of the universe components, $\rho_{\Lambda_*} = 0.7\rho_{tot} = 7.5 \cdot 10^{-30}$ g/cm^3; $\rho_{m*} = 0.3\rho_{tot}$.

Locally measured cosmological constant density [18] $\rho_{\Lambda_0} = (0.44 \div 0.96)\rho_{tot}$, $(2\sigma$ statistics).

$\Lambda = \frac{8\pi G\rho_{\Lambda_*}}{c^2} = 1.40 \cdot 10^{-56}$ cm^{-2}.

Average age of the universe $t_0 = 4.35 \cdot 10^{17}$ s [43].

The ratio of two Hubble constants is $HT_r = \frac{H_l}{H_d} = 1.08$.

Identifying $H_{0\alpha} \equiv H_l$, $H_0 \equiv H_d$, we obtain from (34)

$$\frac{H_{0\alpha}}{H_0} = 1.08 = \frac{\tanh\left(\sqrt{\frac{\Lambda}{3}}\frac{3}{2}ct_0\right)}{\tanh\left(\sqrt{\frac{\Lambda}{3}}\frac{3\alpha}{2(1+\alpha)}ct_0\right)}(z_{rec}+1)^{\frac{3}{2(1+\alpha)}} \tag{35}$$

$$= \frac{\tanh(1.337)}{\tanh\left(1.337\frac{\alpha}{1+\alpha}\right)}(1101)^{\frac{3}{2(1+\alpha)}} = \frac{0.87095}{\tanh\left(1.337\frac{\alpha}{1+\alpha}\right)}(1101)^{\frac{3}{2(1+\alpha)}}.$$

Finally we obtain the equation for α in the form

$$\tanh\left(1.337\frac{\alpha}{1+\alpha}\right) = 0.8064(1101)^{\frac{3}{2(1+\alpha)}}, \quad \alpha \approx 140. \tag{36}$$

Taking $\rho_{mr} = 0.3\rho_{tot}$, we obtain $\rho_{DEV} = 0.0022\rho_{tot}$, and the effective dark energy density at present time should be equal to $\Omega_{0eff} = 0.7022$, which is inside the limits of local measurement of Λ_0, as mentioned above. The local matter density of the flat universe at present time is $\Omega_{0m} = 0.2978$, leading to the dark matter density $\Omega_{0dm} = 0.2578$, for the baryon density $\Omega_{0b} = 0.04$.

5. Discussion

We consider a cosmological model, where the DE input consists of two components: cosmological constant Λ, and small variable part DEV, which is connected with a scalar field, originated in the early universe and creating inflation. The energy density of this variable part is uniquely connected with a matter density, and the matter itself was created during the inflation stage. For simplicity we consider a cosmological model with a linear connection between energy densities as $\alpha\rho_{DEV} = \rho_{DM}$. This model could give a solution to the problem of Hubble Tension, which appears by using an inappropriate extrapolation when finding the present value of the Hubble constant from Planck observational data. We have solved the Friedmann equation in the presence of this relation, and have found the value of α at which HT disappeared. The present DEV density needed for explanation of HT phenomena is very small relative to the cosmological constant Λ. It influences the cosmological expansion at larger redshifts, where the input of the Einstein constant Λ is small. Presently the situation is opposite, $\Lambda \gg \Lambda_V$, because decreasing of matter density during cosmological expansion determines the transition from the quasi-Friedmann to quasi-de Sitter stage. The estimation of the density ρ_{Λ_V} at the present epoch corresponds to $\Omega_{\Lambda_V} \approx 0.0022$, which slightly increases the present dark energy density. In the flat universe it determines decreasing of the dark matter density due to transfer of its energy into DEV in the condition of "equipartition" at constant α.

We have used for estimations a constant ratio of $\rho_{\Lambda_V}/\rho_m = 1/\alpha$ for the universe expansion after recombination, at $z < 1100$, but deviations from this law should not qualitatively change the conclusion that a relatively small average contribution of the variable Ω_{Λ_V} may explain the difference in Hubble constant measured at local and high-z distances.

The present parameters of the LCDM model have been estimated from the analysis of CMB fluctuation measurements in WMAP and PLANCK experiments, having a power-law spectrum of adiabatic scalar perturbations. The procedure is based on a search of extremes in the multidimensional parameter space. The presence of HT (if real) adds an additional restriction to this problem. The universe parameters obtained in this process may be changed with this additional restriction. The computations could be performed in the presence of a variable α. Decreasing of dark matter leads to decreasing of the field amplitude, which may prevent the energy exchange between DM and DEV in the absence of very light DM particles.

In our model the DM should be represented by wide mass spectrum particles, and not by unique mass CDM particles, which are usually considered now. By analogy with CMB, the lowest mass of DM particles should not presently exceed the value $\sim (\Omega_{DEV}/\Omega_{CMB})^{1/4} \times kT_{CMB} \approx 7 \cdot 10^{-4}$ eV, to retain the possibility of an almost constant α.

Funding: This work was supported by the Russian Science Foundation (grant No. 18-12-00378).

Acknowledgments: The author is very grateful to O. Yu. Tsupko for valuable comments.

Conflicts of Interest: The author declares no conflict of interest.

Note

[1] In most equations below it is taken that $c = 1$.

References

1. Gnedin, Y.N.; Krasnikov, S.V. Polarimetric effects associated with the detection of Goldstone bosons in stars and galaxies. *Sov. Phys. JETP* **1992**, *75*, 933–937.
2. Gnedin, Y.N. Resonance magnetic conversion of photons into massless axions and striking feature in quasar polarized light. *Astrophys. Space Sci.* **1997**, *249*, 125–149. [CrossRef]
3. Gnedin, Y.N. Astronomical searches for nonbarionic dark matter. *Astrophys. Space Sci.* **1997**, *252*, 95–106. [CrossRef]
4. Gnedin, Y.N.; Dodonov, S.N.; Vlasyuk, V.V.; Spiridonova, O.I.; Shakhverdov, A.V. Astronomical searches for axions: Observations at the SAO 6-m telescope. *Mon. Not. R. Astron. Soc.* **1999**, *306*, 117–121. [CrossRef]
5. Gnedin, Y.N.; Piotrovich, M.Y.; Natsvlishvili, T.M. PVLAS experiment: Some astrophysical consequences. *Mon. Not. R. Astron. Soc.* **2007**, *374*, 276–281. [CrossRef]

6.   Piotrovich, M.Y.; Gnedin, Y.N.; Natsvlishvili, T.M. Coupling constant for axion and electromagnetic fields and cosmological observations. *Astrophysics* **2009**, *52*, 412–422. [CrossRef]
7.   Zel'dovich, Y.B.; Novikov, I.D. *The Structure and Evolution of the Universe*; Nauka: Moscow, Russia, 1975.
8.   Riess, A.G. The Expansion of the Universe is Faster than Expected. *Nat. Rev. Phys.* **2019**, *2*, 10–12. [CrossRef]
9.   Freedman, W.L. Measurements of the Hubble Constant: Tensions in Perspective. *Astrophys. J.* **2021**, *919*, 16. [CrossRef]
10. Guth, A.H. *The Inflationary Universe*; Perseus Books: Reading, MA, USA, 1998.
11. Linde, A.D. Chaotic inflation. *Phys. Lett. B* **1983**, *129*, 177–181. [CrossRef]
12. Starobinsky, A.A. Dynamics of phase transition in the new inflationary universe scenario and generation of perturbations. *Phys. Lett. B* **1982**, *117*, 175–178. [CrossRef]
13. Peebles, P.J.E. *Principles of Physical Cosmology*; Princeton University Press: Princeton, NJ, USA, 1993.
14. Spergel, D.N.; Verde, L.; Peiris, H.V.; Komatsu, E.; Nolta, M.R.; Bennett, C.L.; Halpern, M.; Hinshaw, G.; Jarosik, N.; Kogut, A.; et al. First-Year Wilkinson Microwave Anisotropy Probe (WMAP) Observations: Determination of Cosmological Parameters. *Astrophys. J. Suppl. Ser.* **2003**, *148*, 175–194. [CrossRef]
15. Ade, P.A.R.; Aghanim, N.; Arnaud, M.; Ashdown, M.; Aumont, J.; Baccigalupi, C.; Banday, A.J.; Barreiro, R.B.; Bartlett, J.G.; Bartolo, N.; et al. Planck Collaboration. Planck 2015 results. XIII. Cosmological parameters. *Astron. Astrophys.* **2016**, *594*, A13.
16. Aghanim, N.; Akrami, Y.; Ashdown, M.; Aumont, J.; Baccigalupi, C.; Ballardini, M.; Banday, A.J.; Barreiro, R.B.; Bartolo, N.; Basak, S.; et al. Planck Collaboration. Planck 2018 results. VI. Cosmological parameters. *arXiv* **2020**, arXiv:1807.06209v3.
17. Riess, A.G.; Filippenko, A.V.; Challis, P.; Clocchiatti, A.; Diercks, A.; Garnavich, P.M.; Gilliland, R.L.; Hogan, C.J.; Jha, S.; Kirshner, R.P.; et al. Observational Evidence from Supernovae for an Accelerating Universe and a Cosmological Constant. *Astron. J.* **1998**, *116*, 1009–1038. [CrossRef]
18. Perlmutter, S.; Aldering, G.; Goldhaber, G.; Knop, R.A.; Nugent, P.; Castro, P.G.; Deustua, S.; Fabbro, S.; Goobar, A.; Groom, D.E.; et al. Measurements of Ω and Λ from 42 High-Redshift Supernovae. *Astrophys. J.* **1999**, *517*, 565–586. [CrossRef]
19. Riess, A.G.; Macri, L.M.; Hoffmann, S.L.; Scolnic, D.; Casertano, S.; Filippenko, A.V.; Tucker, B.E.; Reid, M.J.; Jones, D.O.; Silverman, J.M.; et al. A 2.4% Determination of the Local Value of the Hubble Constant. *Astrophys. J.* **2016**, *826*, 56. [CrossRef]
20. Riess, A.G.; Casertano, S.; Yuan, W.; Macri, L.; Bucciarelli, B.; Lattanzi, M.G.; MacKenty, J.W.; Bowers, J.B.; Zheng, W.; Filippenko, A.V.; et al. Milky Way Cepheid Standards for Measuring Cosmic Distances and Application to Gaia DR2: Implications for the Hubble Constant. *Astrophys. J.* **2018**, *861*, 126. [CrossRef]
21. Riess, A.G.; Casertano, S.; Yuan, W.; Macri, L.M.; Scolnic, D. Large Magellanic Cloud Cepheid Standards Provide a 1% Foundation for the Determination of the Hubble Constant and Stronger Evidence for Physics beyond ΛCDM. *Astrophys. J.* **2019**, *876*, 85. [CrossRef]
22. Wong, K.C.; Suyu, S.H.; Chen, G.C.; Rusu, C.E.; Millon, M.; Sluse, D.; Bonvin, V.; Fassnacht, C.D.; Taubenberger, S.; Auger, M.W.; et al. H0LiCOW XIII. A 2.4% measurement of H_0 from lensed quasars: 5.3σ tension between early and late-Universe probes. *Mon. Not. R. Astron. Soc.* **2020**, *498*, 1420–1439. [CrossRef]
23. Yuan, W.; Riess, A.G.; Macri, L.M.; Casertano, S.; Scolnic, D.M. Consistent Calibration of the Tip of the Red Giant Branch in the Large Magellanic Cloud on the Hubble Space Telescope Photometric System and Implications for the Determination of the Hubble Constant. *Astrophys. J.* **2019**, *886*, 61. [CrossRef]
24. Verde, L.; Treu, T.; Riess, A.G. Tensions between the Early and the Late Universe. *arXiv* **2019**, arXiv:1907.10625.
25. Karwal, T.; Kamionkowski, M. Early dark energy, the Hubble-parameter tension, and the string axiverse. *Phys. Rev. D* **2016**, *94*, 103523. [CrossRef]
26. Moertsell, E.; Dhawan, S. Does the Hubble constant tension call for new physics? *arXiv* **2018**, arXiv:1801.07260.
27. Poulin, V.; Smith, T.L.; Karwal, T.; Kamionkowski, M. Early Dark Energy Can Resolve The Hubble Tension. *Phys. Rev. Lett.* **2019**, *122*, 221301. [CrossRef]
28. Yang, W.; Pan, S.; Di Valentino, E.; Nunes, R.C.; Vagnozzi, S.; Mota, D.F. Tale of stable interacting dark energy, observational signatures, and the H0 tension. *J. Cosmol. Astropart. Phys.* **2018**, *1809*, 019. [CrossRef]
29. Vagnozzi, S. New physics in light of the H0 tension: An alternative view. *arXiv* **2019**, arXiv:1907.07569.
30. Di Valentino, E.; Melchiorri, A.; Mena, O.; Vagnozzi, S. Interacting dark energy after the latest Planck, DES, and H0 measurements: An excellent solution to the H0 and cosmic shear tensions. *arXiv* **2019**, arXiv:1908.04281.
31. Umiltà, C.; Ballardini, M.; Finelli, F.; Paoletti, D. CMB and BAO constraints for an induced gravity dark energy model with a quartic potential. *J. Cosmol. Astropart. Phys.* **2015**, *1508*, 017. [CrossRef]
32. Ballardini, M.; Finelli, F.; Umiltà, C.; Paoletti, D. Cosmological constraints on induced gravity dark energy models. *J. Cosmol. Astropart. Phys.* **2016**, *1605*, 067. [CrossRef]
33. Rossi, M.; Ballardini, M.; Braglia, M.; Finelli, F.; Paoletti, D.; Starobinsky, A.A.; Umilta, C. Cosmological constraints on post-Newtonian parameters in effectively massless scalar-tensor theories of gravity. *Phys. Rev. D* **2019**, *100*, 103524. [CrossRef]
34. Knox, L.; Millea, M. The Hubble Hunter's Guide. *Phys. Rev. D* **2020**, *101*, 043533. [CrossRef]
35. Bengaly, C.A.P.; Clarkson, C.; Maartens, R. The Hubble constant tension with next generation galaxy surveys. *arXiv* **2019**, arXiv:1908.04619.
36. Di Valentino, E.; Mena, O.; Pan, S.; Visinelli, L.; Yang, W.; Melchiorri, A.; Mota, D.F.; Riess, A.G.; Silk, J. In the realm of the Hubble tension—A review of solutions. *Class. Quantum Grav.* **2021**, *38*, 153001. [CrossRef]

37. Einstein, A. Kosmologische Betrachtungen zur allgemeinen Relativitätstheorie. In *Sitzungsberichte der Königlich Preussischen Akademie der Wissenschaften Berlin—Part 1*; 1917; pp. 142–152.
38. Mukhanov, V.F.; Chibisov, G.V. Vacuum energy and large-scale structure of the Universe. *Sov. Phys. JETP* **1982**, *56*, 258–265.
39. Arun, K.; Gudennavar, S.B.; Sivaram, C. Dark matter, dark energy, and alternate models: A review. *Adv. Space Res.* **2017**, *60*, 166–186. [CrossRef]
40. Zhao, G.B.; Raveri, M.; Pogosian, L.; Wang, Y.; Crittenden, R.G.; Handley, W.J.; Percival, W.J.; Beutler, F.; Brinkmann, J.; Chuang, C.H.; et al. Dynamical dark energy in light of the latest observations. *Nat. Astron.* **2017**, *1*, 627–632. [CrossRef]
41. Samart, D.; Phongpichit, P. Unification of inflation and dark matter in the Higgs–Starobinsky model. *Eur. Phys. J. C* **2019**, *79*, 347. [CrossRef]
42. Turner, M.S. Windows on the axion. *Phys. Rep.* **1990**, *197*, 67–97. [CrossRef]
43. Age of the Universe. Available online: https://en.wikipedia.org/wiki/Age_of_the_universe (accessed on 28 October 2021).

Article

Dark Matter Axions, Non-Newtonian Gravity and Constraints on Them from Recent Measurements of the Casimir Force in the Micrometer Separation Range

Galina L. Klimchitskaya [1,2] **and Vladimir M. Mostepanenko** [1,2,3,*]

[1] Central Astronomical Observatory at Pulkovo of the Russian Academy of Sciences, 196140 Saint Petersburg, Russia; g_klimchitskaya@mail.ru

[2] Institute of Physics, Nanotechnology and Telecommunications, Peter the Great Saint Petersburg Polytechnic University, 195251 Saint Petersburg, Russia

[3] Kazan Federal University, 420008 Kazan, Russia

[*] Correspondence: vmostepa@gmail.com

Abstract: We consider axionlike particles as the most probable constituents of dark matter, the Yukawa-type corrections to Newton's gravitational law and constraints on their parameters following from astrophysics and different laboratory experiments. After a brief discussion of the results by Prof. Yu. N. Gnedin in this field, we turn our attention to the recent experiment on measuring the differential Casimir force between Au-coated surfaces of a sphere and the top and bottom of rectangular trenches. In this experiment, the Casimir force was measured over an unusually wide separation region from 0.2 to 8 µm and compared with the exact theory based on first principles of quantum electrodynamics at nonzero temperature. We use the measure of agreement between experiment and theory to obtain the constraints on the coupling constant of axionlike particles to nucleons and on the interaction strength of a Yukawa-type interaction. The constraints obtained on the axion-to-nucleon coupling constant and on the strength of a Yukawa interaction are stronger by factors of 4 and 24, respectively, than those found previously from gravitational experiments and measurements of the Casimir force but weaker than the constraints following from a differential measurement where the Casimir force was nullified. Some other already performed and planned experiments aimed at searching for axions and non-Newtonian gravity are discussed, and their prospects are evaluated.

Keywords: dark matter axions; non-Newtonian gravity; measurements of the Casimir force; hypothetical particles

Citation: Klimchitskaya, G.L.; Mostepanenko, V.M. Dark Matter Axions, Non-Newtonian Gravity and Constraints on Them from Recent Measurements of the Casimir Force in the Micrometer Separation Range. *Universe* **2021**, *7*, 343. https://doi.org/10.3390/universe7090343

Academic Editor: Gerald B. Cleaver

Received: 16 August 2021
Accepted: 9 September 2021
Published: 12 September 2021

1. Introduction

The problem of dark matter has a long history [1]. As found by J. Oort in 1932 when studying stellar motion in the neighborhood of a galaxy, the galaxy mass must be well over than that of its visible constituents [2]. A year later, F. Zwicky [3] applied the virial theorem to the Coma cluster of galaxies in order to determine its mass. The obtained mass value turned out to be much larger than that found from the number of observed galaxies belonging to the Coma cluster multiplied by their mean mass. An excess mass, which reveals itself only gravitationally, received the name *dark matter*.

According to current concepts, dark matter contributes to approximately 27% of the energy of the universe although its physical nature remains unknown. There are many approaches to this problem based on the role of some hypothetical particles, such as axions, arions, massive neutrinos, weakly interacting massive particles (WIMP), barionic dark matter, modified gravity, etc. (see [1,4–8] for a review).

The model of dark matter that finds support from astrophysics and cosmology is referred to as *cold dark matter*. According to this model, dark matter consists of light hypothetical particles that are produced in the early universe and become nonrelativistic already at the first stages of its evolution. The most popular particle of this kind is an axion, i.e., a pseudoscalar Nambu–Goldstone boson introduced to solve the problem of strong CP violation in Quantum Chromodynamics (QCD) [9–11]. It has been known that the gauge invariant QCD vacuum depends on an angle θ, and that this dependence violates the CP invariance of QCD. However, experiments suggest that strong interactions are CP invariant and that the electric dipole moment of a neutron is equal to zero up to a high degree of accuracy. An introduction of the Peccei–Quinn symmetry and axions, which are connected with its violation, helps to solve this problem.

Axions and other axionlike particles that arise in many extensions to the Standard Model can interact both with photons and with fermions (electrons and nucleons). These interactions are used for an axion search and for constraining axion masses and coupling constants from observations of numerous astrophysical and cosmological processes as well as from various laboratory experiments (see reviews [1,4–8,12–20] of already obtained bounds on the axion mass and coupling constants to photons, electrons and nucleons).

Prof. Yuri N. Gnedin obtained many important results investigating the interaction of dark matter axions with photons in astrophysics and cosmology. He proposed [21] to employ the polarimetric methods for a search for axions and arions (i.e., the axions of zero mass [22]) in the emission from pulsars, X-ray binaries with low-mass components, and magnetic white dwarfs. For this purpose, he used the conversion process of photons into axions in the magnetic field of compact stars and in the interstellar and intergalactic space (i.e., the Primakoff effect). Next, Prof. Gnedin demonstrated an appearance of the striking feature in the polarized light of quasistellar objects due to the resonance magnetic conversion of photons into massless axions [23].

Using these results, Prof. Yu. N. Gnedin organized the axion search by the 6 m telescope at the Special Astronomical Observatory in Russia. Both the Primakoff effect and the inverse process of an axion decay into two photons were searched for in the integalactic light of clusters of galaxies and in the brightness of night sky due to axions in the halo of our galaxy [24]. Although no evidences of axions were found, it was possible to find the upper limit on the photon-to-axion coupling constant from the polarimetric observations of magnetic chemically peculiar stars of spectral type A possessing strong hydrogen Balmer absorption lines [24]. The above results, as well as the ground-based cavity experiments searching for galactic axions, searches of an axion decay in the galactic and extragalactic light, for the solar and stellar axions, and the limits obtained on the coupling constant of axions to photons, were discussed in a review [25].

In their further research on dark matter axions, Yu. N. Gnedin and his collaborators analyzed [26] the intermediate results of PVLAS experiments interpreted [27] as arising due to a conversion of photons into axions with a coupling constant to photons of the order of 4×10^{-6} GeV^{-1}. By considering the astrophysical and cosmological constraints, they have shown [26] that this result is in contradiction with the data on stellar evolution that exclude the standard model of QCD axions.

Using the cosmic orientation of the electric field vectors of polarized radiation from distant quasars, Yu. N. Gnedin and his collaborators placed a rather strong limit on the coupling constant of axions to an electric field [28]. Numerous results related to the processes of axion decay into two photons, the transformation of photons into axions in the magnetic fields of stars and of interstellar or intergalactic media, and the transformation of axions generated in the cores of stars into X-ray photons were discussed in a review [29].

It has been known that the coupling constant of axionlike particles to fermions can be constrained in the laboratory experiments measuring the Casimir force between two closely spaced test bodies. This force is caused by the zero-point and thermal fluctuations of the electromagnetic field. It acts between any material surfaces—metallic, dielectric or semiconductor [30,31]. A constraint on the electron-arion coupling constant from old measurements of the Casimir force [32] was obtained in [33]. The competitive constraints on the coupling constants of axionlike particle to nucleons from different experiments measuring the Casimir interaction were obtained in [34–41]. All experiments used for obtaining these constraints have been performed in the separation range below 1 µm.

Starting from 1982 [42], measurements of the van der Waals and Casimir forces were also used for constraining the Yukawa-type corrections to Newton's law of gravitation. These corrections arise due to an exchange of light scalar particles between atoms of two closely spaced macrobodies [43] and in the extra-dimensional unification schemes with a low-energy compactification scale [44–47]. A review of the most precise measurements of the Casimir interaction and constraints on non-Newtonian gravity obtained from them can be found in [20,48].

In this paper, we obtain new constraints on the coupling constant of axionlike particles to nucleons and on the Yukawa-type corrections to Newtonian gravity following from recent experiment measuring the differential Casimir force between two Au-coated bodies spaced at separations from 0.2 to 8 µm [49]. This experiment was performed by Prof. R. S. Decca by means of a micromechanical torsional oscillator. The differential Casimir force was measured between an Au-coated sapphire sphere and the top and bottom of Au-coated deep Si trenches. The measurement results were compared with the exact theory using the scattering approach and were found to be in good agreement with it over the entire measurement range with no fitting parameters under the condition that the relaxation properties of conduction electrons are not included in the computations. Another theoretical approach that takes into account the relaxation properties of conduction electrons was excluded by the measurement data over the range of separations from 0.2 to 4.8 µm [49].

We calculate additional forces arising in the experimental configuration due to an exchange of two axionlike particles between nucleons of the test bodies as well as due to the Yukawa-type correction to Newton's gravitational potential. Taking into account that no extra force was observed, the constraints on the masses and coupling constants of axions and on the strength and interaction range of the Yukawa interaction were found from the extent of agreement between the measured and calculated Casimir forces.

According to our results, the obtained constraints on axionlike particles are up to a factor of 4 stronger than those from other measurements of the Casimir force. The new constraints on the Yukawa-type interaction are up to a factor of 24 stronger than those obtained previously when measuring the Casimir force. Stronger constraints on an axion [38] and non-Newtonian gravity [50] were obtained only from the experiment by R. S. Decca [50], where the Casimir force was completely nullified.

The paper is organized as follows. In Section 2, we consider the effective potentials due to an exchange of pseudoscalar and scalar particles. Section 3 is devoted to a brief discussion of recent experiments on measuring the differential Casimir force in the micrometer range. In Section 4, we calculate additional forces due to an exchange of two axionlike particles between nucleons and obtain constraints on the axion mass and coupling constant. The constraints on the Yukawa-type correction to Newtonian gravity are found in Section 5. In Sections 6 and 7, the reader can find a discussion of the results obtained and our conclusions, respectively.

We use a system of units with $\hbar = c = 1$.

2. Effective Potentials Due to Exchange of Pseudoscalar and Scalar Particles

An interaction of the field of originally introduced QCD axions $a(x)$ [9–11], which describes the Nambu–Goldstone bosons, with the fermionic field $\psi(x)$ is given by the pseudovector Lagrangian density [12,15]

$$\mathcal{L}_{\mathrm{pv}}(x) = \frac{g}{2m_a} \bar{\psi}(x)\gamma_5\gamma_\mu\psi(x)\partial^\mu a(x), \tag{1}$$

where m_a is the mass of an axion, γ_n with $n = 0, 1, 2, 3, 4, 5$ are the Dirac matrices, and g is the dimensionless coupling constant of the axions to fermionic particles of spin $1/2$ (in our case a proton or a neutron).

The axionlike particles introduced in Grand Unified Theories (GUT) interact with fermions by means of the pseudoscalar Lagrangian density [12,15,51]

$$\mathcal{L}_{\mathrm{ps}}(x) = -ig\bar{\psi}(x)\gamma_5\psi(x)a(x). \tag{2}$$

On a tree level, the Lagrangian densities (1) and (2) result in the same action and common effective potential due to an exchange of one axion or axionlike particle between two nucleons of mass m spaced at a distance $r = |\vec{r}_1 - \vec{r}_2|$ [52,53]

$$V_{an}(r; \vec{\sigma}_1, \vec{\sigma}_2) = \frac{g^2}{16\pi m^2}\left[(\vec{\sigma}_1 \cdot \hat{r})(\vec{\sigma}_2 \cdot \hat{r})\left(\frac{m_a^2}{r} + \frac{3m_a}{r^2} + \frac{3}{r^3}\right)\right.$$
$$\left. -(\vec{\sigma}_1 \cdot \vec{\sigma}_2)\left(\frac{m_a}{r^2} + \frac{1}{r^3}\right)\right], \tag{3}$$

where $\hat{r} = (\vec{r}_1 - \vec{r}_2)/r$ is the unit vector directed along the line connecting the two nucleons and $\vec{\sigma}_1$, $\vec{\sigma}_2$ are their spins. Here, we assume that the coupling constants of an axion to a neutron and a proton are equal and denote m as their mean mass.

Taking into account that the effective potential (3) is spin-dependent, the resulting additional force averages to zero when integrated over the volumes of unpolarized test bodies used in experiments on measuring the Casimir force. Therefore, the process of one-axion exchange cannot be used for the axion search in measurements of the Casimir force performed until now [31,54] (the proposed measurement of the Casimir force between two test bodies possessing the polarization of nuclear spins [55] is, however, quite promising for this purpose).

A process of the two-axion exchange between two nucleons deserves particular attention. If the pseudovector Lagrangian density (1) is used, the respective effective potential is still unknown. This is because the actual interaction constant $g/(2m_a)$ is dimensional, and the resulting quantum field theory is nonrenormalizable (the current status of this problem is reflected in [56]). However, in the case of the pseudoscalar Lagrangian density (2) describing the interaction of axionlike particles with fermions, the effective potential due to a two-axion exchange is spin-independent and takes the following simple form [43,57,58]:

$$V_{aan}(r) = -\frac{g^4}{32\pi^3 m^2}\frac{m_a}{r^2}K_1(2m_a r), \tag{4}$$

where $K_1(z)$ is the modified Bessel function of the second kind.

The effective potential (4) can be summed over all pairs of nucleons belonging to the test bodies V_1 and V_2, leading to their total interaction energy

$$U_{aan}(z) = -\frac{m_a g^4}{32\pi^3 m^2}n_1 n_2 \int_{V_1} d\vec{r}_1 \int_{V_2} d\vec{r}_2 \frac{K_1(2m_a r)}{r^2}, \tag{5}$$

where n_1 and n_2 are the numbers of nucleons per unit volume of the first and second test bodies, respectively, and z is the closest separation distance between them. Finally, from (5), one arrives at the additional force acting between the test bodies due to the two-axion exchange

$$F_{aan}(z) = -\frac{\partial U_{aan}(z)}{\partial z}. \tag{6}$$

Equations (5) and (6) can be used to search for axionlike particles in experiments measuring the Casimir force and for constraining their parameters.

A similar situation takes place for the Yukawa-type corrections to Newton's gravitational law, which arise due to an exchange of one scalar particle of mass m_s between two pointlike particles (atoms or nucleons) with masses m_1 and m_2 separated by a distance r. It is conventional to notate the coupling constant of the Yukawa interaction as αG, where G is the Newtonian gravitational constant and α is the proper dimensionless Yukawa strength. Then, the Yukawa-type correction to Newtonian gravity takes the form

$$V_{\text{Yu}}(r) = -\frac{Gm_1 m_2}{r}\alpha e^{-r/\lambda},\tag{7}$$

where the interaction range $\lambda = 1/m_s$ represents the Compton wavelength of a scalar particle. As was mentioned in Section 1, the same correction to the Newton's gravitational potential also arises in extra-dimensional physics with a low-energy compactification scale [44–47].

The Yukawa-type interaction energy between two macrobodies V_1 and V_2 arising due to the potential (7) is given by

$$U_{\text{Yu}}(z) = -\alpha G\rho_1\rho_2 \int_{V_1} d\vec{r}_1 \int_{V_2} d\vec{r}_2\, \frac{e^{-r/\lambda}}{r},\tag{8}$$

where ρ_1 and ρ_2 are the mass densities of the first and second test bodies, respectively. In this case, the additional force acting between two test bodies is equal to

$$F_{\text{Yu}}(z) = -\frac{\partial U_{\text{Yu}}(z)}{\partial z}.\tag{9}$$

Both hypothetical forces (6) and (9) act on the background of the Casimir force measured at separations below a few micrometers. Note that the Newtonian gravitational force

$$F_{\text{gr}}(z) = G\rho_1\rho_2\frac{\partial}{\partial z} \int_{V_1} d\vec{r}_1 \int_{V_2} d\vec{r}_2\, \frac{1}{r}\tag{10}$$

calculated within the same range of separations is much less than the error in measuring the Casimir force and can be neglected (see below).

3. Measurements of the Casimir Force in the Micrometer Separation Range

All experiments measuring the Casimir force between two macrobodies used for obtaining constraints on axionlike particles [35–41] were performed at separations below a micrometer. A single direct measurement of the Casimir force between two macrobodies at separations up to 8 µm was reported quite recently and compared with theory with no fitting parameters [49]. Below, we briefly elucidate the main features of this experiment needed for obtaining new constraints on the axionlike particles and Yukawa-type corrections to Newtonian gravity.

In the experiment [49], the micromechanical torsional oscillator was used to measure the differential Casimir force between an Au-coated sapphire sphere of $R = 149.7$ µm radius and the top and bottom of Au-coated rectangular silicon trenches of $H = 50$ µm depth. Such a large depth of the trenches was chosen in order for the Casimir force between a sphere and a trench bottom (and all the more additional hypothetical forces) to be equal to zero. This means that the actually measured Casimir force $F_C(z)$ acts between a sphere and a trench top. In so doing, however, the differential measurement scheme used allowed us to reach a rather low total experimental error equal to $\Delta F_C = 2.2$ fN at the separation distance of $z = 3$ µm. Similar schemes of differential force measurements were previously used in [50,59–61].

The thicknesses of Au coatings on the sphere and the trench surfaces equal to $d_{\text{Au}}^{(s)} = 250$ nm and $d_{\text{Au}}^{(t)} = 150$ nm, respectively, were thick enough in order for Au-coated test bodies to be considered made from solid Au when calculating the Casimir force. For technological reasons, before depositing Au coatings, the sapphire sphere and silicon trench surfaces were also covered with Cr layers of thickness $d_{\text{Cr}} = 10$ nm. These layers do not influence the Casimir force but should be accounted for in calculations of the additional forces due to two-axion exchange and the Yukawa-type potential in Sections 4 and 5, respectively.

The Casimir force between an Au-coated sphere and an Au-coated trench top was measured over the separation region 0.2 μm $\leqslant z \leqslant 8$ μm and compared with an exact theory developed in the sphere-plate geometry using the scattering approach in the plane-wave representation [62–64] and the derivation expansion [65–68]. In doing so, the contribution of patch potentials to the measured signal was characterized by Kelvin probe microscopy. An impact of surface roughness was taken into account perturbatively [31,54,69,70] and was found to be negligible. It was shown [49] that the measurement data are in agreement with theory to within the total experimental error ΔF_C over the entire measurement range if the relaxation properties of conduction electrons are not included in computations. An inclusion of the relaxation of conduction electrons (i.e., using the dissipative Drude model at low frequencies) results in a strong contradiction between experiment and theory over the range of separations from 0.2 to 4.8 μm. These results are in line with previous precise experiments on measuring the Casimir interaction at shorter sphere-plate separations [61,71–81] and discussions of the so-called Casimir puzzle [82–84] (see also recent approaches to the resolution of this problem [85–87]).

It is necessary to stress that calculation of the Casimir force between the first test body (sphere) and the flat top of rectangular trenches in [49] was based on first principles of quantum electrodynamics at nonzero temperature and did not use any approximate methods, such as the proximity force approximation applied in the most of previously performed experiments [31,54] or fitting parameters. As for the total experimental errors ΔF_C, they were found in [49] at the 95% confidence level as a combination of both random and systematic errors.

Notice that, according to [88,89], the same experiment cannot be used to exclude an alternative model and to constrain the fundamental forces. This claim was made in 2011 when some of the background effects in the Casimir force (such as the role of patch potentials) and theoretical uncertainties (such as an error in the proximity force approximation) were not completely settled. As was, however, immediately objected in the literature [90,91], such a difference between the excluded and confirmed models of the Casimir force is of a form quite different from a correction to the fundamental force. This fact allows us to constrain the parameters of the latter. After the seminal experiment by R. S. Decca performed in 2016 [61], where the theoretical predictions of two models for the Casimir force differed by up to a factor of 1000 and one of them was conclusively rejected whereas another one was confirmed, it became amply clear that a comparison with the confirmed model can be reliably used for constraining the fundamental forces in all subsequent experiments.

When discussing the measure of agreement between experiment and theory, it is necessary also to take into account the contribution of differential Newtonian gravitational force between the sphere and the top and bottom of rectangular trenches. In [49] this force was assumed to be negligibly small. Taking into account that in Sections 4 and 5 the measure of agreement between experiment and theory is used for constraining the parameters of axionlike particles and non-Newtonian gravity, below we estimate the role of Newton's gravitational force more precisely.

The second test body, which is concentrically covered by the rectangular trenches, is a $D = 25.4$ mm diameter Si disc (schematic of the experimental setup is shown in Figure 1 of [49]). As mentioned above, both of the test bodies are coated by layers of Cr and Au. In our estimation of the upper bound for the Newton's gravitational force in this experiment, we assume that both the sphere and the disc of diameter D and thickness H, where H is the trench depth, are all-gold. By choosing the disc thickness equal to H, we disregard the Si substrate underlying trenches because it does not contribute to the differential gravitational force between the top and bottom of the trenches (unlike the Casimir and additional hypothetical forces, the more long-range gravitational force from the trench bottom cannot be considered equal to zero).

Taking into account that the disc radius $D/2$ is much larger that the sphere radius R, the gravitational force between them is given by [92]

$$F_{\rm gr} = -\frac{8\pi^2}{3}G\rho_{\rm Au}^2 HR^3,\tag{11}$$

where $\rho_{\rm Au} = 19.28$ g/cm^3 is the density of gold. Substituting the values of all parameters in (11), one obtains $F_{\rm gr} = 0.11$ fN. It is seen that the upper bound for the gravitational force is much less than the experimental error in measuring the Casimir force, which is equal to $\Delta F_{\rm C} = 2.2$ fN at $z = 3$ μm (the variations in gravitational force depending on whether the sphere is above the top or bottom of the trenches are well below 0.11 fN). This confirms that the contribution of Newton's gravitational force in this experiment is very small and can be neglected.

4. Constraints on Axionlike Particles

The results of the experiment [49] measuring the Casimir force in the micrometer seperation range allow us to obtain new constraints on the mass of axionlike particles and their coupling constants to nucleons. As explained in Section 2, measurements of the Casimir force between unpolarized test bodies can be used for constraining the parameters of axionlike particles in which interaction with fermions is described by the Lagrangian density (2). For this purpose, one should use the process of two-axion exchange between nucleons of the laboratory test bodies, which leads to the effective potential (4) and force (6). Thus, it is necessary to calculate this force in the experimental configuration of [49].

According to Section 3, this configuration reduces to a sapphire (Al$_2$O$_3$) sphere of radius R (the sapphire density is $\rho_{\rm Al_2O_3} = 4.1$ g/cm^3) coated by the layers of Cr and Au of thicknesses $d_{\rm Cr}$ and $d_{\rm Au}^{(s)}$, respectively, interacting with a Si plate ($\rho_{\rm Si} = 2.33$ g/cm^3) in which the thickness exceeds H and can be considered infinitely large when calculating the hypothetical interactions rapidly decreasing with separation. The plate is also coated by the layer of Cr of thickness $d_{\rm Cr}$ and by an external layer of Au of thickness $d_{\rm Au}^{(t)}$ indicated in Section 3.

The additional force arising in this configuration due to a two-axion exchange can be calculated by (5) and (6). The results of this calculation for a homogeneous sphere and plate are presented in [41]. The impact of two metallic layers is easily taken into account in perfect analogy to [38,40]. Finally, the additional force due to a two-axion exchange in the configuration of the experiment [49] takes the form

$$F_{\rm aan}(z) = -\frac{\pi}{2m_a m^2 m_{\rm H}^2}\int_1^\infty du\,\frac{\sqrt{u^2-1}}{u^2}e^{-2m_a uz}X_t(m_a u)X_s(m_a u),\tag{12}$$

where $m_{\rm H}$ is the mass of atomic hydrogen and the following expressions for the functions $X_t(x)$ and $X_s(x)$ are obtained:

$$X_t(x) = C_{\rm Au}\left(1 - e^{-2xd_{\rm Au}^{(t)}}\right) + C_{\rm Cr}e^{-2xd_{\rm Au}^{(t)}}\left(1 - e^{-2xd_{\rm Cr}}\right) + C_{\rm Si}e^{-2x(d_{\rm Au}^{(t)}+d_{\rm Cr})},$$

$$X_s(x) = C_{\rm Au}\left[\Phi(R,x) - e^{-2xd_{\rm Au}^{(s)}}\Phi(R - d_{\rm Au}^{(s)},x)\right]$$

$$+ C_{\rm Cr}e^{-2xd_{\rm Au}^{(s)}}\left[\Phi(R - d_{\rm Au}^{(s)},x) - e^{-2xd_{\rm Cr}}\Phi(R - d_{\rm Au}^{(s)} - d_{\rm Cr},x)\right]$$

$$+ C_{\rm Al_2O_3}e^{-2x(d_{\rm Au}^{(s)}+d_{\rm Cr})}\Phi(R - d_{\rm Au}^{(s)} - d_{\rm Cr},x),\tag{13}$$

where the function Φ is defined as

$$\Phi(r,x) = r - \frac{1}{2x} + e^{-4rx}\left(r + \frac{1}{2x}\right).\tag{14}$$

The numerical coefficients C in (13) are specific for any material. They are calculated by the following equation:

$$C = \rho \frac{g^2}{4\pi} \left(\frac{Z}{\mu} + \frac{N}{\mu} \right),$$

(15)

where Z and N are the number of protons and the mean number of neutrons in an atomic nucleus of a given substance, and $\mu = M/m_H$ with M being the mean atomic or molecular mass.

The values of (15) for Au, Cr, Si and Al_2O_3, i.e., C_{Au}, C_{Cr}, C_{Si} and $C_{Al_2O_3}$, are proportional to g^2 with the factors found using the densities of Au, Si and Al_2O_3 indicated above and $\rho_{Cr} = 7.15$ g/cm^3. The values of Z/μ for Au, Cr, Si and Al_2O_3 are equal to 0.40422, 0.46518, 0.50238 and 0.49422, respectively, and those of N/μ are equal to 0.60378, 0.54379, 0.50628 and 0.51412 for the same respective materials [43].

Taking into consideration that, in the experiment [49], no additional interaction was observed within the total experimental error $\Delta F_C(z)$, the interaction (12) should satisfy the inequality

$$|F_{aan}(z)| \leqslant \Delta F_C(z).$$

(16)

Using Equations (12)–(14), one can constrain the possible values of the axionlike particles mass m_a and their coupling constant to nucleons g from (16).

The numerical analysis of (16) taking into account (12)–(14) and the values of ΔF_C at different z [49] shows that the strongest constraints on m_a and g follow at $z = 3$ μm, where $\Delta F_C = 2.2$ fN (see Section 4). These constraints are presented by the line labeled "new" in Figure 1. For comparison purposes, in the same figure, we show the constraints obtained earlier from the Cavendish-type experiment [93] (line "gr$_1$" [94]), from measuring the smallest gravitational forces using the torsional oscillator [95–97] (line "gr$_2$" [56]), from measuring the effective Casimir pressure [73,74] (line 1 [35]), from the experiment using a beam of molecular hydrogen [98] (line "H$_2$" [99]) and from the differential measurement where the Casimir force was completely nullified [59] (the dashed line [38]). Note that the figure field above each line is excluded and that below each line is allowed by the results of respective experiment.

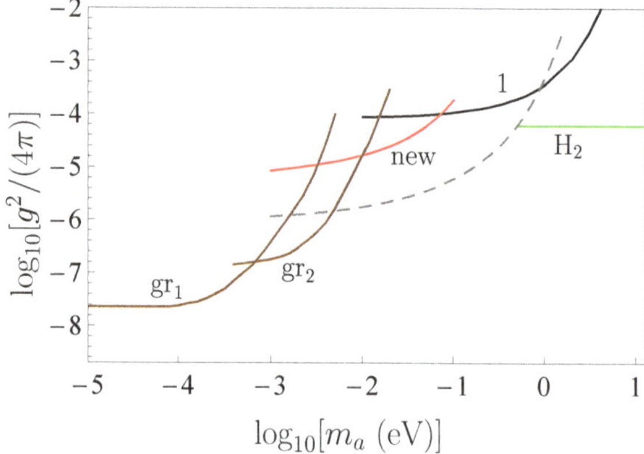

Figure 1. The strongest constraints on the coupling constant of axionlike particles to nucleons obtained from different experiments. Lines labeled "gr$_1$" and "gr$_2$" are found from the Cavendish-type experiments and from measuring the minimum gravitational forces, respectively. Lines labeled "new" and "H$_2$" follow from the recent experiment on measuring the Casimir force in the micrometer separation range and from the experiment using a beam of molecular hydrogen. Line 1 and the dashed line are obtained from measuring the effective Casimir pressure and from the experiment nullifying the Casimir force. The regions above each line are excluded, and those below are allowed.

As is seen in Figure 1, in the region of axion masses from 10 to 74 meV the constraints on g obtained from the experiment [49] (line "new") are stronger than those following from measuring the effective Casimir pressure (line 1) and the smallest gravitational forces (line "gr$_2$"). The maximum strengthening up to a factor of 4 is reached for the axion mass $m_a \approx 16$ meV. Thus, in the region of axion masses indicated above, this experiment leads to stronger constraints than the previous experiments measuring the Casimir force [20,48]. The stronger constraints shown by the dashed line follow only from the experiment [59] where the Casimir force was completely nullified.

5. Constraints on Non-Newtonian Gravity

As mentioned in Section 1, the results of the experiment [49] measuring the Casimir force in the micrometer separation range can be also used for constraining the parameters of Yukawa-type corrections to Newtonian gravity. For this purpose, one should calculate the Yukawa-type force (8), (9) in the experimental configuration using its geometrical characteristics and densities of all constituting materials presented in Sections 3 and 4.

The calculation results using (8) and (9) for a homogeneous sphere above a homogeneous plate are presented in [41]. Here, we modify them in perfect analogy to [38,40] to take into account the contributions of Cr and Au layers covering both a sphere and a trench in this experiment. The result is

$$F_{\mathrm{Yu}}(z) = -4\pi^2 G\alpha\lambda^3 e^{-z/\lambda} Y_t(\lambda) Y_s(\lambda), \tag{17}$$

where

$$Y_t(\lambda) = \rho_{\mathrm{Au}}\left(1 - e^{-d_{\mathrm{Au}}^{(t)}/\lambda}\right) + \rho_{\mathrm{Cr}} e^{-d_{\mathrm{Au}}^{(t)}/\lambda}\left(1 - e^{-d_{\mathrm{Cr}}/\lambda}\right) + \rho_{\mathrm{Si}} e^{-(d_{\mathrm{Au}}^{(t)}+d_{\mathrm{Cr}})/\lambda},$$

$$Y_s(\lambda) = \rho_{\mathrm{Au}}\left[\Psi(R,\lambda) - e^{-d_{\mathrm{Au}}^{(s)}/\lambda}\Psi(R - d_{\mathrm{Au}}^{(s)},\lambda)\right] \tag{18}$$

$$+ \rho_{\mathrm{Cr}} e^{-d_{\mathrm{Au}}^{(s)}/\lambda}\left[\Psi(R - d_{\mathrm{Au}}^{(s)},\lambda) - e^{-d_{\mathrm{Cr}}/\lambda}\Psi(R - d_{\mathrm{Au}}^{(s)} - d_{\mathrm{Cr}},\lambda)\right]$$

$$+ \rho_{\mathrm{Al_2O_3}} e^{-(d_{\mathrm{Au}}^{(s)}+d_{\mathrm{Cr}})/\lambda}\Psi(R - d_{\mathrm{Au}}^{(s)} - d_{\mathrm{Cr}},\lambda),$$

and the function Ψ is defined as

$$\Psi(r,\lambda) = r - \lambda + (r + \lambda)e^{-2r/\lambda}. \tag{19}$$

By virtue of the fact that the Yukawa-type force (17) was not observed within the limits of the measurement error, it should satisfy the inequality

$$|F_{\mathrm{Yu}}(z)| \leqslant \Delta F_C(z). \tag{20}$$

Similar to the case of an additional interaction due to a two-axion exchange between nucleons, the strongest constraints on the parameters of the Yukawa-type interaction α and λ follow from (20) at $z = 3$ μm where $\Delta F_C = 2.2$ fN.

The constraints obtained are shown by the line labeled "new" in Figure 2. For comparison purposes, in this figure we also show the constraints following from measuring of the effective Casimir pressure [73,74] (line 1), of the normal Casimir force between sinusoidally corrugated surfaces of a sphere and a plate at different orientation angles of corrugations [100,101] (line 2 [102]) and of the lateral Casimir force force between the sinusoidally corrugated surfaces of a sphere and a plate [103,104] (line 3 [105]). Note that somewhat stronger constraints than those shown by line 3 were obtained [106] from the experiment measuring the Casimir force between crossed cylinders [107] on an undefined confidence level [31,54]. However, the strongest constraints on the parameters of a Yukawa-type interaction in the region below $\lambda = 10^{-8}$ m follow from the experiments on neutron scattering. They are shown by the line labeled "n" [108,109]. Similar to Figure 1, the regions above each line are excluded by the results of respective experiment, whereas the regions below each line are allowed.

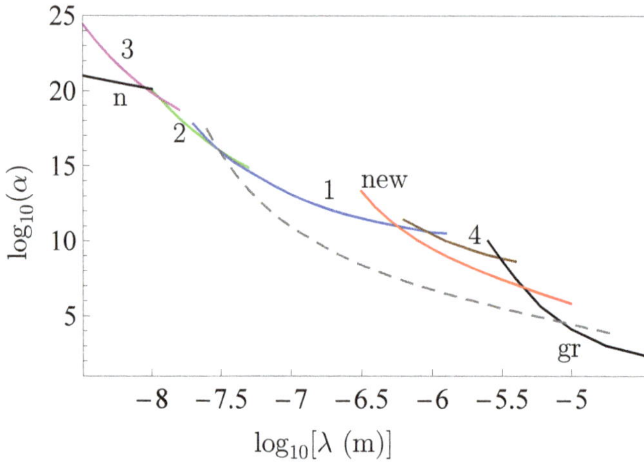

Figure 2. The strongest constraints on the interaction constant and interaction range of the Yukawa-type interaction obtained from different experiments. Lines 1, 2 and 3 are found from measuring the effective Casimir pressure between two parallel plates, and normal and lateral Casimir forces between the sinusoidally corrugated surfaces of a sphere and a plate, respectively. The line labeled "n" is found from the experiments on neutron scattering. The lines labeled "new" and 4 follow from the experiment measuring the Casimir force in the micrometer separation range and from the experiment using a torsion pendulum. The line labeled "gr" and the dashed line are obtained from the Cavendish-type experiments and from the experiment nullifying the Casimir force. The regions above each line are excluded, and those below each line are allowed.

In the region of Figure 2 with a larger λ, the constraints shown by line 4 are obtained from measuring the Casimir force by means of the torsion pendulum [110]. The line labeled "gr" indicates the constraints on the Yukawa-type interaction following from the Cavendish-type experiments [93,94,111]. As for the dashed line, which covers the largest interaction range, it is obtained [59] from the differential force measurement where the Casimir force was completely nullified (compare with the dashed line in Figure 1 found from the same experiment).

As seen in Figure 2, the constraints labeled "new" are stronger than the constraints of lines 1 and "gr" following from measurements of the effective Casimir pressure and from the Cavendish-type experiments over the interaction range from 550 nm to 4.4 μm. The maximum strengthening by up to a factor of 24 is reached at $\lambda = 3.1$ μm. The constraints obtained are weaker only as compared with those following from the experiment where the Casimir force was nullified [59].

6. Discussion

In this article, we considered the problems of dark matter axions, non-Newtonian gravity and the constraints on their parameters. As discussed in Section 1, axions and axionlike particles have gained much recognition as the most probable constituents of dark matter. An active search for axions using their interactions with photons, electrons, and nucleons is under way in many laboratories all over the world. A major contribution to the investigation of interactions between axions and photons in different astrophysical processes has been made by Prof. Yu. N. Gnedin, who suggested several prospective possibilities for the observation of axionlike particles and for constraining their parameters.

Here, we obtain new constraints on the coupling constant of axionlike particles to nucleons, which follow from the recently performed measurement of the differential Casimir force between Au-coated surfaces of the sphere and the top and bottom of rectangular trenches [49]. The differential character of this experiment allowed one to reach a very high precision and to obtain meaningful data up to a very large separation distance of 8 μm. The measure of agreement between the obtained data and the theoretical predictions based on first principles of quantum electrodynamics at nonzero temperature allowed us to find rather strong constraints on the axionlike particles and non-Newtonian gravity of Yukawa type.

The obtained constraints on the coupling constants of axionlike particles to nucleons are stronger by up to a factor of 4 than the previously known ones derived [56] from the gravitational experiments and from measuring the effective Casimir pressure [36,73,74]. This strengthening holds in the range of axion masses m_a from 10 to 74 meV. We have also shown that the same experiment on measuring the differential Casimir force in the micrometer separation range [49] results in up to a factor of 24 stronger constraints on the interaction constant of Yukawa-type interactions compared with the ones found previously from measuring the effective Casimir pressure [73,74], an experiment using the torsion pendulum [110] and the Cavendish-type experiments [93,94,111]. In this case, the strengthening holds in the interaction range from $\lambda = 550$ nm to $\lambda = 3.1$ μm.

Although the constraints obtained are not the strongest ones (the strongest constraints on both the axionlike particles and the Yukawa-type interaction in the interaction ranges indicated above were obtained from the experiment [59] where the Casimir force was nullified), they complement the results found from previous measurements of the Casimir interaction and can be considered as their additional confirmation.

The obtained results fall into intensive studies of axionlike particles, non-Newtonian gravity and constraints on their parameters. In addition to the literature already discussed in Section 1, it is pertinent to mention a haloscope search for dark matter axions that excludes some range of the axion–photon couplings in models of invisible axions [112] and another haloscope experiment for the search of galactic axions using a superconducting resonant cavity [113]. The first results of the promising experiment for searching the dark matter axions with masses below 1 μeV are reported in [114]. Two more haloscope experiments are performed for the search of dark matter axions using their interaction with electronic spins [115] and photons [116].

A few prospective experiments for constraining the parameters of axionlike particles and non-Newtonian gravity are suggested in the literature but have not yet been performed. Here, one should mention an experiment measuring the Casimir pressure between parallel plates at up to 25–30 μm separations (Cannex) [117–120]. An approach for searching for dark matter axions with $m_a < 1$ μeV using a superconducting radio frequency cavity was proposed in [121]. Several possibilities for probing the non-Newtonian gravity in a submillimeter interaction range by means of temporal lensing [122], molecular spectroscopy [123] and neutron interferometry [123,124] are also discussed. Finally, very recently, the possibility to detect the axion-nucleon interaction in the Casimir-less regime by means of levitated system was proposed [125]. According to the authors, their approach provides a possibility to strengthen the current constraints on g by several orders of magnitude in the wide region of axion masses from 10^{-10} eV to 10 eV.

7. Conclusions

To conclude, the search for dark matter axions, non-Newtonian gravity and constraints on their parameters is a multidisciplinary problem of elementary particle physics, quantum field theory, gravitational theory, astrophysics and cosmology. At the moment, neither axions nor corrections to Newton's gravitational law, other than that predicted by General Relativity theory, are observed, but more and more stringent constraints on them have been obtained. Keeping in mind that there are very serious reasons for creating axions at the very early stages of the evolution of the universe and for existence of deviations from Newton's law of gravitation at very short separations, as predicted by the extended Standard Model, Supersymmetry, Supergravity and String theory, one may hope that these predictions will find experimental confirmation in the not too distant future.

Author Contributions: Investigation, both authors; writing, both authors. All authors have read and agreed to the published version of the manuscript.

Funding: This work was supported by the Peter the Great Saint Petersburg Polytechnic University in the framework of the Russian state assignment for basic research (project N FSEG-2020-0024). V.M.M. was also partially funded by the Russian Foundation for Basic Research grant number 19-02-00453 A.

Institutional Review Board Statement: Not applicable.

Informed Consent Statement: Not applicable.

Data Availability Statement: All necessary data are contained in this paper.

Acknowledgments: V.M.M. is grateful for partial support by the Russian Government Program of Competitive Growth of Kazan Federal University.

Conflicts of Interest: The authors declare no conflict of interest.

References

1. Bertone, G.; Hooper, D. Hystory of Dark Matter. *Rev. Mod. Phys.* **2018**, *90*, 045002. [CrossRef]
2. Oort, J.H. The force exerted by the stellar system in the direction perpendicular to the galactic plane and some related problems. *Bull. Astron. Inst. Neth.* **1932**, *6*, 249–287.
3. Zwicky, F. Die Rotverschiebung von extragalaktischen Nebeln. *Helv. Phys. Acta* **1933**, *6*, 110–127.
4. Overduin, J.M.; Wesson, P.S. Dark Matter and Background Light. *Phys. Rep.* **2004**, *402*, 267–406. [CrossRef]
5. Bertone, G.; Hooper, D.; Silk, J. Particle dark matter: Evidence, candidates and constraints. *Phys. Rep.* **2005**, *405*, 279–390. [CrossRef]
6. Sanders, R.H. *The Dark Matter Problem: A Historical Perspective*; Cambridge University Press: Cambridge, UK, 2010.
7. Matarrese, S.; Colpi, M.; Gorini, V.; Moshella, U. (Eds.) *Dark Matter and Dark Energy*; Springer: Dordrecht, The Netherlands, 2011.
8. Chadha-Day, F.; Ellis, J.; Marsh, D.J.E. Axion Dark Matter: What is it and Why Now? *arXiv* **2021**, arXiv:2105.01406
9. Peccei, R.D.; Quinn, H.R. CP Conservation in the Presence of Pseudoparticles. *Phys. Rev. Lett.* **1977**, *38*, 1440–1443. [CrossRef]
10. Weinberg, S. A New Light Boson? *Phys. Rev. Lett.* **1978**, *40*, 223–226. [CrossRef]
11. Wilczek, F. Problem of Strong P and T Invariance in the Presence of Instantons. *Phys. Rev. Lett.* **1978**, *40*, 279–283. [CrossRef]
12. Kim, J.E. Light pseudoscalars, particle physics and cosmology. *Phys. Rep.* **1987**, *150*, 1–177. [CrossRef]
13. Adelberger, E.G.; Heckel, B.R.; Stubbs, C.W.; Rogers, W.F. Searches for new Macroscopic forces. *Annu. Rev. Nucl. Part. Sci.* **1991**, *41*, 269–320. [CrossRef]
14. Rosenberg, L.J.; van Bibber, K.A. Searches for invisible axions. *Phys. Rep.* **2000**, *325*, 1–39. [CrossRef]
15. Raffelt, G.G. Axions—Motivation, limits and searches. *J. Phys. A Math. Theor.* **2007**, *40*, 6607–6620. [CrossRef]
16. Antoniadis, I.; Baessler, S.; Bücher, M.; Fedorov, V.V.; Hoedl, S.; Lambrecht, A.; Nesvizhevsky, V.V.; Pignol, G.; Protasov, K.V.; Reynaud, S.; et al. Short-range fundamental forces. *C. R. Phys.* **2011**, *12*, 755–778. [CrossRef]
17. Kawasaki, M.; Nakayama, K. Axions: Theory and Cosmological Role. *Annu. Rev. Nucl. Part. Sci.* **2013**, *63*, 69–95. [CrossRef]
18. Ivastorza, I.G.; Redondo, J. New experimental approaches in the search for axion-like particles. *Progr. Part. Nucl. Phys.* **2018**, *102*, 89–159. [CrossRef]
19. Safronova, M.S.; Budker, D.; DeMille, D.; Jackson Kimball, D.F.; Derevianko, A.; Clark, C.W. Search for new physics with atoms and molecules. *Rev. Mod. Phys.* **2018**, *90*, 025008. [CrossRef]
20. Klimchitskaya, G.L. Constraints on Theoretical Predictions beyond the Standard Model from the Casimir Effect and Some Other Tabletop Physics. *Universe* **2021**, *7*, 47. [CrossRef]
21. Gnedin, Y.N.; Krasnikov, S.V. Polarimetric effects associated with the detection of Goldstone bosons in stars and galaxies. *Sov. Phys. JETP* **1992**, *75*, 933–937; Translated in *Zh. Eksp. Teor. Fiz.* **1992**, *102*, 1729–1738.
22. Anselm, A.A.; Uraltsev, N.G. A second massless axion? *Phys. Lett. B* **1982**, *114*, 39–41. [CrossRef]
23. Gnedin, Y.N. Magnetic Conversion of Photons into Massless Axions and Striking Feature in Quasar Polarized Light. *Astrophys. Space Sci.* **1997**, *249*, 125–129. [CrossRef]
24. Gnedin, Y.N.; Dodonov, S.N.; Vlasyuk, V.V.; Spiridonova, O.I.; Shakhverdov, A.V. Astronomical searches for axions: Observation at the SAO 6-m telescope. *Mon. Not. R. Astron. Soc.* **1999**, *306*, 117–121. [CrossRef]
25. Gnedin, Y.N. Current status of modern dark matter problem. *Int. J. Mod. Phys. A* **2002**, *17*, 4251–4260. [CrossRef]
26. Gnedin, Y.N.; Piotrovich, M.Y.; Natsvlishvili, T.M. PVLAS experiment: Some astrophysical consequences. *Mon. Not. R. Astron. Soc.* **2007**, *374*, 276–281. [CrossRef]

27. PVLAS Collaboration. Experimental Observation of Optical Photons Generated in Vacuum by a Magnetic Field. *Phys. Rev. Lett.* **2007**, *96*, 110406.
28. Piotrovich, M.Y.; Gnedin, Y.N.; Natsvlishvili, T.M. Coupling constants for axions and electromagnetic fields and cosmological observations. *Astrophysics* **2009**, *52*, 412–422. [CrossRef]
29. Gnedin, Y.N.; Piotrovich, M.Y.; Natsvlishvili, T.M. New results in searching for axions by astronomical methods. *Int. J. Mod. Phys. A* **2016**, *31*, 1641019. [CrossRef]
30. Casimir, H.B.G. On the attraction between two perfectly conducting plates. *Proc. Kon. Ned. Akad. Wet.* **1948**, *51*, 793–795.
31. Bordag, M.; Klimchitskaya, G.L.; Mohideen, U.; Mostepanenko, V.M. *Advances in the Casimir Effect*; Oxford University Press: Oxford, UK, 2015.
32. Derjaguin, B.V.; Abricosova, I.I.; Lifshitz E.M. Direct measurement of molecular attraction between solids separated by a narrow gap. *Quat. Rev.* **1956**, *10*, 295–329. [CrossRef]
33. Mostepanenko, V.M.; Sokolov, I.Y. Casimir effect leads to new restrictions on long-rang force constant. *Phys. Lett. A* **1987**, *125*, 405–407. [CrossRef]
34. Bezerra, V.B.; Klimchitskaya, G.L.; Mostepanenko, V.M.; Romero, C. Constraints on the parameters of an axion from measurements of the thermal Casimir-Polder force. *Phys. Rev. D* **2014**, *89*, 035010. [CrossRef]
35. Bezerra, V.B.; Klimchitskaya, G.L.; Mostepanenko, V.M.; Romero, C. Stronger constraints on an axion from measuring the Casimir interaction by means of a dynamic atomic force microscope. *Phys. Rev. D* **2014**, *89*, 075002. [CrossRef]
36. Bezerra, V.B.; Klimchitskaya, G.L.; Mostepanenko, V.M.; Romero, C. Constraining axion-nucleon coupling constants from measurements of effective Casimir pressure by means of micromachined oscillator. *Eur. Phys. J. C* **2014**, *74*, 2859. [CrossRef]
37. Bezerra, V.B.; Klimchitskaya, G.L.; Mostepanenko, V.M.; Romero, C. Constraints on axion-nucleon coupling constants from measuring the Casimir force between corrugated surfaces. *Phys. Rev. D* **2014**, *90*, 055013. [CrossRef]
38. Klimchitskaya, G.L.; Mostepanenko, V.M. Improved constraints on the coupling constants of axion-like particles to nucleons from recent Casimir-less experiment. *Eur. Phys. J. C* **2015**, *75*, 164. [CrossRef]
39. Klimchitskaya, G.L.; Mostepanenko, V.M. Constraints on axionlike particles and non-Newtonian gravity from measuring the difference of Casimir forces. *Phys. Rev. D* **2017**, *95*, 123013. [CrossRef]
40. Klimchitskaya, G.L. Recent breakthrough and outlook in constraining the non-Newtonian gravity and axion-like particles from Casimir physics. *Eur. Phys. J. C* **2017**, *77*, 315. [CrossRef]
41. Klimchitskaya, G.L.; Kuusk, P.; Mostepanenko, V.M. Constraints on non-Newtonian gravity and axionlike particles from measuring the Casimir force in nanometer separation range. *Phys. Rev. D* **2020**, *101*, 056013. [CrossRef]
42. Kuzmin, V.A.; Tkachev, I.I.; Shaposhnikov, M.E. Restrictions imposed on light scalar particles by measurements of van der Waals forces. *Pis'ma v Zh. Eksp. Teor. Fiz.* **1982**, *36*, 49–52; Translated: *JETP Lett.* **1982**, *36*, 59–62.
43. Fischbach, E.; Talmadge, C.L. *The Search for Non-Newtonian Gravity*; Springer: New York, NY, USA, 1999.
44. Antoniadis, I.; Arkani-Hamed, N.; Dimopoulos, S.; Dvali, G. New dimensions at a millimeter to a fermi and superstrings at a TeV. *Phys. Lett. B* **1998**, *436*, 257–263. [CrossRef]
45. Arkani-Hamed, N.; Dimopoulos, S.; Dvali, G. Phenomenology, astrophysics, and cosmology of theories with millimeter dimensions and TeV scale quantum gravity. *Phys. Rev. D* **1999**, *59*, 086004. [CrossRef]
46. Floratos, E.G.; Leontaris, G.K. Low scale unification, Newton's law and extra dimensions. *Phys. Lett. B* **1999**, *465*, 95–100. [CrossRef]
47. Kehagias, A.; Sfetsos, K. Deviations from $1/r^2$ Newton law due to extra dimensions. *Phys. Lett. B* **2000**, *472*, 39–44. [CrossRef]
48. Mostepanenko, V.M.; Klimchitskaya, G.L. The State of the Art in Constraining Axion-to-Nucleon Coupling and Non-Newtonian Gravity from Laboratory Experiments. *Universe* **2020**, *6*, 147. [CrossRef]
49. Bimonte, G.; Spreng, B.; Maia Neto, P.A.; Ingold, G.-L.; Klimchitskaya, G.L.; Mostepanenko, V.M.; Decca, R.S. Measurement of the Casimir Force between 0.2 and 8 μm: Experimental Procedures and Comparison with Theory. *Universe* **2021**, *7*, 93. [CrossRef]
50. Chen, Y.J.; Tham, W.K.; Krause, D.E.; López, D.; Fischbach, E.; Decca, R.S. Stronger Limits on Hypothetical Yukawa Interactions in the 30–8000 Nm Range. *Phys. Rev. Lett.* **2016**, *116*, 221102. [CrossRef]
51. Moody, J.E.; Wilczek, F. New macroscopic forces? *Phys. Rev. D* **1984**, *30*, 130–139. [CrossRef]
52. Bohr, A.; Mottelson, B.R. *Nuclear Structure, Volume 1*; Benjamin: New York, NY, USA, 1969.
53. Adelberger, E.G.; Fischbach, E.; Krause, D.E.; Newman, R.D. Constraining the couplings of massive pseudoscalars using gravity and optical experiments. *Phys. Rev. D* **2003**, *68*, 062002. [CrossRef]
54. Klimchitskaya, G.L.; Mohideen, U.; Mostepanenko, V.M. The Casimir force between real materials: Experiment and theory. *Rev. Mod. Phys.* **2009**, *81*, 1827–1885. [CrossRef]
55. Bezerra, V.B.; Klimchitskaya, G.L.; Mostepanenko, V.M.; Romero, C. Constraining axion coupling constants from measuring the Casimir interaction between polarized test bodies. *Phys. Rev. D* **2016**, *94*, 035011. [CrossRef]
56. Aldaihan, S.; Krause, D.E.; Long, J.C.; Snow, W.M. Calculations of the dominant long-range, spin-independent contributions to the interaction energy between two nonrelativistic Dirac fermions from double-boson exchange of spin-0 and spin-1 bosons with spin-dependent couplings. *Phys. Rev. D* **2017**, *95*, 096005. [CrossRef]
57. Ferrer, F.; Nowakowski, M. Higg- and Goldstone-boson-mediated long range forces. *Phys. Rev. D* **1999**, *59*, 075009. [CrossRef]
58. Drell, S.D.; Huang, K. Many-Body Forces and Nuclear Saturation. *Phys. Rev.* **1953**, *91*, 1527–1543. [CrossRef]

59. Decca, R.S.; López, D.; Chan, H.B.; Fischbach, E.; Klimchitskaya, G.L.; Krause, D.E.; Mostepanenko, V.M. Precise measurements of the Casimir force and first realization of the "Casimir-less" experiment. *J. Low Temp. Phys.* **2004**, *135*, 63–74. [CrossRef]

60. Decca, R.S.; López, D.; Chan, H.B.; Fischbach, E.; Krause, D.E.; Jamell, C.R. Constraining New Forces in the Casimir Regime Using the Isoelectronic Technique. *Phys. Rev. Lett.* **2005**, *94*, 240401. [CrossRef]

61. Bimonte, G.; López, D.; Decca, R.S. Isoelectronic determination of the thermal Casimir force. *Phys. Rev. B* **2016**, *93*, 184434. [CrossRef]

62. Spreng, B.; Hartmann, M.; Henning, V.; Maia Neto, P.A.; Ingold, G.-L. Proximity force approximation and specular reflection: Application of the WKB limit of Mie scattering to the Casimir effect. *Phys. Rev. A* **2018**, *97*, 062504. [CrossRef]

63. Henning, V.; Spreng, B.; Hartmann, M.; Ingold, G.-L.; Maia Neto, P.A. Role of diffraction in the Casimir effect beyond the proximity force approximation. *J. Opt. Soc. Am. B* **2019**, *36*, C77–C87. [CrossRef]

64. Spreng, B.; Maia Neto, P.A.; Ingold, G.-L. Plane-wave approach to the exact van der Waals interaction between colloid particles. *J. Chem. Phys.* **2020**, *153*, 024115. [CrossRef] [PubMed]

65. Fosco, C.D.; Lombardo, F.C.; Mazzitelli, F.D. Proximity force approximation for the Casimir energy as a derivative expansion. *Phys. Rev. D* **2011**, *84*, 105031. [CrossRef]

66. Bimonte, G.; Emig, T.; Kardar, M. Material dependence of Casimir force: Gradient expansion beyond proximity. *Appl. Phys. Lett.* **2012**, *100*, 074110. [CrossRef]

67. Bimonte, G.; Emig, T.; Jaffe, R.L.; Kardar, M. Casimir forces beyond the proximity force approximation. *Europhys. Lett.* **2012**, *97*, 50001. [CrossRef]

68. Bimonte, G. Going beyond PFA: A precise formula for the sphere-plate Casimir force. *Europhys. Lett.* **2017**, *118*, 20002. [CrossRef]

69. Bordag, M.; Klimchitskaya, G.L.; Mostepanenko, V.M. Casimir force between plates with small deviations from plane parallel geometry. *Int. J. Mod. Phys. A* **1995**, *10*, 2661–2681. [CrossRef]

70. van Zwol, P.J.; Palasantzas, G.; De Hosson, J.T.M. Influence of random roughness on the Casimir force at small separations. *Phys. Rev. B* **2008**, *77*, 075412. [CrossRef]

71. Decca, R.S.; Fischbach, E.; Klimchitskaya, G.L.; Krause, D.E.; López, D.; Mostepanenko, V.M. Improved tests of extra-dimensional physics and thermal quantum field theory from new Casimir force measurements. *Phys. Rev. D* **2003**, *68*, 116003. [CrossRef]

72. Decca, R.S.; López, D.; Fischbach, E.; Klimchitskaya, G.L.; Krause, D.E.; Mostepanenko, V.M. Precise comparison of theory and new experiment for the Casimir force leads to stronger constraints on thermal quantum effects and long-range interactions. *Ann. Phys. (N. Y.)* **2005**, *318*, 37–80. [CrossRef]

73. Decca, R.S.; López, D.; Fischbach, E.; Klimchitskaya, G.L.; Krause, D.E.; Mostepanenko, V.M. Tests of new physics from precise measurements of the Casimir pressure between two gold-coated plates. *Phys. Rev. D* **2007**, *75*, 077101. [CrossRef]

74. Decca, R.S.; López, D.; Fischbach, E.; Klimchitskaya, G.L.; Krause, D.E.; Mostepanenko, V.M. Novel constraints on light elementary particles and extra-dimensional physics from the Casimir effect. *Eur. Phys. J. C* **2007**, *51*, 963–975. [CrossRef]

75. Chang, C.-C.; Banishev, A.A.; Castillo-Garza, R.; Klimchitskaya, G.L.; Mostepanenko, V.M.; Mohideen, U. Gradient of the Casimir force between Au surfaces of a sphere and a plate measured using an atomic force microscope in a frequency-shift technique. *Phys. Rev. B* **2012**, *85*, 165443. [CrossRef]

76. Banishev, A.A.; Chang, C.-C.; Klimchitskaya, G.L.; Mostepanenko, V.M.; Mohideen, U. Measurement of the gradient of the Casimir force between a nonmagnetic gold sphere and a magnetic nickel plate. *Phys. Rev. B* **2012**, *85*, 195422. [CrossRef]

77. Banishev, A.A.; Klimchitskaya, G.L.; Mostepanenko, V.M.; Mohideen, U. Demonstration of the Casimir Force between Ferromagnetic Surfaces of a Ni-Coated Sphere and a Ni-Coated Plate. *Phys. Rev. Lett.* **2013**, *110*, 137401. [CrossRef] [PubMed]

78. Banishev, A.A.; Klimchitskaya, G.L.; Mostepanenko, V.M.; Mohideen, U. Casimir interaction between two magnetic metals in comparison with nonmagnetic test bodies. *Phys. Rev. B* **2013**, *88*, 155410. [CrossRef]

79. Xu, J.; Klimchitskaya, G.L.; Mostepanenko, V.M.; Mohideen, U. Reducing detrimental electrostatic effects in Casimir-force measurements and Casimir-force-based microdevices. *Phys. Rev. A* **2018**, *97*, 032501. [CrossRef]

80. Liu, M.; Xu, J.; Klimchitskaya, G.L.; Mostepanenko, V.M.; Mohideen, U. Examining the Casimir puzzle with an upgraded AFM-based technique and advanced surface cleaning. *Phys. Rev. B* **2019**, *100*, 081406(R). [CrossRef]

81. Liu, M.; Xu, J.; Klimchitskaya, G.L.; Mostepanenko, V.M.; Mohideen, U. Precision measurements of the gradient of the Casimir force between ultraclean metallic surfaces at larger separations. *Phys. Rev. A* **2019**, *100*, 052511. [CrossRef]

82. Mostepanenko, V.M.; Bezerra, V.B.; Decca, R.S.; Fischbach, E.; Geyer, B.; Klimchitskaya, G.L.; Krause, D.E.; López, D.; Romero, C. Present status of controversies regarding the thermal Casimir force. *J. Phys. A Math. Gen.* **2006**, *39*, 6589–6600. [CrossRef]

83. Bezerra, V.B.; Decca, R.S.; Fischbach, E.; Geyer, B.; Klimchitskaya, G.L.; Krause, D.E.; López, D.; Mostepanenko, V.M.; Romero, C. Comment on "Temperature dependence of the Casimir effect". *Phys. Rev. E* **2006**, *73*, 028101. [CrossRef]

84. Mostepanenko, V.M. Casimir Puzzle and Casimir Conundrum: Discovery and Search for Resolution. *Universe* **2021**, *7*, 84. [CrossRef]

85. Klimchitskaya, G.L.; Mostepanenko, V.M. An alternative response to the off-shell quantum fluctuations: A step forward in resolution of the Casimir puzzle. *Eur. Phys. J. C* **2020**, *80*, 900. [CrossRef]

86. Hannemann, M.; Wegner, G.; Henkel, C. No-Slip Boundary Conditions for Electron Hydrodynamics and the Thermal Casimir Pressure. *Universe* **2021**, *7*, 108. [CrossRef]

87. Klimchitskaya, G.L.; Mostepanenko, V.M. Casimir entropy and nonlocal response function to the off-shell quantum fluctuations. *Phys. Rev. D* **2021**, *103*, 096007. [CrossRef]

88. Lambrecht, A.; Reynaud, S. Casimir and short-range gravity tests. In *Gravitational Waves and Experimental Gravity*; Augé, E., Dumarchez, J., Vân, J.T.T., Eds.; Thê Gioi: Hanoi, Vietnam, 2011; pp. 199–206.

89. Lambrecht, A.; Canaguier-Durand, A.; Guérout, R.; Reynaud, S. Casimir effect in the scattering approach: Correlations between material properties, temperature and geometry. In *Casimir Physics*; Dalvit, D.A.R., Milonni, P.W., Roberts, D.C., Rosa, F.S.S., Eds.; Springer: Heidelberg, Germany, 2011; pp. 97–127.

90. Mostepanenko, V.M.; Bezerra, V.B.; Klimchitskaya, G.L.; Romero, C. New constraints on Yukawa-type interactions from the Casimir effect. *Int. J. Mod. Phys. A* **2012**, *27*, 1260015. [CrossRef]

91. Klimchitskaya, G.L.; Mohideen, U.; Mostepanenko, V.M. Constraints on non-Newtonian gravity and light elementary particles from measurements of the Casimir force by means of a dynamic atomic microscope. *Phys. Rev. D* **2012**, *86*, 065025. [CrossRef]

92. Decca, R.S.; Fischbach, E.; Klimchitskaya, G.L.; Krause, D.E.; López, D.; Mostepanenko, V.M. Application of the proximity force approximation to gravitational and Yukawa-type forces. *Phys. Rev. D* **2009**, *79*, 124021. [CrossRef]

93. Kapner, D.J.; Cook, T.S.; Adelberger, E.G.; Gundlach, J.H.; Heckel, B.R.; Hoyle, C.D.; Swanson, H.E. Tests of the Gravitational Inverse-Square Law below the Dark-Energy Length Scale. *Phys. Rev. Lett.* **2007**, *98*, 021101. [CrossRef]

94. Adelberger, E.G.; Heckel, B.R.; Hoedl, S.; Hoyle, C.D.; Kapner, D.J.; Upadhye, A. Particle-Physics Implications of a Recent Test of the Gravitational Inverse-Square Law. *Phys. Rev. Lett.* **2007**, *98*, 131104. [CrossRef]

95. Long, J.C.; Chan, H.W.; Churnside, A.B.; Gulbis, E.A.; Varney, M.C.M.; Price, J.C. Upper limits to submillimetre-range forces from extra space-time dimensions. *Nature* **2003**, *421*, 922–925. [CrossRef]

96. Yan, H.; Housworth, E.A.; Meyer, H.O.; Visser, G.; Weisman, E.; Long, J.C. Absolute measurement of thermal noise in a resonant short-range force experiment. *Class. Quant. Grav.* **2014**, *31*, 205007. [CrossRef]

97. Long, J.C.; Kostelecký, V.A. Search for Lorentz violation in short-range gravity. *Phys. Rev. D* **2015**, *91*, 092003. [CrossRef]

98. Ramsey, N.F. The tensor force between two protons at long range. *Phys. A* **1979**, *96*, 285–289. [CrossRef]

99. Ledbetter, M.P.; Romalis, M.V.; Jackson Kimball, D.F. Constraints on Short-Range Spin-Dependent Interactions from Scalar Spin-Spin Coupling in Deuterated Molecular Hydrogen. *Phys. Rev. Lett.* **2013**, *110*, 040402. [CrossRef] [PubMed]

100. Banishev, A.A.; Wagner, J.; Emig, T.; Zandi, R.; Mohideen, U. Demonstration of Angle-Dependent Casimir Force between Corrugations. *Phys. Rev. Lett.* **2013**, *110*, 250403. [CrossRef] [PubMed]

101. Banishev, A.A.; Wagner, J.; Emig, T.; Zandi, R.; Mohideen, U. Experimental and theoretical investigation of the angular dependence of the Casimir force between sinusoidally corrugated surfaces. *Phys. Rev. B* **2014**, *89*, 235436. [CrossRef]

102. Klimchitskaya, G.L.; Mohideen, U.; Mostepanenko, V.M. Constraints on corrections to Newtonian gravity from two recent measurements of the Casimir interaction between metallic surfaces. *Phys. Rev. D* **2013**, *87*, 125031. [CrossRef]

103. Chiu, H.C.; Klimchitskaya, G.L.; Marachevsky, V.N.; Mostepanenko, V.M.; Mohideen, U. Demonstration of the asymmetric lateral Casimir force between corrugated surfaces in the nonadditive regime. *Phys. Rev. B* **2009**, *80*, 121402(R). [CrossRef]

104. Chiu, H.C.; Klimchitskaya, G.L.; Marachevsky, V.N.; Mostepanenko, V.M.; Mohideen, U. Lateral Casimir force between sinusoidally corrugated surfaces: Asymmetric profiles, deviations from the proximity force approximation, and comparison with exact theory. *Phys. Rev. B* **2010**, *81*, 115417. [CrossRef]

105. Bezerra, V.B.; Klimchitskaya, G.L.; Mostepanenko, V.M.; Romero, C. Advance and prospects in constraining the Yukawa-type corrections to Newtonian gravity from the Casimir effect. *Phys. Rev. D* **2010**, *81*, 055003. [CrossRef]

106. Mostepanenko, V.M.; Novello, M. Constraints on non-Newtonian gravity from the Casimir force measurements between two crossed cylinders. *Phys. Rev. D* **2001**, *63*, 115003. [CrossRef]

107. Ederth, T. Template-stripped gold surfaces with 0.4-nm rms roughness suitable for force measurements: Application to the Casimir force in the 20–100-nm range. *Phys. Rev. A* **2000**, *62*, 062104. [CrossRef]

108. Nesvizhevsky, V.V.; Pignol, G.; Protasov, K.V. Neutron scattering and extra short range interactions. *Phys. Rev. D* **2008**, *77*, 034020. [CrossRef]

109. Kamiya, Y.; Itagami, K.; Tani, M.; Kim, G.N.; Komamiya, S. Constraints on New Gravitylike Forces in the Nanometer Range. *Phys. Rev. Lett.* **2015**, *114*, 161101. [CrossRef]

110. Masuda, M.; Sasaki, M. Limits on Nonstandard Forces in the Submicrometer Range. *Phys. Rev. Lett.* **2009**, *102*, 171101. [CrossRef]

111. Smullin, S.J.; Geraci, A.A.; Weld, D.M.; Chiaverini, J.; Holmes, S.; Kapitulnik, A. Constraints on Yukawa-type deviations from Newtonian gravity at 20 microns. *Phys. Rev. D* **2005**, *72*, 122001. [CrossRef]

112. Du, N.; Force, N.; Khatiwada, R.; Lentz, E.; Ottens, R.; Rosenberg, L.J.; Rybka, G.; Carosi, G.; Woollett, N.; Bowring, D.; et al. Search for Invisible Axion Dark Matter with the Axion Dark Matter Experiment. *Phys. Rev. Lett.* **2018**, *120*, 151301. [CrossRef] [PubMed]

113. Alesini, D.; Braggio, C.; Carugno, G.; Crescini, N.; D'Agostino, D.; Di Gioacchino, D.; Di Vora, R.; Falferi, P.; Gallo, S.; Gambardella, U.; et al. Galactic axions search with a superconducting resonant cavity. *Phys. Rev. D* **2019**, *99*, 101101. [CrossRef]

114. Ouellet, J.L.; Salemi, C.P.; Foster, J.W.; Henning, R.; Bogorad, Z.; Conrad, J.M.; Formaggio, J.A.; Kahn, Y.; Minervini, J.; Radovinsky, A.; et al. First Results from ABRACADABRA-10 cm: A Search for Sub-μeV Axion Dark Matter. *Phys. Rev. Lett.* **2019**, *122*, 121802. [CrossRef]

115. Crescini, N.; Alesini, D.; Braggio, C.; Carugno, G.; D'Agostino, D.; Di Gioacchino, D.; Falferi, P.; Gambardella, U.; Gatti, C.; Iannone, G.; et al. Axion Search with a Quantum-Limited Ferromagnetic Haloscope. *Phys. Rev. Lett.* **2020**, *124*, 171801. [CrossRef] [PubMed]

116. Lee, S.; Ahn, S.; Choi, J.; Ko, B.R.; Semertzidis, Y.K. Axion Dark Matter Search around 6.7 μeV. *Phys. Rev. Lett.* **2020**, *124*, 101802. [CrossRef]
117. Klimchitskaya, G.L.; Mostepanenko, V.M.; Sedmik, R.I.P.; Abele, H. Prospects for Searching Thermal Effects, Non-Newtonian Gravity and Axion-Like Particles: CANNEX Test of the Quantum Vacuum. *Symmetry* **2019**, *11*, 407. [CrossRef]
118. Klimchitskaya, G.L.; Mostepanenko, V.M.; Sedmik, R.I.P. Casimir pressure between metallic plates out of thermal equilibrium: Proposed test for the relaxation properties of free electrons. *Phys. Rev. A* **2019**, *100*, 022511. [CrossRef]
119. Sedmik, R.I.P. Casimir and non-Newtonian force experiment (CANNEX): Review, status, and outlook. *Int. J. Mod. Phys. A* **2020**, *35*, 2040008. [CrossRef]
120. Sedmik, R.I.P.; Pitschmann, M. Next Generation Design and Prospects for Cannex. *Universe* **2021**, *7*, 234. [CrossRef]
121. Berlin, A.; D'Agnolo, R.T.; Ellis, S.A.; Nantista, C.; Neilson, J.; Schuster, P.; Tantawi, S.; Toro, N.; Zhou, K. Axion dark matter detection by superconducting resonant frequency conversion. *J. High Energy Phys.* **2020**, *2020*, 88. [CrossRef]
122. Faizal, M.; Patel, H. Probing Short Distance Gravity using Temporal Lensing. *Int. J. Mod. Phys. A* **2021**, *36*, 2150115. [CrossRef]
123. Banks, H.; McCullough, M. Charting the fifth force landscape. *Phys. Rev. D* **2021**, *103*, 075018. [CrossRef]
124. Rocha, J.M.; Dahia, F. Neutron interferometry and tests of short-range modifications of gravity. *Phys. Rev. D* **2021**, *103*, 124014. [CrossRef]
125. Chen, L.; Liu, J.; Zhuy, K. Constraining Axion-to-Nucleon interaction via ultranarrow linewidth in the Casimir-less regime. *arXiv* **2021**, arXiv:2107.08216.

MDPI

St. Alban-Anlage 66

4052 Basel

Switzerland

Tel. +41 61 683 77 34

Fax +41 61 302 89 18

www.mdpi.com

Universe Editorial Office

E-mail: universe@mdpi.com

www.mdpi.com/journal/universe